A HISTORY OF TACTICS.

A

HISTORY OF TACTICS

CAPT. H. M. JOHNSTONE, R.E.

The Naval & Military Press Ltd

Published by
The Naval & Military Press Ltd
Unit 10, Ridgewood Industrial Park,
Uckfield, East Sussex,
TN22 5QE England
Tel: +44 (0) 1825 749494
Fax: +44 (0) 1825 765701
www.naval-military-press.com

PREFACE.

AFTER any study of the development of Tactics, one is soon aware of the fact that it has two natural branches. Of these one embraces the question of the actual formations of the troops—companies, battalions, brigades, squadrons—when near the enemy; the other, which older writers like Jomini called Grand Tactics, bears upon the question of the *directions* imposed upon the portions of an army by the higher commanders. The two branches are quite distinct. The former is allied to the questions of drill, discipline and training; the latter borders upon the domain, in many cases, of strategy, and depends greatly on the genius and temperament of the commander who gives the order.

But just as successful tactics is imitated and produces a *fashion* in tactics, so also is it with successful Grand Tactics. Frederick the Great's tactics, based on rigid and continuous drill, discipline and training, became a "fashion" and was imitated; so also his flank marches and Oblique Order of Battle—his Grand Tactics—became a "fashion" and were imitated.

Thus both branches lend themselves naturally to treatment as the subject of a History, and can be dealt with side by side without detriment to clearness and reasonable continuity, just as the many branches of the development of a nation are dealt with by the ordinary historian in a single work.

The author confesses to some diffidence in publishing this work, under its rather ambitious title; a real and complete *History* would be the work of a lifetime, and would require qualities which he lays no claim to. But he claims to have done his best to present the *stages* of the development, to ensure decent accuracy in the short accounts of the battles he utilizes, and to make his maps as clear and simple as is consistent with their showing the movements described in the text.

He has, of course, borrowed all of the facts and many of the opinions, but his hope is that he has arranged them in such a way as to render the subject readable to British officers and all others interested in the Art of War.

Holding that description of a fight, without a plan of it, leaves but a vague impression of the Science or Art of it, he has in every case drawn the map and arranged the troops on it at that hour which would best illustrate the matter under discussion. In some cases, where a fairly long and complicated series of operations is described, to give one map showing "position of troops at dawn," for example, would throw on the reader an extra labour in following the movements; to show many different stages on one map, without producing confusion, requires a profuse use of colours—and then often leaves confusion. In many cases, therefore, the author has shown one position only, choosing a critical moment or one that seems to him to show some typical action in offence or defence—a position near enough to the beginning of the fight for the stages that led up to it to be easily followed, and near enough to the supreme crisis for the further stages to be readily understood.

The sources from which the maps were drawn are numerous; the following table shows the matter approximately:—

PLATE	I.—	Jomini.
"	II.—	Adapted from Jomini.
"	III.—	Collected from Grivet.
"	IV.—	Collected and adapted from Grivet.
"	V.—	Adapted from various sources.
"	VI.—	Collected and adapted from Grivet.
"	VII.—	Napier's.
"	VIII.—	Adapted from U.S. Records, and General Doubleday.
"	IX.—	U.S. Records' Atlas.
"	X.—	Topography from Prussian Official; positions of troops from various.
"	XI.—	" " " "
"	XII.—	Adapted from French Official and Prussian Official.
"	XIII.—	" " "
"	XIV.—	" " "
"	XV.—	Prussian Official.
"	XVI.—	Adapted from Prussian Official.
"	XVII.—	Prussian Official.
"	XVIII.—	Adapted from Prussian Official.
"	XVIIIA.—	" " , and a Bavarian account.
"	XIX.—	" " , " "
"	XX.—	" Mouzaffer, Greene, and others.
"	XXI.—	" a French account of Skobeleff; the "Mutiny" action is from an Indian Government Official.
"	XXII.—	Adapted from Shadbolt, and from others.
"	XXIII.—	British Intelligence Department.
"	XXIV.—	Colville.
"	XXV.—	Adapted from German Official, *Times'* History, etc.
"	XXVI.—	" " "
"	XXVII.—	" Russian Official, Richmond Smith, *Times'* History, etc.

TABLE OF CONTENTS.

A HISTORY OF TACTICS.

CHAPTER I.

INTRODUCTION.

THE chief matter with which we are concerned in the History of Tactics is the development of methods of actual fighting, both in attack and defence. The changes from time to time were due sometimes essentially to the improvement of weapons, both for infantry, artillery and cavalry. These changes naturally brought about in some cases modifications in some of the other branches of Tactics, such as orders of march and the details of Military Engineering. Usually the full effect of an improved weapon was not understood until one belligerent or another proved it on the field of battle. In this case the corresponding tactical development may be said to have been due to the very practical experience gained.

Importance of Tactics and Discipline.

When one studies history from the military point of view, one is at once struck by the fact that the victorious side was nearly always superior to its adversaries on account of its tactics or its discipline, two things that are not the same, but that are at the same time closely allied. Without the discipline the tactics are bound to be fitful and desultory; no reliance can be placed upon a good method of fighting unless the troops are under the control of their officers, though, to be sure, we found in the Boer war the very greatest trouble in defeating an adversary in whose case what we look upon as discipline was not very conspicuous. In this case, and in similar cases, the place of discipline was taken by a quite extraordinary appreciation, on the part of individuals, of the right thing to do—an appreciation arising from the natural aptitude of the Boer, fostered by generations of continuous use of the rifle in attack and defence, combined with the extreme mobility obtained by having every man mounted. Their style of fighting, most effective in such a country as theirs and against any enemy less mobile than themselves, proved itself much more effective when things were going favourably than when the opposite was the case. It is in reverse, and in those cases where the issue hangs long in doubt, and when manœuvring in the presence of the enemy is essential—it is in such cases that strict discipline is of so great effect. A badly disciplined force, however great the capacity of its individuals may be, is apt to go to pieces and become a mob the moment it experiences a reverse.

Discipline, too, enables the commander to be sure that his orders will be carried out, even when the troops so ordered are not under his own eye.

It should, then, be clearly understood, that a mere theory of tactics, however good, depends on discipline for its effective application. The idea that 10,000 intelligent men, skilled with the rifle and quite understanding the use of individual cover, are for these reasons alone an army, is quite a mistake. The discipline of combined action, both in marching, attacking and defending, is required in addition. There is probably little doubt that the failure of the Boers to make full use of their initial successes in the manner that the Germans (say) would have done, was due in a great degree to the lack of discipline; for we hear of commandos refusing to push on at the orders of Joubert, when they thought it risky.

With the successful nations all through history we find that their armies, in their combined movements as well as in the smaller manœuvres of each section, had been drilled in methods superior to those of their enemies.

The Greek phalanx vanquished the Persian cavalry, but it disappeared under the blows of the Roman Legion, which was as well disciplined as itself, and had the extra advantage of a formation more mobile than the Phalanx. It was by its tactics that the Legion conquered the world.

Strategy is a great and important matter, but tactics puts the coping stone on the edifice. Manœuvre with what skill you please, it is tactics that must settle the question at issue.

For a moment the Roman Legion was arrested in its career of victory. Hannibal saw the use of independent reserves, to be thrown into the fight by a C.-in-C. who has been carefully watching the vicissitudes of the struggle. But when the Romans learned it from him, they in their turn won. Cæsar perfected the legionary tactics, keeping always under his own hand a reserve of both infantry and cavalry, and became the undisputed master of the world.

It is, then, perfection of manœuvres and their application to each field of battle that give victory. The great strategic movements that precede battle lead only to disaster if one is not tactically superior to one's enemy.

The study of the tactical means employed by Frederick the Great would be doubly interesting if one traced, in previous commanders, the successive developments and improvements which these earlier chiefs introduced; for the most successful leaders in tactics have been those who have made use, as Frederick did, of the experience gained in all former campaigns, while rejecting what had become obsolete through improvements in arms, increase of size of armies, increase of national wealth, and who by much thought and study endeavoured to make use of the new facilities and powers put into their hands by moral and material progress.

As an instance of what I imply by the effect of general increase of wealth, I may mention increased facility of strategic manœuvre by the larger number and better quality of roads, the latter improvement making it gradually possible to take the field with heavier artillery than before.

But as the thorough investigation of the pre-Frederick development would take much time, I shall do little more than mention the leaders who laid the foundation of Frederick's system.

Stage of Supremacy of Cavalry.

After the fall of the Roman Empire, infantry began to lose its preponderance on the field of battle, and cavalry became the masters. But gradually English archers and infantry, and then Swiss infantry, came again to the top. In the long wars between England and France the former for the most part won, thanks to their tactics. Creçy was won by the archers; and Edward had introduced discipline and tactics in his army of a kind unknown on the Continent. His nobles fought at the head of infantry, and the English bow proved itself superior to the cross-bow of the French, having twice the rapidity of fire; also the archer could carry 24 arrows to 18 of the cross-bow man. Edward also taught his archers—*i.e.*, his artillery—to shelter themselves with palisades from the charges of the French knights, thus bringing elementary field engineering into its proper relation as a branch of tactics. Edward in his battles, just as Wellington at Talavera, Busaco, Albuera, Waterloo, showed a fine instinct in choosing positions suitable to the number, quality and armament of their men. But it must not for a moment be supposed that these two great men were so often victorious from skill alone in the choosing of positions.

One hundred years after Creçy there emerges the beginnings of standing armies. In France Louis XI. dominates the great vassals, creates a permanent army and organizes the artillery. The Swiss infantry begins to appear on the scene, brave, compact and highly disciplined, and gives the *coup de grâce* to the lingering remnants of chivalry.

Swiss Infantry.

Necessity and experience produced this remarkable force, and instructed them in tactics. Having to cope with the abounding cavalry of Austria, they massed themselves into bodies bristling with halberds, and easily held the mounted enemy in the narrow fields of battle of their native land. Victorious at Morgarten in 1315, at Sempach in 1386, they brought back faith in the compact formation of the phalanx.

Charles the Bold (1476), the last of those who trusted to chivalry, was beaten by this vigorous infantry. All the powers of Europe started to enroll bodies of Swiss as mercenaries; in fact, Swiss tactics became the fashion.

Artillery.

Artillery improved in mobility about this time; its previous sluggishness of movement, combined with its shortness of range and the erraticness of the flight of its projectiles, had caused it to be neglected and scorned. One historian refers to it as rather useful for frightening the enemy's horses. Francis I. won at Marignan (1515) against the Swiss through his artillery, and this arm returned to favour.

Maurice of Nassau (1600) formed his infantry in three lines on the Roman system; and in the field he made the great advance of broadening the front by diminishing the depth, thus winning by getting more fire from his men.

Gustavus.

A new era opens with Gustavus Adolphus (1611). Deeply learned in war, he knew most thoroughly all that was to be learned from the late wars, and had studied most carefully the arms, equipments, organization and tactics of all his immediate predecessors and his contemporaries. Having put himself to the test in his fights with the Danes and the Russians, he brought to the succour of the Protestant League in Germany a brave army, hardened to blows and privations, admirably disciplined and full of confidence in its chief. His enemy's troops were without discipline; the infantry strong and heavy, but without much "go"; the cavalry, clad in steel, were a redoubtable foe, his artillery strong, but lacking in mobility. The Germans employed very deep formations, forming their infantry in battalions up to 45 men deep in some cases, two lines of pikes in front with a third of musketeers.

Now these Germans, at the same time, were no mean adversaries; their generals were well-proved men, who understood the use of the three arms, and the employment of reserves; but Gustavus hit off the necessity of the case when he understood that he must improve on his opponent's tactics—not by strategy only, or even chiefly, was success going to be won. He based his whole organization on two eternal principles, discipline and mobility. He freed his armies from the swarms of camp-followers then customary, and reduced baggage to the smallest possible limits. He instituted regular uniforms, and promoted purely by merit. Finding that a formation of great depth simply wasted the men in rear, and that cannon shots produced great ravages, he arranged his infantry in the field with a maximum depth of six; and as soon as the fight was joined, these became three deep only. His infantry he divided into regiments of 1,008 men, in eight companies, each company having 72 musketeers and 54 pikemen. Two regiments formed a brigade. This was the beginning of the organization that led up to modern ways.

His cavalry he arranged in three ranks; it was to charge, the first two ranks firing a pistol at close range, and immediately to draw swords; the 3rd rank did not fire, but had each man a pair of pistols for the *mêlée*. They were divided into squadrons of 120 horses. The intervals of the squadrons were partially filled with companies of 80 to 200 musketeers, picked men from the best regiments.

The Prussians with their deeper formations could only show as broad a front of battle as Gustavus by putting all their troops into the 1st line, thus yielding to their adversary the advantage of having a supporting and a reserve line.

Gustavus greatly improved the musket, and introduced the cartridge, both for muskets and cannon. He supplied regiments with a few short, light cannon, which fired not balls, but handfuls of bullets—the first beginnings of the battalion machine-gun.

But Gustavus' great merit lay in his appreciation of the advantage of so organizing his troops that they could be moved about freely and without confusion on the field of battle. By this means unexpected developments could be more easily met, and full advantage could be taken of gaps or weak points that might appear in the enemy's front, and of signs of wavering at any point of that front. It is not meant that anything like modern mobility was achieved, but a great step in advance was made from the utter immobility of previous methods. Gustavus' methods, improved upon by Frederick, were far ahead of those of their adversaries, just as the tactical methods of the French under Napoleon were a vast advance upon Frederick; and the advance lay in the provision of greater power of manœuvring on the field. At each stage we find *that* belligerent beaten whose troops, from their training and formation, were more tied down to positions than the adversary's.

About Gustavus it has been said that "no one ever equalled him in leading his army against an enemy, or in conducting a retreat without loss, or in camping his troops or in surrounding his camp with hasty entrenchments; it was impossible to know better than he did the business of fortifying, of attacking and defending. No one knew better than he how to estimate his enemy and to act in the various vicissitudes of war. Taking in a moment his resolution from the appearance of the foe and profiting by every chance, it was impossible to equal him in his manner of ranging his troops for battle."

In fact, his genius saw what ought to be done, and the previous training he had given his men and the formation he had imposed upon them enabled the thing to be done without delay and without confusion.

Turenne.

Turenne comes next as a great tactician, great attention being paid to orders of march which would render quick and easy the conversion from march order to battle array. From this time dates the two lines for the order of battle, with a reserve, at first of cavalry, and later of all three arms.

CHAPTER II.

FREDERICK THE GREAT.

AFTER Louis XIV. (1715) all study of, and improvement in, tactics, ceased for a time; and after Saxe, who defeated the Duke of Cumberland at Fontenoy in 1745, the French armies fell into a state of decay. The Seven Years' War commenced, and Frederick appears on the scene as one of the most remarkable figures of all time. He inherited an army 70,000 strong, of whom 26,000 were foreign mercenaries, the whole being in a state of perfect discipline.

He was no blind reformer; he interfered with nothing that was good, but, seeing himself surrounded by possible enemies, and having a country without any natural strength of frontier, he set himself to work to achieve a position of strength in which he could take the offensive on all occasions. One may suppose that, asking himself the question of the secret of victory, he early arrived at the just conclusion that to win you must contrive to be stronger than the enemy at the decisive point. Pursuing the matter into the question of how to ensure being able to achieve this, he concluded that he must so drill his men that they could be manœuvred on the field of battle with certainty and with a celerity surpassing that of other armies. The sluggishness and inertia of his enemies on the field of battle was something that is to us inconceivable; and in his hands a system of tactics, that went down long after his time before the still superior mobility of the French Revolutionary Armies, gave him the means of achieving the most brilliant victories. In his time, the rest of the armies of Europe had become most cumbrous machines; when they got drawn up laboriously in position anywhere, it was as if they were afraid to move again. They would remain thus in position while a more mobile enemy filed past them in full view for hours, seeking a good point of attack.

Frederick made Prussia into an armed camp, and worked without pause at the perfecting of his forces in every particular, exactly as modern Germany has done. Each campaign taught him faults, which were mended in time for the next one. One of his enemies, Austria, stood still the while, improving only to some extent its artillery; another, Russia, was tactically and administratively in a state of infancy. The French Army became indescribably slack, ignorant and undisciplined. Frederick's discipline, on the contrary, was most severe; in his generals he punished even misfortune, his officers were kept with a tight hand, but everything, short of relaxation of discipline, was done to ameliorate the condition of the soldier.

In the matter of arms, he improved the musket so that it was not necessary to prime the touch-hole for each discharge, and so that rapidity of loading was increased to the point of being able to fire six times in a minute.

His battalion, like his predecessors', had eight companies of fusiliers and two of grenadiers; but the others had brigaded, for administrative purposes, all their grenadier companies into separate battalions, thus making the tactical unit different from the administrative, a thing which he properly saw to be absurd. The battalion was drawn up in three ranks. The drill aimed at accurate alignment and extreme skill in handling the weapons. The former produced a stiffness in tactics which was very marked when the system came to be opposed to the French of Napoleon's time; but it was so vast an improvement upon the tactics and discipline of Frederick's own opponents that it gave him the inestimable advantage of superior mobility.

The line system of advancing to the attack gave a great amount of fire as compared with that obtained from an equal number of men in the deeper formations previously in vogue; and in theory, fire was the thing to be depended upon. But on the field of battle Frederick's men generally settled the dispute with the bayonet. To march to a flank, the line wheeled into the required direction by companies, and preserved company distance, so that line could be rapidly re-formed. Their perfect drill enabled them to march considerable distances on a straight front of several battalions, without confusion or loss of alignment; but it is evident that the success of the whole system depends greatly on the ground being open and unobstructed. As we shall find, the next development, which destroyed Frederick's system, produced a system free from this objection.

The rapid and accurate forming of square was an important part of Frederick's drill; but a very weak point in the whole of the drill was the total lack of any system of skirmishing. In this point also, the next, or French, system proved its superiority.

Frederick's cavalry regiments had each five to ten squadrons, each divided into two troops of 70 horses. He reduced their order from three ranks to two. Understanding that its chief use was shock, he instilled into them the rule of charging without hesitation, and without any halting to fire, while leaving them in possession of their short muskets and pistols, which might often be useful, as on outpost duty. So perfect was their drill that under the great leader Seidlitz they could manœuvre and charge in masses with strict precision. Great care was taken also to exercise the cavalry in all varieties of ground.

Frederick took some trouble to produce a horse artillery, but on the whole he laid chief stress on perfection of infantry and cavalry.

The order of battle was in two lines, infantry in the centre, cavalry on the wings. The 1st line was usually rather stronger in numbers than the 2nd; both had their battalions deployed, the front line having small gaps of eight paces between the units. Between the two lines there would be a few battalions here and there, doubtless as a first support to any part of the front line that showed signs of wavering. The cavalry would also be in two lines deployed on both flanks of the infantry, and at the outer extremities there would often be a few squadrons in column at whole distance, so that an outward wheel by squadrons would form a line looking to the flank. The artillery was massed in great batteries in front of the infantry.

What strikes one most in the whole thing is the extreme rigidity of the system. Compared to the later French system, and to modern ways, it is terribly cumbrous. Imagine four columns, each a mile long, marching in parallel lines for several miles. The two centre columns are infantry, the outer cavalry. The forward half of each column belongs to the front line of battle, the rear half to the 2nd line. The distance

BATTLE OF HOHENFRIEDBERG. 4th June 1745. Pl. I.

Rohnstock
Kauder
Pilgramshain
Hausdorf
Hohenfriedberg
STRIGAU
R
R
R
R

Scale.

1 0 1 2 3 4 5 6 7 8 Miles

Du Moulin's 1st Position	King's 1st Line of Battle
Weissenfel's " "	do. 2nd do.
do. 2nd "	Austrian Main Line, 1st Position
Austrian Reserve R R R	do. do. , 2nd "

H.M.J.

BATTLE OF SOOR – 30th Sept. 1745.

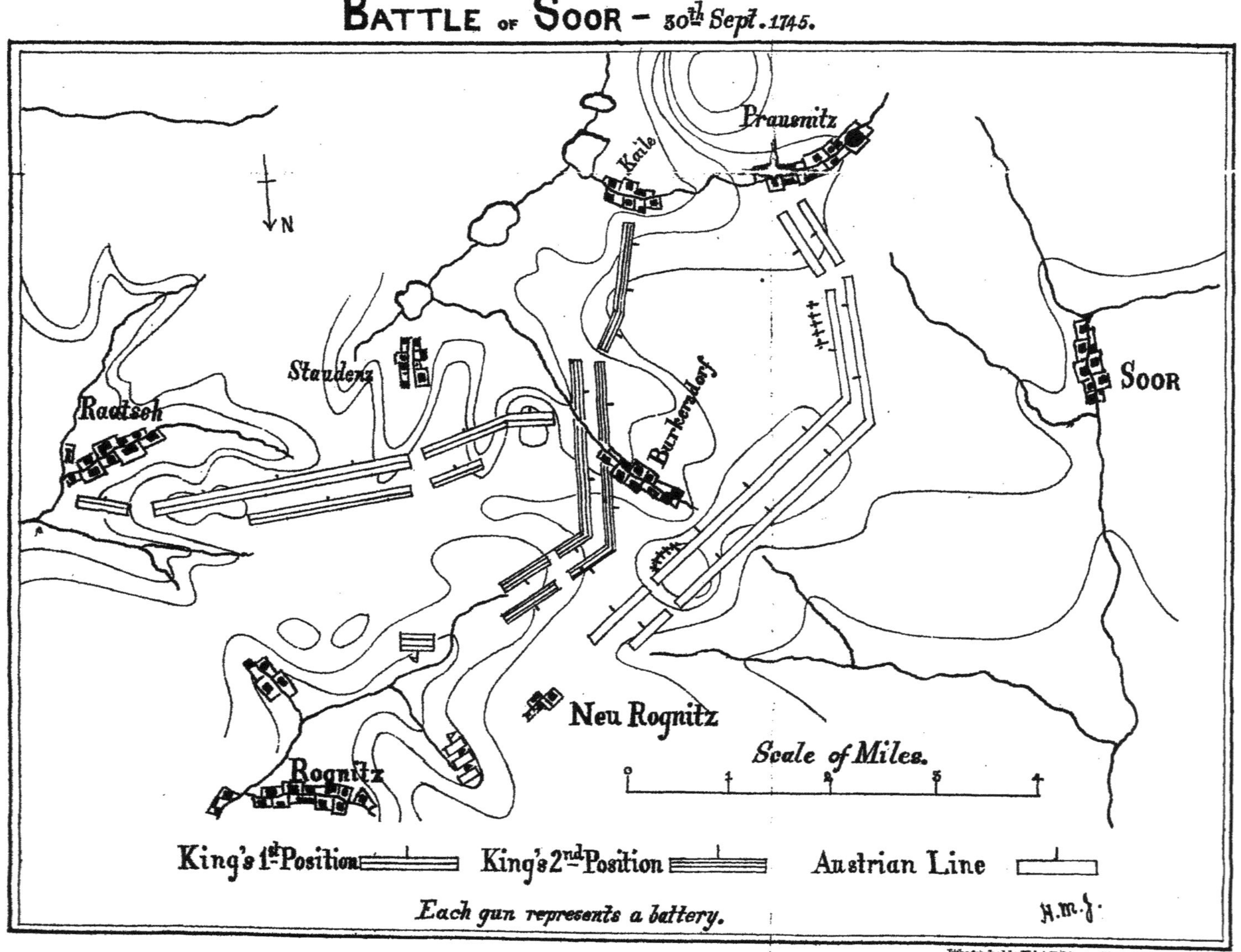

apart of the columns is the exact front that the forward half of each will present when in line of battle; and if, on forming line of battle, there are no gaps and no troops are to be crowded out, this distance must be rigidly preserved. The difficulty of achieving this feat even on an unobstructed plain is obviously great, and could only be surmounted by the most perfectly trained and disciplined troops. On rough and intersected ground, the movement must either end in dislocation and loss of distance, or there must be very frequent halting of portions to let those who have been delayed or diverted regain their places. When the pre-arranged place was reached at which line of battle was to be formed, the leading company of each half of each column wheeled to right or left as ordered, followed by the companies in rear of it. As soon as the companies were all facing in the new direction, the order was given to wheel into line. Imagine the confusion if the distances apart of the long columns had not been kept!

Frederick's aim was to reach the enemy in oblique order of battle, *i.e.*, the line when formed would have its centre roughly opposite the hostile flank, the line itself standing obliquely to the enemy's front. An advance in this way upon an immobile and sluggish enemy who motionlessly awaited attack would effect the required object, viz., bringing a preponderance of force against a decisive point.

One is always tempted to ask why men were not in those days taught to fight more individually. I suppose the reason is twofold. In the first place, the private soldier or trooper was usually an utterly uneducated and ignorant man. The best you could expect of him was that he would obey orders if he had previously had plenty of drill and discipline under the fear of punishment. In the second place, though firearms had been long in use, their range and accuracy were still so limited that the arbiter of the fight was still the bayonet and the sword. For this part of the fight, close order in well-kept lines without gaps was the important thing, handed down as a perfectly correct legacy from the times of the Roman Legion. But already, as early as the Seven Years' War, there were signs that the firearm was going to be easily the arbiter of the fight; for in France we find Mesnil-Durand advocating skirmishers, and expecting to so far distract and demoralize the enemy by their use that the mere appearance at close range of the solid masses in support would move the enemy to give way. And we find the Austrians in the Seven Years' War itself employing a few of their light troops as skirmishers, with the object of harassing (be it noted) by fire alone.

But the really efficient use of skirmishers had to wait till the French put their Revolution armies into the field; and we shall see the reason of their special efficiency when we come into the next stage of the history of tactics.

Hohenfriedberg or Strigau. *v. Plate* I.

Of Frederick's many battles, I shall choose first that of Hohenfriedberg, won by him on 4th June, 1745, *i.e.*, during the war of the Austrian Succession. Frederick had coveted Silesia and at first obtained it; but by April, 1745, things military and political had taken such a turn that Frederick found himself isolated, with every appearance that the retention of Silesia was going to be no easy task, for the number of his enemies, the Austrians, had been reinforced by the whole Saxon Army. In this strait he asked from France, his ally, that she should operate on the Danube and thus take some of the pressure off Prussia. But Louis XV. remembered how Frederick had left him in the lurch in 1742, and prepared to push the war against the English in Flanders, where Saxe had on 10th May gained the battle of Fontenoy.

At the end of April the Prussian Army was much scattered; the bulk of it, and the King's headquarters, being between Breslau and Neisse. By the 24th May it was concentrated in camp at Frankenstein, to the number of 65,000. Austrian cavalry and irregulars had been about the country for some time, and had been severely handled by Frederick's men. Their army, Austrians and Saxons combined, had passed through the mountains at Landshut, and numbered 85,000.

Frederick spread among the enemy the false information that the Prussians were afraid and were about to retreat out of Silesia; in reality he was preparing the very position where he intended to fight a victorious battle.

On the 29th May he was in camp at Reichenbach, a short march from Schweidnitz; on the 1st June he passed through this fortress. The advanced guard, of considerable strength, moved on Strigau; the army encamped in the plain behind.

The hostile advanced guard appeared on the heights of Freibourg. The Prince of Lorraine had marched the allied army from Landshut with the intention of descending into the plain of Freibourg, Hohenfriedberg and Kauder. The King made a careful reconnaissance of the whole surroundings, and employed three days in road-making, having, as Wellington often did, fixed the battlefield and thus prepared for the struggle.

On the 2nd June, the allied generals had a council of war. Their troops had, so far, discovered of the Prussians nothing but a few small detachments, Frederick hiding his men away in ravines in order to lure the enemy on. The Prince of Lorraine accordingly resolved to act boldly, and ordered the advanced guard to rush Schweidnitz, and to pursue towards Breslau, whither he supposed the Prussians to be retreating; a division was to take Strigau, and move on to blockade Glogau. The Prince had no idea that he had on his hands an enemy 60,000 strong, who had no intention of yielding a foot of ground.

On the 3rd June, from a height in front of his main camp, Frederick sees eight columns on the march; their right on the burn, their left (the Saxons) stretching to Pilgramshain. The Prussian Army now executed a remarkable night march. Camp was struck at 8 p.m., and in two columns the 60,000 men marched. This performance, executed in such silence that the enemy never discovered it, speaks volumes both for the perfect reconnaissance of the ground and for the excellent mobility of the Prussian Army under Frederick. It reached the bridges of Strigau at midnight, and there rested for two hours. Frederick called together the C.O.'s and gave them the following verbal orders:—"The army will march off continuously by the right, on two lines, to cross the stream; the cavalry will form up opposite the enemy's left, which is about Pilgramshain, Dumoulin's Corps covering its right. The infantry will range itself to the left of the cavalry, facing the woods of Rohnstock. The cavalry of the left wing will protect its left on the burn of Strigau, well away from the town. The 10 squadrons of dragoons and the 20 of hussars, forming the cavalry reserve, will be stationed behind the centre of the 2nd line, to be employed as may be required. A regiment of hussars will form itself into a 3rd line to each wing of cavalry, to cover its flank and help in the pursuit. The cavalry will charge impetuously, will strike at the enemies' faces, and will make no prisoners in the heat of the action. When it has overturned and dispersed the hostile cavalry, it will return at once against the infantry, and take it in flank and rear. The Prussian infantry will advance at the double, and as far as circumstances permit, it will attack with the bayonet; if firing becomes necessary, it will only fire at 150 paces. When any villages are come across which the enemy are not in possession of, the Prussian generals should occupy them and line their edges with infantry. This may sometimes help in gaining the hostile flank. But no troops are to be placed in houses or yards, in order that nothing may tend to prevent rapid advance or pursuit."

The army now put itself in motion, Dumoulin crossing first. His division became at once aware that hostile infantry were on a height in front of them. These proved to be the Saxon corps of the Duc de Weissenfels that had been ordered to take Strigau. Dumoulin sheered off to the right and climbed the high ground, getting thus on the Saxon left flank, completely surprising them. The King at the same time took up a battery of six heavy guns, and confusion was at once produced in the Saxon ranks. The Duke, thinking he had to do with a mere detachment, pushed up all his men; but the unexpected artillery fire, an attack in front by the rest of Dumoulin's Division, and two charges by the cavalry of the right wing, put the whole Saxon 1st line to flight. A rapid advance of the Prussian masses brought them on the 2nd line, which they cut to pieces with great slaughter. Thus the whole Saxon corps was dispersed before the Prussian line of battle was completely formed. The Prince of Lorraine was at Hausdorf, and heard the firing, but took it to be the capture of Strigau by the Saxons. But finding presently that the Saxons were in flight all over the countryside, he ordered a general advance. Between Rohnstock and the Strigau is a plain with copses here and there and small ditches. In the plain some hard fighting took place, the Austrians making intelligent use of the ground, but the Prussians, with bayonets fixed, were not to be denied. The right wing, having disposed of the Saxons, was to endeavour to get round the enemy's left, but the marshy ground delayed it badly, and meanwhile the left was gaining ground; but ten squadrons of the Prussian left, before they had time to form after passing the brook, were badly handled. They should have crossed under cover of their infantry, for cavalry caught forming up are an easy prey. However, the ten squadrons succeeded in rallying, and took a good part in the battle.

The left wing was now all across; one regiment managed to slip forward into a village near the stream, and got the Austrians in flank. At the same time a cavalry brigade from the 2nd line poured through an interval in the Prussian infantry, and charged impetuously; 4,000 Austrians laid down their arms to this charge.

By this time the right had got well on to the Austrian left flank, and the battle was decided. The rout was complete. The Austrians lost 4 generals, 200 officers and 7,000 men taken prisoner, 60 guns captured, and 8,000 men killed or wounded. The reserve of the Austrians took no part in the fight; it was very strongly posted on the heights of Hohenfriedberg. The King occupied the heights of Kauder, but his men had had enough of fighting after their night march; he let the reserve escape into Bohemia.

Frederick's tactics in this fight do not thus show the great formality of the regular linear system, and it is not a very marked case of the oblique order. But his method is as ever. He saw, from personal reconnaissance, the opportunity an unexpected night march would give him of falling upon the enemy's left wing in force before it could be supported. Having disposed of that, raising the spirits of his men and depressing the enemy in proportion, he was more than equal in number to the rest of the foe. Moreover, he had planned all this, while the enemy was advancing blindly, not knowing whether he had the whole Prussian Army to meet or not.

Battle of Soor. *v. Plate* I.

On the 30th September of the same year, 1745, Frederick beat the Austrians again in the battle of Soor. After Hohenfriedberg, he had slowly and cautiously followed the defeated Austrians into Bohemia; but the detachments he had felt compelled to make had reduced his force below 25,000 men. Some conferences about a general peace occupied three months, and then Frederick began to see that Austria, at least, would not make peace. Threatened by greatly superior forces, he retreated in September towards Silesia, and was, on the 30th of that month, encamped to the south of Trautenau. The Austrian Army, of twice his strength, came up with him, and he thought it better to attack than to be chased through the defiles of the Bohemian mountains.

His troops had received orders the night before to march northwards at 10.0 next morning; but at 4 a.m. word came in that a great body of Austrian cavalry was at hand, followed by infantry that was already deploying. This was a genuine surprise, but Frederick was fit for the emergency. When he went forward to reconnoitre in person, he found that the enemy would deploy across his right flank, if he let them.

v. Map.

The position was as follows:—Frederick's right was between Prausnitz and Soor, facing southwards; but this part of the force seems to have joined the main body before the Austrians had interfered. The position of the rest was extended from a hill above Bürkersdorf to a very steep ravine near Radtsch, also facing south. The Austrians gradually got into position from near Prausnitz to a point not far from Neu Rognitz, the line forming a pretty deep re-entering angle. Evidently Frederick would experience the effectiveness of the oblique order, if he remained as he was.

The Austrian commander fully expected the enemy, caught at such a disadvantage by much superior numbers, to retreat precipitately, when he hoped at least to become owner of most of his baggage. But in spite of the danger of manœuvring in presence of an enemy already almost in order of battle, Frederick ordered a change of front to the right. The splendid discipline and drill of the Prussians enabled this delicate manœuvre to be carried through with incredible speed and precision; before the Austrians could complete their formations, Frederick was facing them on a line almost parallel with theirs; but he could only approach theirs in length by extending on a single line, and keeping back only a very small reserve. The right wing, which had to do a "right turn, right wheel," was the whole time under the fire of a powerful battery of 28 Austrian cannon, but not a man left the ranks of his own accord. The Austrian cavalry of their left wing had posted itself, in three lines, in a position so cramped that the lines were not more than 20 paces apart. Frederick, seeing this, ordered Buddenbrock to charge; the Prussian cavalry had only just arrived in position, and was apparently still in column; it charged as it was, without firing, received the customary discharge of the Austrian cavalry, drove its 1st line on to its 2nd almost before the Austrian troopers had time to draw their swords, drove these on to the 3rd line, and cut them up badly. They numbered 50 squadrons.

Frederick's right flank brigade, cheered by the sight, went at the hostile guns in a frontal attack, a most rash proceeding, and were driven back with some loss; but the little reserve of five battalions arriving in the nick of time, the five repulsed battalions re-formed beside them, and by a joint effort the ten battalions captured the height and the guns.

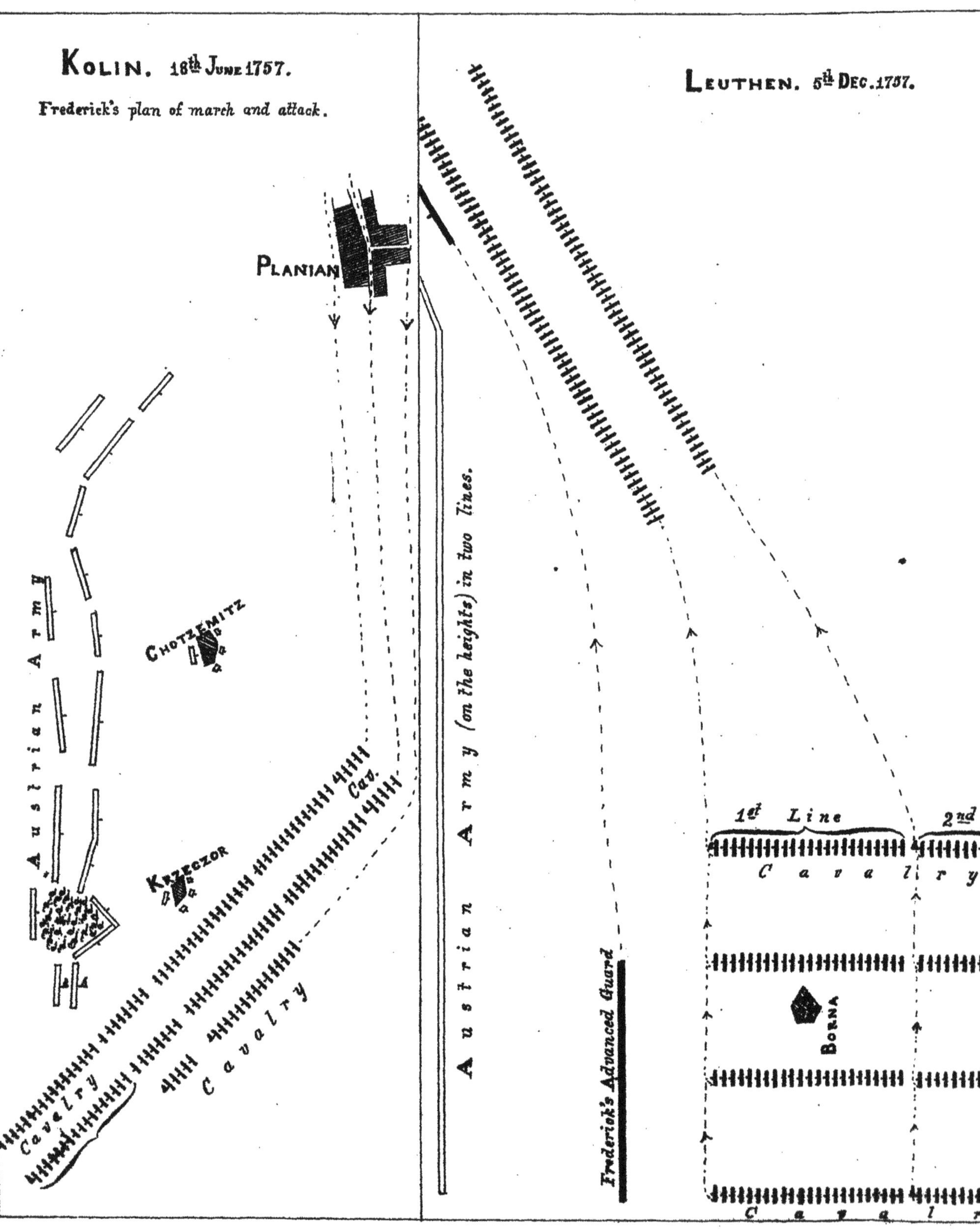
KOLIN. 18th JUNE 1757.
Frederick's plan of march and attack.
PLANIAN
Austrian Army
CHOTZEMITZ
KRZECZOR
Cav.
Cavalry
Cavalry
LEUTHEN. 5th DEC. 1757.
Austrian Army (on the heights) in two lines.
Frederick's Advanced Guard
1st Line
2nd
Cavalry
BORNA
Cavalr

Pl. II.

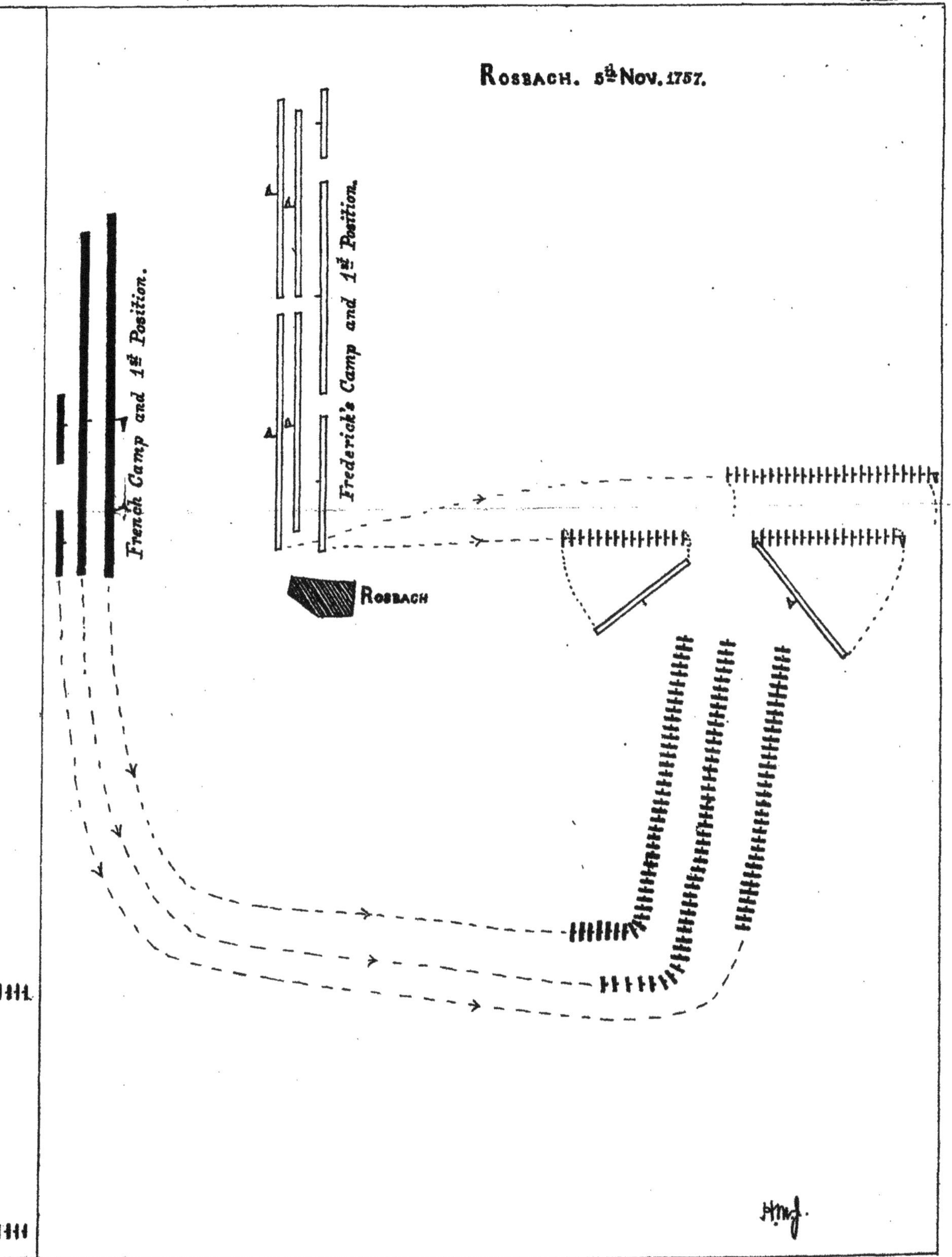

Lithographed by W. & A. K. Johnston, Limited, Edinburgh & London.

A great column of Austrians was now perceived coming down from their right on the village of Bürkersdorf. Like lightning, Frederick anticipated them by doubling a battalion into it. The clumsy column moved slowly, giving Frederick's men time to set fire to the houses along the far side; sheltered behind this novel obstacle, the Prussian left moved down and plied the column with musketry fire, dispersing it.

Meantime, the Prussian cavalry on the right had driven the Austrian horse into impossible ground; leaving two regiments to support the right wing, 20 squadrons were moved across to threaten the hostile right.

At the same time, the infantry of the Prussian right had, as we have seen, gained a footing on the Austrian heights at their left flank; they proceeded to roll up the hostile line. As soon as this movement was well in hand, the Guard regiment at the Prussian centre made frontal attack on the enemy's centre, where they held a steep and wooded hill.

The Austrian line was broken and their flank forced. Bit by bit they gave ground, fighting at every opportunity among the tumbled ravines and heights which make up the landscape; but the Prussians would not be denied, and the retreat gradually degenerated into a route. The cavalry, now on the Prussian left, caught and made prisoners of 1,700 men; but the Austrian cavalry of this wing had retired early beyond Soor, into the depths of the great forest beyond. At that village Frederick called off his men; 20,000 had defeated 40,000, with a loss to themselves of 3,000. The Austrians lost 22 guns, 30 officers and 2,000 prisoners, 6,000 killed and wounded.

This is one of Frederick's most remarkable battles; the plan shows that there was something in it of the oblique order. Threatened by an enemy with an attack perpendicularly on his right flank, he turned the tables completely, thanks to the grand discipline and training of his army. Unable to get his "oblique order" in perfection, he produced the same effect by attacking in échelon from the right, and placing there the small reserve he had available. At Hohenfriedberg, he also had advanced in *échelon* from the right and overwhelmed the Saxons before their centre and right could assist. In both cases, he refused his left till the blow of his right took effect. At Soor, the attack from the Austrian right was correct, if it had been done earlier and in better formation; but of course their great fault was the woful pusillanimity with which they allowed their nimble adversary to get 20,000 men wheeled at right angles to their original line without interfering with the movement more actively than by a smooth-bore bombardment at longish range.

The Baron de Jomini, who has left us a very full and careful study of Frederick's battles, lays down certain maxims that he arrives at from the study; these maxims serve to fix in our minds the stage of development at which strategy and tactics had arrived.

He says, after describing the battles of Kolin and Prague and the two of which I have given an account,—"An army, immobile in its position, is susceptible of being turned or overwhelmed at one extremity; and the only means of opposing such a move of the enemy is to manœuvre in the same direction as the enemy, that is to say, offensively and so as to threaten his line."

Plainly, this implies active reconnoitring of his movements from the very outset; and this is what Frederick himself did at Rosbach, when the French, finding him in position and trying against him the artifice of marching past his front and round his

v. Plate II.

flank, found also that he was aware of their march, that he was ready for them at the point where they hoped to gain the advantage, and that he there attacked vigorously the head of their long marching column before it had time to deploy.

Referring then to cases, such as Kolin, when villages lay a short distance in front of the main position, he says :—"An army, posted behind villages, ought to use them to cover its front; their outer edges and their issues should be garnished with infantry and artillery; the main line should be within range of them, to help the holding of them and to cover the retreat of their garrisons at need; these posts, however, should be usually looked upon as advanced posts, not containing too great a proportion of your force, and seldom worth holding *à outrance*." He seems to refer to Blenheim at this point; Marlborough, ignoring two villages in which the French had put 20 battalions, won by attacking at another point—Jomini thinks these battalions, if at the point of danger instead of stuck in the villages, would have turned the tables on Marlborough.

At Kolin, where Frederick was beaten, the Austrians held the villages in front of their line—held them with light troops, who harassed the processional march of the Prussian column, and so disarranged his plan that the oblique position was not achieved and the attack was made piecemeal, and with gaps. If Frederick had done this dangerous part of his manœuvre in the dark, as he did at Hohenfriedberg, he might well have won.

Frederick was not, as a rule, content to get obliquely on to the nearest, and easiest, flank of the enemy; by aiming at the other flank, he often achieved two things—(1), he attacked where the enemy was not expecting him; (2), he made his attack strategic as well as tactical, by threatening their communications and means of retreat; but at the battle of Prague he went for the distant flank, because the near one was resting on the fortifications of the town, and was very strong also in front.

From a consideration of Kolin, Jomini lays down the following *dicta* :—"If it be recognized that the most advantageous attacks are those which operate by a concentrated effort on a single extremity of the hostile line, it becomes indispensable to take measures to gain that extremity without unmasking your intention; if you neglect this precaution, the enemy will be able to follow the march of your column, always to present his front to you, or to take your column itself in flank." At Rosbach, Frederick effected the former against the French attempt; at Kolin, the Austrians used the latter from the villages against him.

"You must then conceal your march, either by means of darkness" (as Frederick did at Hohenfriedberg), "or by the help of the contour of the ground" (as Frederick at Leuthen), "or by a feint against the hostile front, for the purpose of fixing his attention there." This last was effected to some extent by the threatening attitude of Frederick's advanced guard after it passed through Borna; and then the march shown in the sketch went on unmolested, covered partly by the nature of the ground and partly by the fog which prevailed. The troops marched in columns of companies at wheeling distance; the leading companies of the four columns of the 1st line wheeled simultaneously to the right, and were followed by each company as it reached the same point, so that the whole four columns presently became a single column. The advanced guard, moving along parallel to this great column, became flank guard for the whole movement. The heads of the columns of the 2nd line wheeled at the same time as the 1st line, and

v. Plate II.

marched as shown. It is said that Frederick experimented in this fight with some new long cannon he had manufactured, and found them very effective.

The Prussian cavalry attacks at Kolin and Reichenberg show the danger of taking cavalry near woods without being sure of what the woods contain; and generally, the danger of imperfect reconnaissance. The Prussian cavalry succeeded in its charges against the Austrian horse, but immediately after suffered severely, and was completely checked, by the fire of infantry along the edges of woods.

In the matter of sieges, several of which Frederick undertook—among them, Prague and Olmütz—there are some important lessons. Jomini says:—"The initiative being the surest guarantee of victory, an army covering a siege (*i.e.*, preventing the enemy's field army from interfering with your siege operations) should never let itself be attacked by the enemy; it ought, on the contrary, to anticipate attack, for it is by beating the succouring army that you will hasten the fall of the fortress; if the enemy presents himself in such strength that you cannot attack him without using your siege troops, you must raise the siege without hesitation, and unite your forces. When you have beaten the field army, pursue it far enough to be free of it, and take up the siege again, but do not draw back your covering army to the vicinity of the place; the covering army should always meet the enemy as far away from the place as possible."

Napoleon's proceedings at Mantua in 1796 are a model in this connection.

The following maxims suggest themselves from the battles of Rosbach and Soor:—"When an army foresees an attack" (*i.e.*, an attack from a dangerous direction), "it must never wait for the enemy; it must on the contrary anticipate it without a moment's loss of time, and attack vigorously; if the enemy tries a turning movement on the field of battle, an active reply is necessary."

But it may be said that Wellington awaited the enemy at Waterloo, that he would quite possibly have been beaten, if he had quitted his position and attacked. It was with this and similar cases in view that I added to Jomini the words "an attack from a dangerous direction," such as a flank. Napoleon made no such attack; doubtless, Wellington would have known how to meet it if he had.

We have now obtained a good idea of Frederick's system of tactics. Stiff and formal as his method was, and depending for success on the still stiffer immobility of his adversaries, it was a great advance on former methods. There was a clear understanding that the attack should be made at a decisive point., *i.e.*, the flank, and that the force brought to bear at that point should be superior to anything the enemy could mass there in time to oppose it. It was necessary, therefore, for success that the march into the required oblique position opposite the hostile flank should be either made unseen by the enemy, or, if that was impossible, then that the enemy should be content to remain still and await attack. The experience of previous wars had taught Frederick that he could usually depend upon this quiescence on the part of the enemy. So poor was the drill and training of the hostile forces compared to his that they were afraid to move from the position they had taken up as soon as the Prussian King hove in sight. But at Kolin, and still more at Rosbach, the inherent danger of the stiff linear tactics of Frederick was clearly shown. When, as in these cases, the defender attacked during the flank march of the attacker, disaster occurred to the latter. And the reason of this lay in the unwieldy stiffness of the linear system. Its success depended on an extreme accuracy of distance-keeping, on freedom from molestation by the enemy, and on the most perfect discipline and obedience to orders. If these three conditions prevailed, the army reached the chosen position, formed accurate line by a wheel of the units of the great columns, and success was assured. But a double line, of approximately equal strength at every point, was specially strong nowhere. If the columns be caught on their march, and compelled to form line parallel to the enemy, then the whole condition of tactical success is lost; for we found that that condition is to be in special strength at a decisive point. But when the march was allowed to complete itself without molestation, and the great line came down obliquely on the hostile flank, the necessary superiority of force at a decisive point was achieved; for although the line may not have been specially weighted at the part opposite the hostile flank, its advance enables it to bring a converging and superior fire, and then a converging and superior charge of bayonets. When the enemy, as at the battle of Prague, drew back their right wing at right angles to his original front, their position was no better than before; the salient angle in their line was a very weak point, and the whole of the folded-back wing, as well as a good deal of that part of the line that had kept its original position, were open to enfilade by the attacker's artillery. This latter result, to be sure, depended altogether upon the lie of the ground; but the weakness of the right-angle salient remained. And further, with the dense formations of those days, a retreat of the troops at the angle could hardly be effected without mingling and confusion.

Moreover, if a line of Frederick's kind be suddenly attacked during its march, and be forced to form front parallel to the enemy, a dash made by a strong column on any part of it is likely to effect a breach in the line. And we shall find that the next development of tactics, by the French namely, was eminently adapted for this purpose.

MESNIL DURAND.

Pl. III.

FIG. 1

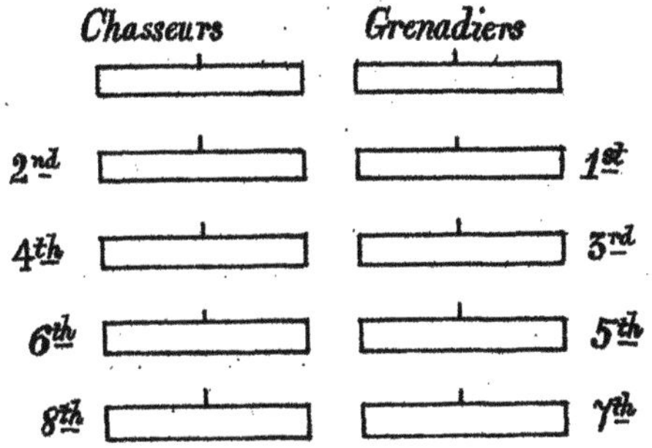

Chasseurs *Grenadiers*

2nd 1st

4th 3rd

6th 5th

8th 7th

Battalion in Line.

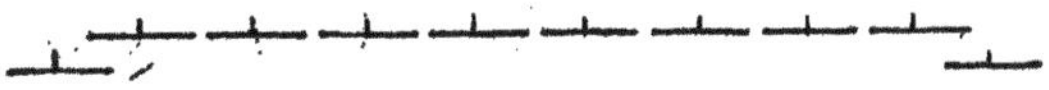

FIG. 2

Brigade of 6 Battalions deploying into line of Battalion-Columns at full distance.

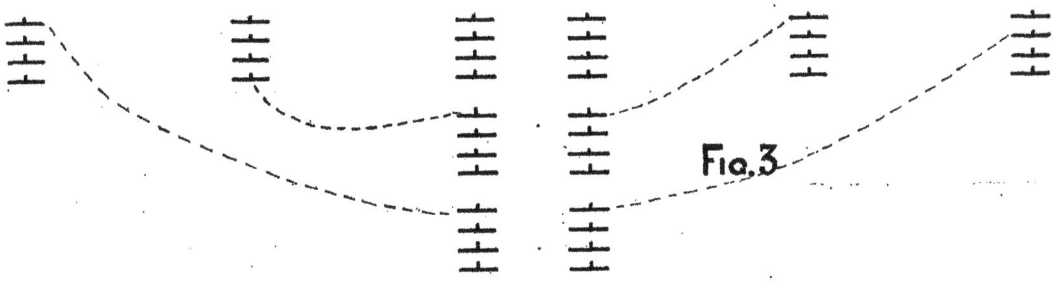

FIG. 3

FIG. 4

Ordre demi-plein.

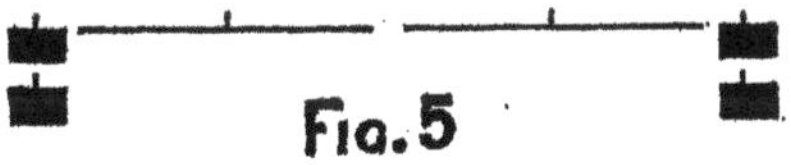

FIG. 5

Brigade in ordre demi-plein.

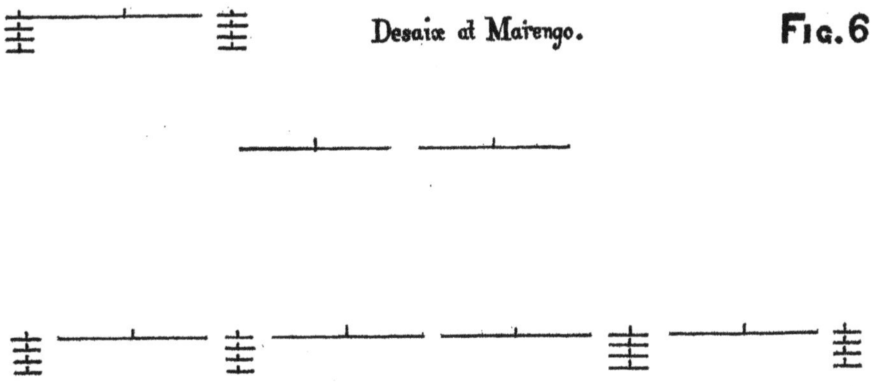

Desaix at Marengo.

FIG. 6

W. & A. K. Johnston, Limited, Edinburgh & London

CHAPTER III.

THE FRENCH REVOLUTION, AND NAPOLEON.

WHEN Louis XVI. mounted the French throne in 1774, the Prussian Army was the model for the whole of Europe, just as, since 1870, the German organization has been copied all over the world. The details of Frederick's system were searched by military men as a means to success. But in France there was a soldier, Mesnil-Durand, who saw that the linear system of Frederick could not be the last word in tactics and who set to work to produce something better. He saw the weak points in it—the excessive stiffness and formality of the evolutions, and the ease with which a manœuvre on the Prussian system could be upset, whether by accidents of ground or by active work on the part of an enemy.

About 1774 Mesnil-Durand presented his system of battalion drill. The ordinary formation of a battalion of ten companies was to be the column of double companies in close order. The leading double company was composed of chasseurs and grenadiers, who skirmished. When the battalion was to form line, it deployed outwards, on both flanks of the 2nd double company—obviously a very rapid method. The leading double company separated into two, and took post on the wings and slightly in rear of them. Movements in advance or retreat were made in column, the centre double company of the line setting off and the rest, in pairs, forming behind it. The skirmishing companies doubled to the head of the column, and led the advance.

Plate III. *Figs.* 1 *and* 2.

In the same way a brigade of four battalions moved in a single column, and deployed into a line of battalion columns to right or left as ordered. The regular brigade of two regiments, or six battalions, moved in two columns of three battalions each and deployed outwards; or moved in a single column, and deployed outwards by alternate battalions. It will be observed that these movements were planned so as to be executed in the least possible time—an important matter, if within range of the hostile artillery.

Fig. 3.

The deployment was always such that the battalions reached their position at full distance so that line could be formed, if required.

For the attack there were to be battalion columns; and squares were to be formed of single battalions. The intervals between the columns were filled with swarms of skirmishers supplied by the chasseur and grenadier companies, who would clear rapidly out of the way when the battalions had formed into line, or when the gaps were about to be utilized to deliver a charge of cavalry, or to bring up guns.

Such a line of battalion columns, with its skirmishers and its artillery, would advance upon the enemy in this formation; on getting within 150 yards or so, the commander would judge whether the fire of the guns and of the skirmishers had shaken the enemy sufficiently to render an immediate charge likely to succeed; if he thought it had, the columns would fix bayonets and charge by order at that part of the hostile front where success seemed most likely, or where success would be most decisive on the fate of the rest of his front. And observe here the tactical advantage of this method; the battalion columns, by a slight wheel, could be all brought towards the same part of the hostile front, thereby bringing preponderance of force, without delay, suddenly against a chosen part. Now if those battalions had been advancing in line, no such manœuvre could have been performed until battalions had been dislocated from their places in the line and brought behind the one that was facing in the required direction—an operation that would take time, that would have to be performed under fire and that would demand much discipline from the troops.

If, on the contrary, the enemy were showing a staunch front, and the success of a charge looked doubtful, each battalion, or at least the one opposite the chosen point, would be deployed rapidly into line and commence firing. A crisis would shortly arise, and soon one side or the other would establish a superiority. The following are Mesnil-Durand's own words:—"Columns bring together great force on a small front; they alone can, by reason of the smallness of their front and of the relative size of the intervals, preserve the necessary freedom of movement for the passage of the columns of cavalry that support them, and for the employment of a numerous artillery; but these columns must be connected together by thick chains of skirmishers."

We must remember, of course, the short range and the inaccuracy of the weapons of the day; it is stated by a French officer, writing early last century, that the 1777 model of French musket, in the hands of the best shot of the company, missed a 6-foot square target almost every time at 150 yards and only made 50 per cent. of hits at 75 yards.

In the statement I have just quoted there is a clause that incidentally shows the great gap between the tactics of those days and of modern times—it is that the cavalry support the infantry. It is as well to understand quite clearly the reason of the change. Assuming that a cavalry charge of those days was made at the same rate of gallop as now, the cavalry would be under the fire of the hostile infantry for only about ⅛th of the time during which it would have nowadays to endure it—unless, of course, the ground favoured concealment; and the rapidity of modern fire is such that it is no exaggeration to say that each trooper in charging now would have to run the risk of 50 bullets for every one he risked in 1796.

Mesnil-Durand's proposals were, of course, combatted even in France by the believers in Frederick's system. In arguing with them he further says:—"Infantry has two weapons to fight with, and ought to have two distinct formations in consequence; the deployed order is the better for fire, the order in column for manœuvring and attacking the enemy. In every case, without exception, we must employ that order which is most advantageous for the circumstances of the moment."

He proceeds to repeat that the order in line of battalion columns is the "primitive order," as he calls it, meaning that the army should come on to the field of battle and approach in that formation to the verge of being within range. Then circumstances are to be the guide as to the method of the further advance. He argues that the thin and shallow line system, following its natural tendency, is apt to induce the men to halt and burn powder, while the order in columns has the tendency to get the men to advance to close quarters. The deployed order in two ranks is weak everywhere, chiefly on the flanks, and is incapable of the least manœuvre. The cavalry and artillery can seldom support it effectively, except on the extremities of the flanks. In the French system, on the contrary, the front is, or can be quickly made, strong anywhere; it threatens the enemy's line by the fire of the skirmishers, the charge of the columns, the support of the cavalry. Observe, however, that we have not yet by any means reached the stage in which the decisive attacking body is a line of riflemen in skirmishing order, as we find it now. Mesnil-Durand's skirmishers are merely a useful adjunct; they serve to keep the enemy's attention occupied while the columns advance to charging distance, or while these columns are deploying into line, in the case where they find it necessary to ply the hostile front with volleys before venturing upon the charge. Thus the skirmishers were not strictly part of the line of battle; they cleared out of the way to make room for cavalry or artillery and also when the column formed line. But when the system was accepted by the French, as it soon was, and came to be tried in the field in actual fighting, the usefulness of the skirmishers and their comparative freedom from loss, on account of their extension and their ability to use the shape of the ground, gradually led to a practical recognition of their importance. And the intelligence of the men who poured into the ranks of the French armies of the Revolution further enhanced the reputation of the skirmishing system. Their opponents, adhering to the stiff formations of Frederick, could make nothing of the lawless activity of these skirmishers, who, tied down to no ranks, scrambled about in the hills and over rough ground and appeared where they were least expected.

We ourselves have been lately surprised and troubled in our South African War by just such another unexpected appearance. I refer, of course, to the astonishing appearance of heavy artillery on the tops of rugged and pathless mountains. That development has also come to stay. The Boers, regardless of what the regular armies of Europe had come to think reasonably possible or impossible, dragged great Creusot guns of 10,000 yards range up to impossible places. Everyone will do it in future; the Russians and Japanese have both been doing it.

To return to Mesnil-Durand. He proceeded to give several varieties of orders of battle, combining line formations with column, among them the "*ordre demi-plein,*" as it was called, an order very frequently used by Napoleon's generals. But Mesnil-Durand himself only admits these modifications as a concession to the prejudices of the time; and it is plain that this "*ordre demi-plein*" militates against the chief advantage that Durand claimed for his system, viz., complete freedom of movement for each battalion column and room for the free and effective support of cavalry and artillery *at any point.*

v. Plate III., *Figs.* 4 *and* 5.

An army assuming his system as its "primitive order" will, he says, march and manœuvre with the greatest ease and rapidity, and will have no need of those eternal alignments which the linear system cannot dispense with. When the battalions are placed in two lines, those of the 2nd line will, in his system, be opposite the middle of the intervals of the 1st line. The base, then, of this system is little battalion masses covered by clouds of skirmishers. In the later wars of the Empire, when France was becoming exhausted and the ranks were full of raw recruits, it was found necessary to enlarge the size of the columns, so as to give more confidence to the untrained soldiers. Brigade columns began to be used; and at Waterloo even a division column. This, of course, was a vicious application of the system and was only useful at all against inferior troops of doubtful pluck and poor discipline. Before *them* the threatening appearance of the vast mass of men was sometimes enough to dissolve the hostile ranks before the great column reached them. But the British infantry, splendidly drilled, brave and under perfect discipline, were not to be frightened by show. They instinctively recognized that the greater part of the mass could do no fighting, and that very few of them indeed could do any firing, so they stood in lines and fired steady volleys. Sometimes as at Waterloo the portion of the lines to right and left of the point at which the head of the column was aiming, wheeled inwards and smote the flanks of the column with volleys. At the first sign of wavering on the part of the column, the British line would fix bayonets and charge ; and so good was their discipline, that, on their best occasions, the success of a charge did not cause them to lose their heads and start a wild and disorganized pursuit, which would have been very dangerous as long as the enemy had reserves in rear. On the contrary they would pursue only a few hundred yards, and return at the command of their officers to their original position. This would go on until the C.-in-C. saw that the time had come for a general advance of the whole line.

The British were thus at a further stage of training and tactics than their French enemies, but the British system was only possible to exceptionally well-drilled and brave troops. It is worth noting, too, that British musketry fire had a great deal to say to our successes. It is allowed on all hands that the British musketeer of that day was a much better shot than any Continental ; and the same is true of the British artillery.

The great credit of Mesnil-Durand's system was its simplicity. A single "primitive order," and simple and rapid movements for two kinds of fighting, fire-fighting and bayonet-fighting. When the second of these was being used, it was easy to direct the shock of several battalions against the same section of the hostile front. And in the system we find for the first time the distinct and deliberate use of skirmishers. Finally and as a last improvement, he wished to use a single company front for his battalion columns, a method that was in the next century adopted by the Prussians and the Swedes, but not by the French themselves in the Revolutionary Wars.

Mesnil-Durand finished by converting to his views the French military authorities, even those who had at first opposed them. In 1776, and more fully in 1791, his system became the official drill in France. Napoleon as a cadet was brought up in it, at his Military School.

At the beginning of the Wars of the French Revolution the French armies had a most difficult task before them. Themselves hurriedly raised and partially trained, they had to fight against the veteran and highly disciplined troops of Prussia, Austria, and England.

French Armies of the Revolution.

"Accordingly," writes Marshal St. Cyr, "it would have been the height of imprudence, at the beginning of 1793, for the Republican armies to engage in what is called *une bataille rangée*, except when they had, as at Jemappes, a great numerical superiority. (This fight took place in Belgium in 1792 against the Austrians, and the French won). What is called *la grande guerre*, that of open countries, was to them interdicted."

We see, in fact, the beginning of the end of the "war of positions." But these young and enthusiastic troops, rapidly trained in the easy drill required by Durand's system, quickly began to distinguish themselves as the exponents of a novel way of fighting. The battalion column became the tactical unit, and skirmishers were pushed out in a thick band. In rugged or intersected country the effect was marvellous ; and the effect was due to mobility.

Napoleon.

But, until Napoleon came on the scene, the full advantage of the new development was not gained. There was still a tendency to have the army in a single long line parallel to the hostile front; each division then fought straight to its front on its own account. It rarely happened that there was a 2nd line, or a reserve proportioned to the size of the army, which the C.-in-C. could wield for the purpose of striking in at a decisive point, or to grapple with some unforeseen event. With such tactics there was also the risk of a general retreat resulting from the defeat of a single column, there being no adequate free force at hand to block the breach in the line.

Napoleon's first command was in the campaign of 1796. Everything was against him; his troops were badly supplied, badly clothed, badly fed and unpaid. They had a great mountain-barrier before them, and a single line of communication running in a narrow strip between the hills and the sea.

On reading a complete account of that campaign you would find that at Montenotte, at Dego, at Lodi, at Lonato and Castiglione, at Rivoli, and at many another fight, the *ordre demi-plein*, or the pure tactics of Durand, carried the day. Always there were swarms of skirmishers who greatly incommoded the unaccustomed Austrians by the boldness and the lawless irregularity of their movements.

When the enemy was in great strength or very densely posted, as at Castiglione, the battalion column became a brigade column on the more important flank, as the readiest manner of strengthening it. As General Duhesme says:—"The manœuvres of the army were simple. Little deploying; brigades, concentrated in mass of three or even six battalions in depth, advanced vigorously upon the enemy in front, while the light infantry, skirmishing, gained his flanks, crowned the heights, carried with its skirmishers trouble and confusion on the enemy's rear, and even often hindered his retreat. We, who were fighting on the Rhine, used to march methodically from one position to another, coming upon the enemy on a front, extended and nearly parallel; we repulsed or were repulsed with a loss often less than 500 or 600 men, and regarded as great trophies 4,000 or 5,000 prisoners; we could not conceive how the Army of Italy was able to make such enormous hauls of prisoners. Occupied with our own fighting, we had no time to study and to realise the sublime perfection of this war of movements that Bonaparte had created there." The skirmishers working round the hostile rear had often a considerable say in the number of prisoners captured.

This kind of tactics requires, as under Napoleon it had, the ability and readiness of all ranks to take responsibility and immediate action thereupon. As a general of the time writes:—

Quality of the Revolutionary Soldiers an Important Point.

"In this army the soldier, no more than the officer, awaited the command to attack; to deploy, to attack in front and flank, to kill or take prisoner, was only an affair of minutes. It looked like an electric communication established between the troops and the generals. The soldiers incited each other, and every advice was well received and instantly followed which looked like leading to the ruin of the enemy. In certain corps peculiar societies of soldiers set themselves to pay particular attention to the Austrian officers. From the start of an action, you would see them (they would be among the skirmishers) disappear in small bodies, and soon the sound of their firing would be carrying terror and disorder right back among the Austrian reserves."

Be it observed, this freedom of movement was conjoined to a very real discipline. It reminds one of the free action of the leading divisions of the German armies, who brought about prematurely the battles of Wörth and Spicheren. But there was in both cases—Napoleon's troops and the Germans—the same whole-souled desire and intelligent co-operation of every unit. This free-skirmishing system proved peculiarly effective in mountain warfare. These troops, who were at first probably hardly a match for their better-drilled opponents in set battles, won the desperate series of fights in the Tyrol in 1796 and 1797 mainly through this new skirmishing system.

1796.

At Lodi, the skirmishers got on to the islands in the stream and materially assisted by their fire the arduous passage of the bridge. As the head of each battalion reached the far end, the men deployed rapidly and instinctively without command and were among the Austrian guns in no time. Now, a force, drilled and trained under the system of Frederick, would have halted on getting clear of the bridge and waited for orders for an elaborate and formal deployment, *i.e.*, if the opposing guns had allowed of it.

The next great fights were Lonato and Castiglione, both of which I shall describe. Arcola was a 3-days' scrimmage, a soldier's battle, a combat of skirmishers. Rivoli, which took place shortly after the battle of Arcola, was an admirable instance of a fight of small columns and skirmishers, and was also a remarkable example of Napoleon's extraordinary genius in seeing in a flash the decisive point. The Austrians had split up their forces into several columns; three of these had united at the head of the plateau of Rivoli, but a fourth, containing the most of the artillery and cavalry, had taken the good road down the right bank of the Adige. Separated as the crow flies, only a couple of miles from the troops on the plateau, it was nevertheless prevented from joining them by the precipitous nature of the valley. It could only join them by a steep and narrow road that zigzagged up into the middle of the plateau. Napoleon seized the head of the ravine in which this road lay, and prevented the junction which would have caused him to be greatly outnumbered in the *bataille rangée* that would have ensued. If you wish an instance of the danger of marching your columns along such routes that they cannot readily combine, there is no better case than the battle of Rivoli.

Egypt.

During his Egyptian campaign Napoleon had to use squares on the march, as we have often had to do in the same regions and for the same reason substantially. His opponents had a vast force of irregular cavalry, which the musket alone could not be depended upon to stop; we had vast numbers of bounding dervishes who got over the ground with astonishing speed. For days on end both of us were watched by hordes of these peculiar warriors waiting for a chance to charge. Napoleon had his great squares six deep, with artillery at the corners, and the cavalry and carts inside. We had less depth, machine-guns at the angles, and small bodies of reserves inside to fill gaps caused by any irregularity in the marching of the square.

Marengo. *Fig.* 6.

When Desaix came so opportunely on the Austrian right flank at Marengo, just at the moment when Napoleon had put his last reserve into the fight and things looked black for the French, he used the new order in the most effective manner.

This attack, assisted by a brilliant charge of Kellermann's cavalry, completely altered the aspect of the battle.

1807.

In 1807 come the battles of Jéna and Auerstädt—a double battle. Napoleon, with the mass of his army, attacked a much inferior Prussian force and beat it without difficulty. Marshal Davoût with his single corps sustained a brilliant combat against

66,000 Prussians, and eventually beat them. This battle is one of the very best examples of the best stage of French tactics.

1809.

In 1809 Napoleon was prepared to invade England. His vast army was splendidly equipped and organized, but much of the fire of the earlier individualism was gone. There was more regularity—something like a going back to the drill-sergeant stage of Frederick the Great. The battalion remained the unit; there were from eight to 12 of them in a division; two, three or four divisions were united under a general, and the Army Corps emerged as a unit. To each division was attached an artillery company of eight guns. A Cavalry Division, an Artillery Reserve (what we call the Corps Artillery), one or two companies of engineers, a corps of gunner artificers, a commissariat and transport corps, all these were attached to each Army Corps. We have evidently come upon the germ of all our modern organization. Huge cavalry divisions were formed, and these took with them horse batteries of six guns. The bulk of the artillery was no longer scattered among the units—we have seen that the division only carried eight guns of its own—and from this time Napoleon made such a use of his artillery as was never seen before.

Army Corps.

Deterioration of French Tactics.

He found it most useful to keep, practically under his own hand, an overpowering mass of artillery to throw into the fight at a critical moment. Thus when Masséna at Wagram had to withdraw and leave a great gap in the line, Napoleon threw 100 guns into the breach and most effectually stopped the Austrian advance.

As this cannonade began to take effect, Napoleon massed MacDonald's Corps into a remarkable column. It had eight battalions in line one behind the other; 13 more battalions formed up in close columns on the wings, and cavalry followed close on the eight battalions. As might be expected, it suffered horribly in its advance; but it broke the Austrians. Jomini, however, thinks it would not have been successful but for the success of Oudinot and Davoût to right and left.

The gradual deterioration of the French infantry tactics is seen in the step-by-step change from the free movements of battalion columns to brigade columns, and even to division columns. Such columns could do no firing at all proportionate to their numbers, and the firing was more and more relegated to huge masses of artillery. The imposing mass of the columns was to strike terror into the enemy, and if necessary to break into his line by sheer weight of numbers.

And yet Napoleon laid down the maxim—*L'arme à feu c'est tout; le reste ce n'est rien.*"

CHAPTER IV.

EXAMPLES OF NAPOLEON'S BATTLES.

As examples of French tactics under Napoleon, I shall take the battles of Castiglione (1796), Austerlitz (1805), Auerstädt (1806), Pultusk (1806), Eylau (1807), describing them sufficiently to show the great variety available under the French training of that epoch.

The next step will naturally be some of the battles in which French tactics met British tactics, and succumbed.

BATTLES OF CASTIGLIONE.

Towards the end of July, 1796, the Austrian general, Würmser, reinforced to nearly 60,000, was in the Tyrol to the north-east of Lake Garda. Napoleon was engaged with the Siege of Mantua, and with the occupation of the conquered territories in North Italy. The siege took 15,300 men; the occupation, 10,000 men, leaving to the French 26,000 to 27,000 as field army.

Würmser divided his force into two columns, one of 20,000 moving down the west side of the lake, and the remainder on the east side towards Mantua. The latter force, under Würmser himself, drove Masséna's Division from the French advanced post at Rivoli, at the same time that the other column began to debouch from the hills at the other side of the lake. Napoleon raised the siege, and concentrated his forces from Lonato to Castiglione, in the plain south of Lake Garda. The head of the smaller Austrian column arrived at Lonato on August 3rd, and the combat of that day took the form shown in *Fig.* 1, *Plate* IV., the French using Durand's Battalion Columns, with skirmishers, the Austrians employing Frederick's line. The latter outflanked the line of columns, partly on account of their superior numbers, and partly because the French kept a reserve in hand of two demi-brigades. When the outer Austrian battalions threatened to lap round the French flanks, the latter were able to meet it by extending their skirmishers further, and wheeling slightly outwards the flank battalion columns. The central columns, meanwhile, advanced steadily and broke the Austrian line. The small cavalry force at Napoleon's disposal took up the pursuit, and would have held the remnants against Lake Garda, but for the timely arrival of a brigade of fresh Austrian troops.

Having arranged for the pushing back of the Austrians into the mountains west of the lake, Napoleon now turned his attention to Würmser, who was moving up slowly from Mantua, which lies to the south-east of the scene of the combat just referred to. His advanced guard under Liptay occupied, on the 4th, the village and heights of Castiglione. Augereau, with a division about 10,000 strong, attacked as shown. Both sides had the same formations as at Lonato. Robert's demi-brigade of three battalions had during the night marched completely round the Austrians and lay concealed behind a wood in their rear. Liptay's force was nearly equal to the French, who kept a reserve of two

Fig. 2, *Plate* IV. August 4th, 1796.

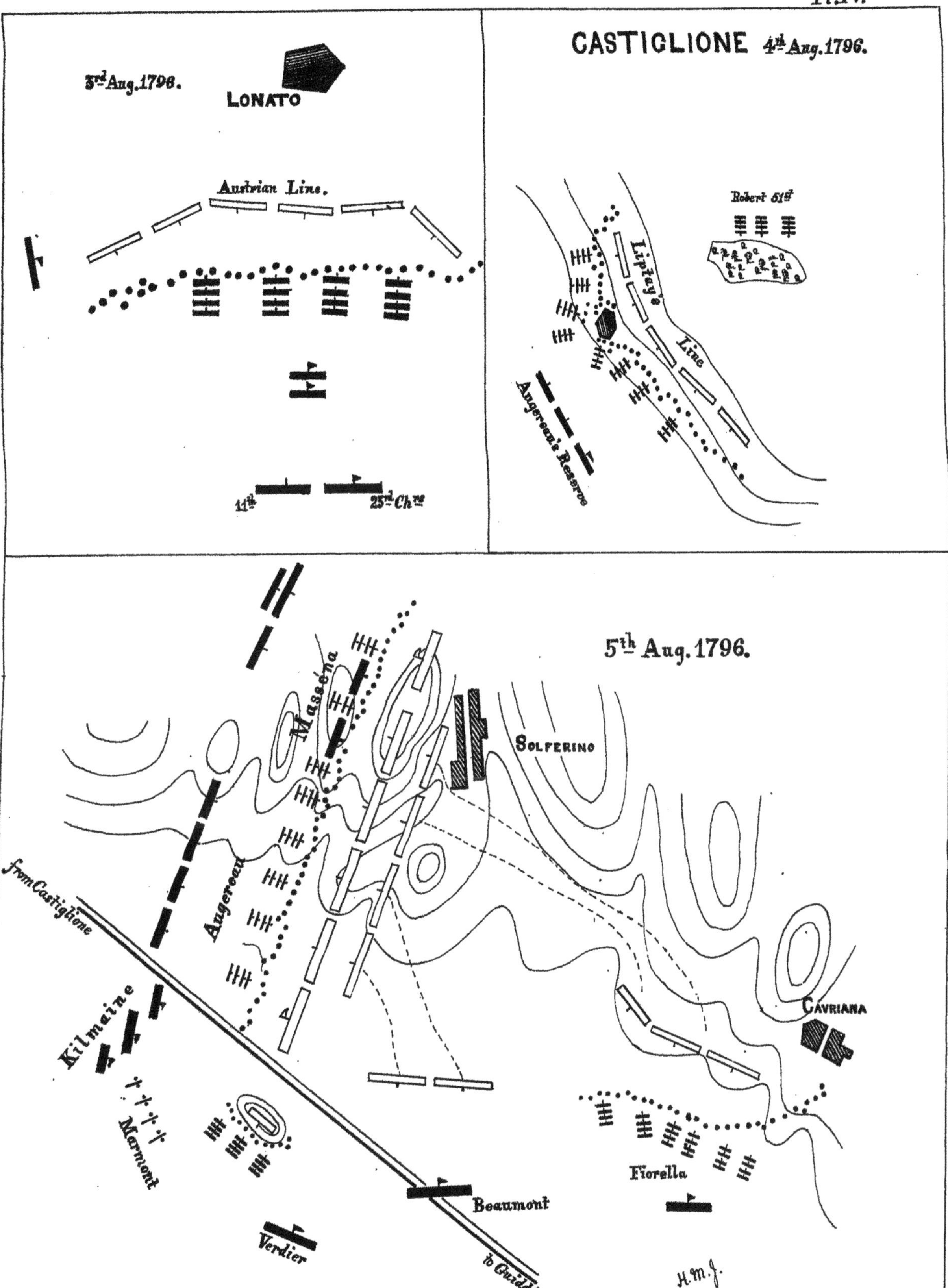

CASTIGLIONE 4th Aug. 1796.
3rd Aug. 1796.
LONATO
Austrian Line.
11th
25th Ch.rs
Robert 51st
Liptay's
Line
Angereau's Reserve
5th Aug. 1796.
Masséna
Solferino
Augereau
from Castiglione
Kilmaine
Marmont
Cavriana
Fiorella
Beaumont
Verdier
to Guiddizzolo
H.M.J.

battalions and a regiment of cavalry. After a brisk fight, the Austrians were driven from the field, Robert falling on their rear as the frontal attack developed.

During that evening and the ensuing night, Napoleon massed all his available strength about Castiglione, Würmser reaching the heights of Solferino, a few miles distant. Serrurier, who had been besieging Mantua with 10,500 men, had previously parted with a portion of his force, to reinforce Augereau. The remainder, 5,000 or 6,000, with some cavalry, were on the 4th a long way to the south. This division had orders to make a night march to Cavriana, on the Austrian rear, and the arrangement was that the sound of its guns would be the signal for Napoleon's attack from the north. Würmser was drawn up in two lines, the right half of his force being on the high ground, the left half on the plain, and the extreme left covered by an earthwork on a mound in the plain. Detaching had brought down his numbers to about 25,000.

Fig. 3, Plate IV. August 5th, 1796.

At daybreak Napoleon moved forward against this line; on his left he had Masséna, in *ordre demi-plein;* in the centre Augereau, in battalion columns at deploying intervals. On the other side of the high road was Marmont with a mass of guns, Verdier's Brigade of Infantry, and the six cavalry regiments of Kilmaine and Beaumont.

Napoleon wished to get round the hostile left, so as to join hands with Fiorella, commanding what had been Serrurier's Division; therefore, on the other flank, Masséna was deployed for fire, rather than for active attack. Masséna and Augereau began the battle with feints, calculated to hold the Austrians in position; but very soon the sound was heard of Fiorella's guns, fighting with the right half of Würmser's 2nd line, which had been brought back in haste to Cavriana when Fiorella's advance was reported. Then Masséna and Augereau began to advance in earnest, the guns on the plain pounded the earthwork, Verdier and a regiment of horse took the mound, Beaumont meanwhile passing round it to behind the Austrian left wing. This move compelled Würmser to detach the rest of his 2nd line, to face Beaumont; and the long Austrian front, now without support, was broken by Augereau, while Masséna cleared the way to Solferino. Würmser gave the order to retreat, having lost 1,000 in prisoners, 2,000 in killed and wounded, and 20 guns. The French troops were incapable of active pursuit, having been marching and fighting continuously for several days.

BATTLE OF AUSTERLITZ, 1805.

Towards the end of November, 1805, Napoleon was at Brünn, in Moravia. He had under his hand, in the immediate neighbourhood, about 50,000 men; Bernadotte was at Iglau, two good marches to the west, with an Army Corps; Davoût had one division at Porlitz, within a short march of the proposed field of battle, another between that and Brünn, while Gudin's Division was at Presburg on the Danube, three long marches distant.

The Allies, Austrians and Russians, were in front of Olmütz; by the arrival of Russian reinforcements their strength was raised to over 90,000. The French would be unable to reach this total. The Allies also excelled in number of guns. Moreover, they had other forces in the field; Archduke Ferdinand was in Bohemia with 20,000; another Austrian Army under Archduke Charles could join in a few weeks from the south, and the Prussians, while for the moment exercising their usual tortuous diplomacy, might at any moment step into the arena with 100,000 men.

Brünn to Olmütz, 35 miles.

Napoleon's aim, therefore, was to induce the immediate enemy to attack, by feigning weakness; and he succeeded with his usual skill. He narrowed his bivouacs, started entrenching, allowed no enterprise against the hostile outposts, and withdrew his own advanced posts at the slightest threat. The Allies fell into the trap. Napoleon wished to suggest to them that they might turn his right and sever him from Vienna; and this was the very plan they took.

Plate V.

Personally reconnoitring towards Austerlitz on November 30th, he came upon the heights of Pratzen, and Soult asked permission to push forward and occupy them. Napoleon replied with these remarkable words:—"If I wished to prevent the enemy from passing, I should certainly place myself here; but the result would only be an ordinary battle. If, on the contrary, I refuse my right by retiring it towards Brünn, and the Russians go beyond these heights (*i.e.*, to the south), they are lost without hope."

French Dispositions and Strength.

On the 1st December, accordingly, the French Army bivouacked as shown on the map. Lannes had 18 battalions and 8 squadrons. The hill, Santon, was entrenched, and supplied with 18 heavy guns, and the regiment there was instructed to hold it to the last man.

Bernadotte, 1st Corps, was of the same strength.

Oudinot had 10 battalions of Grenadiers.

Murat commanded the Cavalry Reserve, 44 squadrons.

Bessières had the Imperial Guard.

Soult's Corps, 31 battalions and 6 squadrons, was much the strongest. With his advanced battalions he held the east exits of the villages from Girzikowitz, through Puntowitz and Kobelnitz, to Sokolnitz; and Telnitz was occupied by a small force. Cavalry reconnoitred towards Pratzen.

Davoût, the extreme right, 10 battalions and 12 squadrons, was refused by being drawn back as shown.

Total in the battle:—97 battalions, 78 squadrons.

The position was thus secured on the left by the mountainous country and the fortified hill, on the right by the large lakes; but on this flank no show of force was made, with the object of luring on the enemy. Between the extremities, the front was in great part covered by the swampy rivulets, which made each village into a veritable defile; and, as we have seen, the defiles were secured.

Allies.

The five columns in which the Allies had been moving are shown on the map, as well as the directions which they took before sunrise on the 2nd. On the previous day's march, Bagration had become the Flank Guard, but was now the right wing.

Kienmayer had 5 battalions, 22 Austrian and 10 Russian squadrons.

Bagration, 12 battalions and 14 squadrons.

Doctorow, 24 Russian battalions (1st Column).

Langeron, 18 Russian battalions (2nd Column).

Przybyszewski, 18 Russian battalions (3rd Column).

{ Kollowrath, 15 Austrian battalions (4th Column).
{ Miloradowich, 12 Russian battalions (4th Column).

Lichtenstein, 82 squadrons (5th Column).

Constantine, 10 battalions, 18 squadrons (Reserve).

Total, 114 battalions, 146 squadrons—or, counting in the gunners, 90,000 men, of whom 15,000 were cavalry.

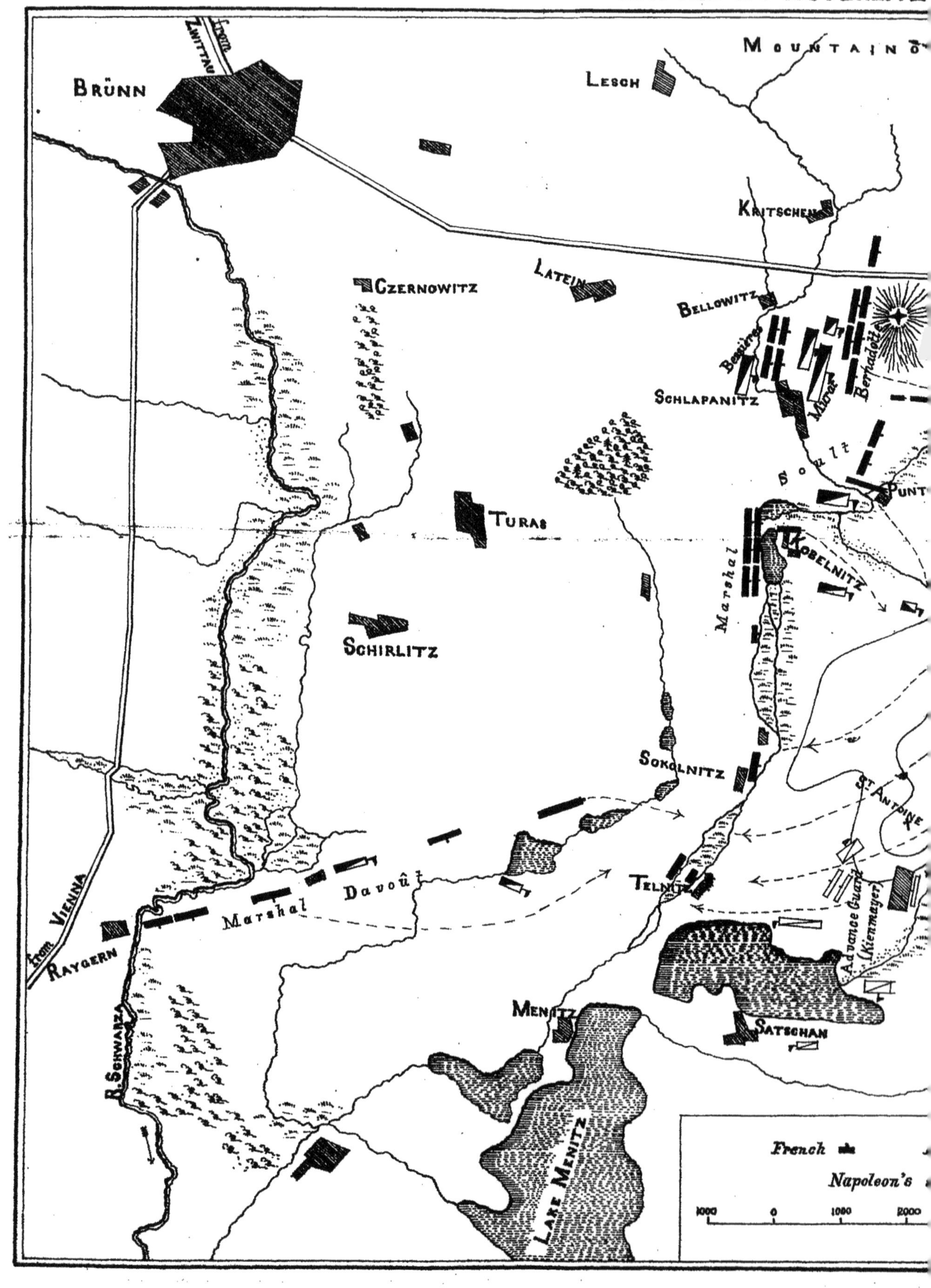

Battle of Austerlitz
from Zwittau
Brünn
Lesch
Kritschen
Latein
Czernowitz
Bellowitz
Bessières
Murat
Bernadotte
Schlapanitz
Soult
Turas
Kobelnitz
Marshal
Schirlitz
Sokolnitz
St Antoine
Telnitz
Marshal Davoût
from Vienna
Raygern
R. Schwarza
Menitz
Satschan
Advance Guard (Kienmayer)
Lake Menitz
French
Napoleon's
1000
0
1000
2000

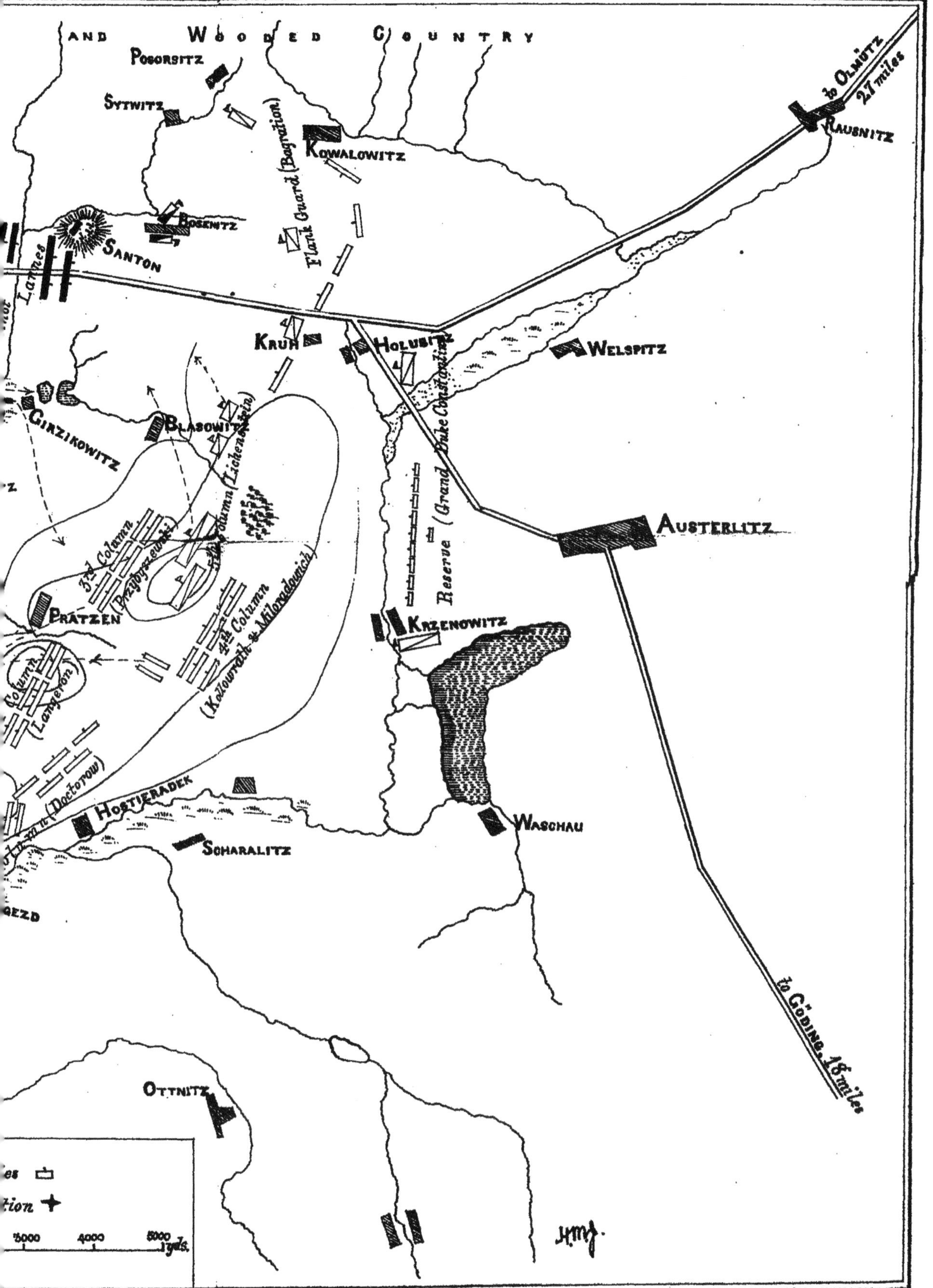

Lithographed by W. & A. K. Johnston, Limited, Edinburgh & London.

Napoleon allowed the first three columns and the advanced guard (Kienmayer) to get well on their way, before himself making any forward move. Thus 65 battalions and 32 squadrons abandoned the heights, and left a serious gap in the allied centre. The 4th Column was following the turning movement, and its leading regiment was passing Pratzen, when Napoleon launched Soult against the heights, a division issuing from Puntowitz and another from Girzikowitz, as shown. A sanguinary struggle took place, and in two hours Soult was firmly established, driving the remains of the 4th Column over to Hostieradek and capturing most of its guns.

Meanwhile the three columns, in pursuance of their orders, had taken Telnitz and Sokolnitz; but Davoût was now to the fore, and, with that tenacity he showed again on the day of Jena, held the great masses that came against him and prevented them from debouching. Learning what was going on at Pratzen, Doctorow sent back a column, but it was too late.

In the meantime, Bagration, Lichtenstein and Constantine were fighting a hard battle with Lannes and Bernadotte, in which some pretty cavalry fighting on a large scale took place. Long before evening, the Allies were beaten in all parts of the field, losing the bulk of their artillery, 11,000 killed and wounded, and 19,000 prisoners.*

Comments.

This battle is cited in this History of Tactics on account of the notable instructions issued by Napoleon beforehand to his generals. In the centre, where the ground was cut up into defiles, each brigade was to have its own definite object to attain, and was to advance, when deployed, in the *ordre demi-plein*.

Napoleon had foreseen that the enemy would throw his splendid and numerous cavalry against Lannes and Bernadotte, that portion of the field being open and free from defiles. He kept, therefore, the bulk of his cavalry on this wing, but was careful to support it with four whole divisions of infantry. Further, he prescribed the joint formation to be observed. Dumas says:—"These divisions were formed in two lines; the first in line, the second in battalion columns; the light cavalry out in front, with its artillery; the heavy cavalry in several lines in rear of the infantry. In this order the arms, lending a mutual support, defied all the efforts of the numerous horsemen of the Allies. If a charge of theirs succeeded at first" (as it several times did), "their half-broken squadrons promptly broke up against the staunch infantry" squares that had formed in the 2nd line; "thrown into disorder, and passing through the intervals of the squares, they were charged in turn and hurled back," and suffered dreadfully each time from the fire poured upon them in retreat.

Napoleon had made the same disposition at Waterloo, when the British heavy cavalry charged, and with the same success.

BATTLE OF AUERSTÄDT, 1806.

Napoleon had, on 14th October, 1806, turned the flank of the Prussian Army, and had even succeeded in placing Davoût's Corps, of three divisions, on the line of retreat which the Prussians intended to use.

* The student who may be desirous of going more fully into the details of the battle is recommended Dumas' "Précis des Evénemens Militaires," Tome XIV., and Stutterheim's pamphlet "Austerlitz."

On the 13th, the King of Prussia had written to his generals:—"The enemy is marching on Leipsic; his marshals, Lannes and Augereau, are covering the movement by taking post opposite Jena; consequently, we shall march to-morrow to Auerstädt" (*i.e.*, northwards along the left bank of the Saale), "and next day to Freiburg. Prince Hohenlohe will cover the movement by holding Jena, without attacking the enemy. Our divisions will march off by the left in a single column at two-hour intervals."

Plate VI. Hassenhausen is a village on the road thus indicated by the King of Prussia; it lies between Jena and the point where Davoût had crossed the Saale. That marshal, on crossing, had sent his cavalry in the direction of Jena; it had been met by Blücher's horsemen and pushed back. Following Blücher came Schmettau's Division, as advanced guard to the 66,000 men marching under the King's orders.

Gudin's Division was leading the French Corps. Its 85th Regiment took possession of Hassenhausen and established itself there, and in the small wood to the right front, posting also some guns to sweep the road. The 25th, 21st and 12th, which made up the rest of the division, posted themselves to the right of the village, the first two in two lines, the 12th in solid column. Thus 12 battalions, or about 10,000 men, were in position to check the Prussian march. Their formation is a model; they cannot expect to attack, and they expect cavalry; so the 85th keep snugly in the village and the wood, the 25th and the 21st deploy for maximum of fire; the 12th keeps in close order, as it may have to receive cavalry attack in flank or infantry in front, and in close column it will be able to form square or deploy for fire with equal celerity. *Fig.* 1.

Schmettau presses forward to drive in the skirmishers, to give a clear field to the cavalry, who will charge towards the French right flank; the 12th forms a regimental square, the right battalions of the 25th and 21st form squares. *Fig.* 2.

The Prussians recoil; and just then Davoût's 2nd Division, Friant, arrives. Friant extends the line to the right, forming up in two lines deployed, while Gudin brings in the 12th to the other side of the village. *Fig.* 3.

The 2nd Prussian Division, Wartensleben, now appears on the scene. It deploys off the road to its right, and is followed pretty closely by a third division as support. Wartensleben advances in échelons from his left, and it looks as if he will gain the French left and rear; Schmettau supports by attacking again towards the village. But Davoût's 3rd Division, Morand, has been hastened up; it comes along at the double, marching in a single close column on a front of a double company, with its artillery on its inner flank. While it doubles, Wartensleben's left, or leading, échelon, of two deployed battalions, with a numerous cavalry on its right, drives back for the moment the 12th, and Schmettau forces the 85th into the interior of the village, but is unable to get possession. Schmettau and the Duke of Brunswick are killed, and a marshal mortally wounded, and the strife continues. A, in *Fig.* 4.

Morand is now close up; his leading battalion comes into the open under a heavy fire from Wartensleben's guns. It advances rapidly, while the rest of the division deploys to the left in line of battalion columns, each at deploying interval from its predecessor. The advance is thus begun in échelon from the right; but the leading échelons are at first compelled to fall back, and presently the whole division is in a single line of battalion columns, covered with a band of skirmishers, and having the artillery divided between the two flanks. B, in *Fig.* 4.

Wartensleben's front is now inclined towards the village, and the fresh advance of

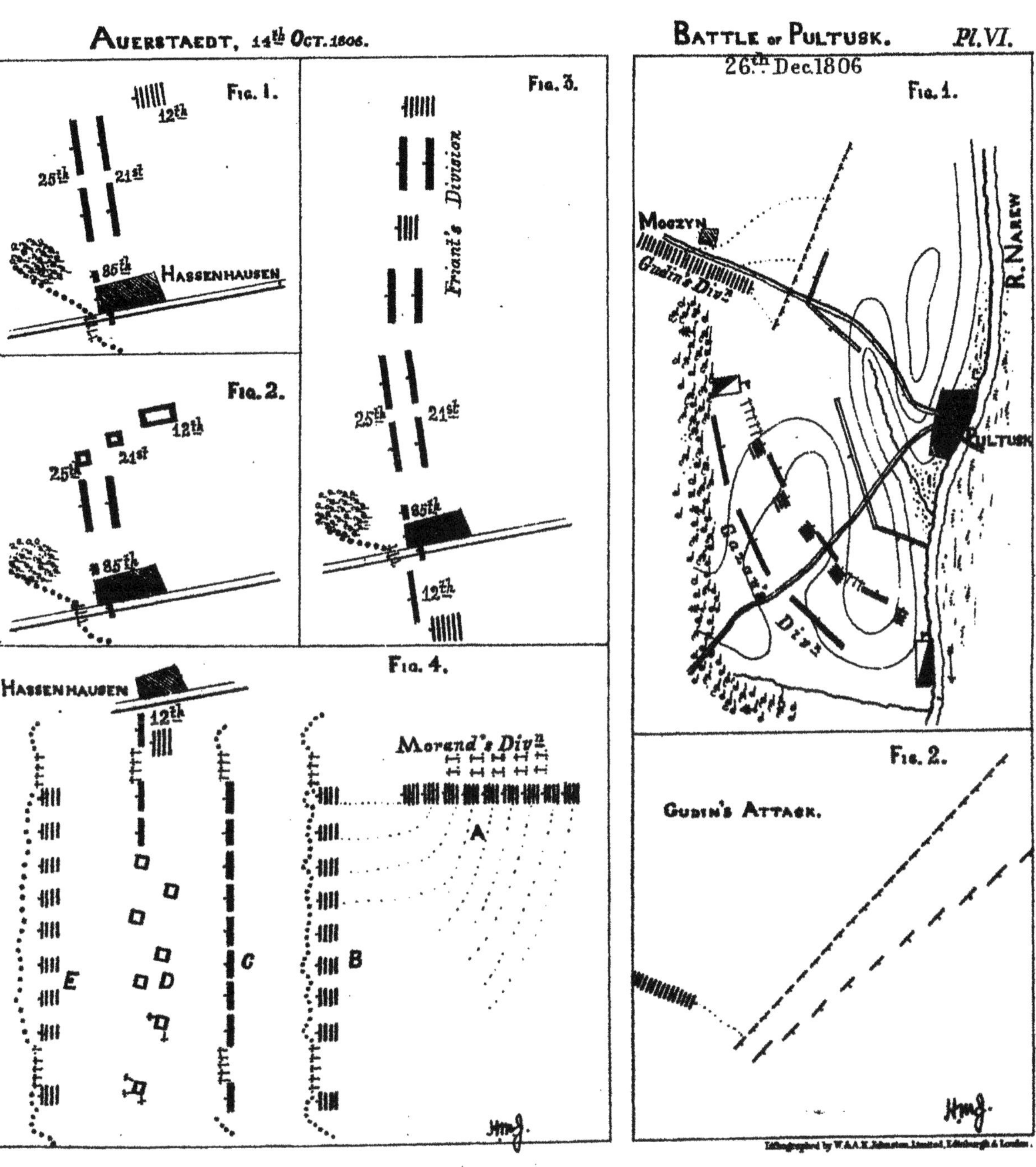

Lithographed by W. & A. K. Johnston, Limited, Edinburgh & London.

Morand is taking it partly in flank; the Prussians recoil, and the village is saved. Wartensleben, brought thus to a halt, begins a continuous fire along his line, and Morand also deploys into line, his skirmishers withdrawing through the gaps. The 12th has recovered its position between the village and Morand's line. C, in *Fig.* 4.

Prussian cavalry, to the number of 14,000, now pass through the intervals of Wartensleben's battalions and hurl themselves on the French in a *frontal* attack. This, of course, is a mistake, for the Austrian line has to cease fire. Seven battalions of Morand's at once form independent squares, the two on the left posting guns at the corners. The Prussian horse make several furious charges, lose frightfully, but break not a single square. D, in *Fig.* 4.

Morand then re-forms his line in battalion columns, skirmishers in front, and advancing to the attack drives back and cuts up Wartensleben. E, in *Fig.* 4.

Meantime, on the right of the village, Friant has been gaining ground, foiling the outflanking attempts of the two Reserve Prussian Divisions. After a lot of fighting, Davoût gains possession of villages to the front, and of a fine artillery position for all his guns. The Prussians are finally beaten, and disperse over the country in flight.

The extreme handiness of the battalion columns in such a fight as this is manifest, and the work done by the skirmishers was most notable. The village of Hassenhausen also played an important part; the 85th demi-brigade, using its shelter tenaciously, saved the French from having their centre completely pierced.

The faulty frontal attack of the cavalry has been commented upon. If the regiments had swept round the Prussian right and lined up for the charge more or less perpendicular to Morand's line, its mere appearance would have compelled the latter's battalions to form squares and his skirmishers to seek their shelter. Prussian guns might then have been advanced to play upon the squares till the cavalry charge fetched home. By this means *all the available force* would have been brought to bear at the *critical* moment.

BATTLE OF PULTUSK, 1806.

After disposing of the Prussian armies and marching through their country, Napoleon was face to face with Russia. On 23rd December, the Narew River, a tributary of the Vistula, was crossed by night in the presence of Russian forces, with a mixture of audacity and skill. Three days later, Lannes found himself near Pultusk in presence of a considerable Russian force. He sent forward his front line in the *ordre demi-plein;* it contained five demi-brigades of infantry, a considerable artillery employed in two masses and a few regiments of cavalry on each wing. The 2nd line, Gazan's Division, made as imposing a show as possible along the edge of the woods, with the general intention of making the enemy expect the whole French advance from that side. The Russians were covering Pultusk in a rough semi-circle. *Plate* VI.

While Lannes advanced with every appearance of confidence, and caused his cavalry to make charges, Gudin's Division arrived in column from the westward, debouching opposite the salient of the Russian position. He wheeled the head of his column slightly to the right, and then deployed rapidly to the left of the leading half-battalion. By advancing now from the left in échelon of half-battalions at 50 paces distance (*v. Fig.* 2), he came well on the left flank of the Russians. This attack decided the contest, and the Russians were driven from the field.

BATTLE OF EYLAU, 1807.

At this battle, only a few months after Jena, we find a French marshal, Augereau, for the first time sinning against the true principles of the French system. He was ordered to take his two divisions against the Russian centre, and he marched them in two solid columns of a division each. During the advance a blizzard stopped all view, but the columns plunged on; they lost direction and found themselves, when the squall was over, facing a battery of 72 Russian guns. Immediate deployment was imperative, and the intention was to form line, with a supporting line of battalion columns; but the 72 guns got in such good work that drill-ground manœuvres were impossible, and the divisions were horribly mauled. The Russian cavalry now came into the gap caused by the deviation of the two great columns, and completed the rout of the corps.

The only excuse is that it was easier to keep the divisions in hand during the advance, and that the snowstorm alone caused the disaster; if this be allowed, it would stand as a parallel case to that of the Highland Brigade at Magersfontein; but the latter had the excuse of darkness during the whole advance. Augereau should have split up into battalion columns and halted the lot during the squall; or, letting them boldly advance, have thus utilized the obscuration of view to get to close quarters without loss. Single battalions would have been easily extricated, if in too dangerous a situation. It was the great merit of the small-column-and-skirmisher system that hostile fire could not be concentrated on all the attackers at once, and that the small columns could manœuvre quickly.

Eventually, the battle was not lost, but it was hardly won. The pre-arranged appearance of Davoût's Corps on the left flank of the enemy restored the fight. This marshal, most intelligently using the ground, extended his men across the hostile flank more widely than was customary at that time. The copses and walls enabled him to ignore the enemy's cavalry, the great terror of skirmishers at that epoch; and by extending he was able to bring to bear a powerful convergent fire. Then, when the moment was ripe, each brigade in battalion columns chose its distinct objective and advanced to a successful attack.

The battle had begun with a terrific artillery combat at short range, Augereau then making his abortive attempt near the centre.

It is said that Napoleon was much impressed by the execution done by the numerous Russian artillery. Shortly after this he set to work to increase his own strength in that arm; and it was subsequently to this period, at Wagram for instance, that he astonished Europe by the telling effect he produced by the employment of guns in great masses.

CHAPTER V.

STATE OF TACTICS IN NAPOLEONIC TIMES, AND AFTER.

A FRENCH writer thus describes a normal fight of those days of Napoleon; but before quoting him I may state that Napoleon's strategy was all along as sound and brilliant as ever. To what extent the deterioration of his infantry tactics was due to the inferior type of conscript now being obtained, or to Napoleon himself, is a difficult question.

Normal Fight by Napoleon.

"Brigades of light cavalry, supported by divisions of dragoons, reconnoitre for the enemy, force him to deploy and to show his strength. The infantry columns arrive towards the end of the day, and prepare for to-morrow's action by occupying outlying villages, hillocks, woods. In the early morning the Emperor reconnoitres in person. If the enemy offers an opportunity for attack, the march-columns, a division each, without waiting to deploy, advance from all sides, under cover of artillery; the cavalry deploys and charges."

Plainly we are not yet in modern times.

"When night stops the fighting, the army bivouacks; the distant corps march on without halt to line up for to-morrow's battle.

"At break of day, the whole are under arms; the army sheltered as much as possible, and the Emperor contenting himself at first with a cannonade. But the moment for attack arrives; the battalions in attacking columns at deploying intervals march to the front, the light cavalry on the wings, the reserve cavalry behind the infantry, the guard more in rear, the artillery in front. Clouds of skirmishers escape from the flanks of the masses of infantry and cover the field with smoke; they profit by all the accidents of the ground to gain cover, and they cover everywhere the movements of the columns. Sometimes the cavalry precedes the infantry and facilitates its action. (I suppose he means by this that the threat of an imminent cavalry charge would compel the hostile infantry to form square and thus greatly diminish its fire). In other cases, it follows close, ready to give support and to profit by the least sign of confusion in the hostile ranks.

"If these stand firm, the French masses deploy and after a short fusillade attack with the bayonet, while the artillery, now left behind, limbers up and comes to closer quarters.

"Unfortunately the heavy and deep columns were not always given time to deploy, as at Busaco and Albuera."

The British had an awkward way of dropping their volley-firing just at this moment, and charging with the bayonet.

When the Austrians, fighting Marshal Masséna at Caldiero in Italy, tried the method of breaking through with an enormous column and failed, an object-lesson was given to the French of the viciousness of the system against good infantry with a fire-arm and a staunch artillery. But the French evidently did not learn the lesson.

Already, in the days of the musket, men were beginning to see that cavalry could not successfully charge unshaken infantry in position. At Waterloo the French found it out to their cost.

Jomini, writing after Waterloo, in 1838 in fact, sums up on the question of artillery thus :—

Jomini sums up in 1838.

"(1). In the offensive, a certain portion of the artillery should concentrate its fire upon the point where a decisive blow is to be struck. Its first use is to shatter the enemy's line, and then to assist with its fire the attack of the infantry and cavalry.

"(2). Several batteries of horse artillery should follow the offensive movement of the columns of attack. Too much foot artillery should not move with an offensive column.

"(3). Half of the horse artillery should be held in reserve, that it may be rapidly moved to any point. For this purpose it should be posted where it can be moved readily in any direction.

"(4). The batteries, whatever may be their general distribution along the line, should give their attention particularly to those points where the enemy would be most likely to approach. The general of artillery should therefore know the decisive strategic and tactical points of the field, as well as the topography of the whole place.

"(5). Artillery placed on level ground or ground sloping gently to the front is most favourably situated either for point-blank or ricochet firing; a converging fire is the best.

"(6). It should be borne in mind that the chief office of all artillery in battles is to overwhelm the enemy's troops, and not to reply to their batteries. It is nevertheless often useful to fire at the batteries, in order to attract their fire. A third of the disposable artillery may be assigned this duty, but two-thirds at least should be directed against the infantry and cavalry..

"(7). If the enemy advance in deployed lines, the batteries should endeavour to cross their fire in order to strike the lines obliquely. If guns can be placed so as to enfilade a line of troops, a most powerful effect is produced.

"(8). When the enemy advance in columns, they may be battered in front. It is advantageous also to attack them in flank and reverse, the moral effect of which is sometimes inconceivable. Ney at the battle of Bautzen was prevented from continuing the well-conceived direction of his advance by the flank fire of a very few pieces of light artillery. It is worth risking guns for such a purpose.

"(9). Batteries must always have supports of infantry or cavalry.

"(10). It is very important that gunners, when threatened by cavalry, preserve their coolness. They should fire first solid shot, next shells and then grape, as long as possible. The infantry supports should form squares in the vicinity, to shelter the horses, and when necessary the gunners. Rocket batteries are very useful to frighten the horses.

"(11). When infantry threatens artillery, the latter should continue its fire to the last moment. The gunners can always be sheltered from infantry attack at short notice, if the battery is properly supported. This is a case for the co-operation of the three arms.

"(12). The proportions of artillery have varied in different wars. Napoleon conquered Italy in 1800 with 40 or 50 pieces, while in 1812 he took 1,000 completely equipped guns into Russia and failed. Usually three pieces to 1,000 combatants are allowed.

"At Eylau Napoleon's troops were badly handled by the Russian artillery, which was extraordinarily numerous. The havoc caused in the French ranks opened his eyes to the necessity of increasing his own. In three months he had, by incredible energy, doubled the *matériel* and *personnel* of his own artillery."

I have collected together these ideas of Jomini, because they show the state of artillery tactics at the end of the Napoleonic wars, and because a comparison of them with the ideas of the present day may be useful.

Comments on Jomini.

Jomini's first dictum is one that is now quite impossible. In attacking an enemy who has guns, you must deal with his artillery first. If unmolested and well served, it can of itself almost stop infantry advancing over open ground. Having got the better of his guns, you can then turn part of yours on to his infantry positions.

His second point, about guns following up your infantry attack, is applicable to the present day.

His third, keeping back an artillery reserve, would never be done now, if your artillery was powerful enough to have any chance of coping with his. If it were not, you could hardly attack him at all, unless the ground was very unfavourable for gunfire, or unless you outnumbered him greatly in infantry and were prepared to accept very heavy loss. If on the defensive and weak in guns, you might quite properly hold back your pieces in reserve at first, so as to have them for use against his infantry attack; at *that* stage you must bring them into action at all hazards. Sometimes a defender, outmatched in guns, may compensate for his weakness by the nature of the ground enabling him to conceal the positions of his guns even when firing, and by providing several alternative emplacements to which a gun could be shifted as soon as the superior hostile artillery discovered its location. The effecting of this plainly indicates a good deal of labour in the making of communications.

But to keep a reserve of guns during a pitched battle, just as you keep a reserve of infantry, would seldom be resorted to, except as just stated. His fourth is plainly applicable to all time.

His fifth, about placing guns on level ground or ground gently sloping to the front, has no sort of connection with modern times. The round cannon-balls of those days did much execution as they bounded along the ground; also the guns ought to see all the ground between themselves and the enemy. Nowadays you can leave the near 1,000 yards—or if the infantry are well in front of the guns, the near 1,500 yards—to be looked after by the rifles and machine-guns during the first part of the fight at least. Later, your guns may have to move to take part in the final stages.

Artillery.

He says "a converging fire is the best"; and this is true always. When field artillery made a start towards its present capacity by becoming rifled, accurate, long-ranged and breech-loading, stages were also gone through in its tactical handling. The precious pieces were never to be risked, the loss of a gun was held a disgrace to the commander of the battery. The guns were accordingly used with timidity, were kept far back in the line of march, were apt to arrive too late on the battle-field, seldom accompanied the advanced guard. Even as late as 1866, the Prussians acted thus; but in 1870 the guns were usually well forward and took their proper *rôle* of opening the battle. Then came the appreciation of the powerful effect produced by being able to direct the fire of many batteries simultaneously on chosen parts of the hostile front. This led to what is called "massing the guns," done frequently by the Prussians with great success in 1870. The fire could not always be called "converging," but the

Massing.

advantage gained was the control of the fire by one commander. The objection sometimes urged against the method, that it afforded a huge target to the enemy, was nonsense in those days, when the great advantage of invisibility was not so important as now. A single battery occupies about 100 yards of front; and a target of this width is quite enough for any decent gunners to keep within, as a matter of lateral deviation. If massing was brought about by putting several batteries one close behind the other at slightly different levels, then the objection of a huge target would apply.

Latest Artillery Improvements

Then comes the stage of smokeless powder, quick-fire, greater range and accuracy than ever. Concealment from view now becomes of supreme importance, and 100 guns side by side can hardly be hidden. Each gun if possible, each battery certainly, must be allowed to find a position in which it can hope to fire for some time without its exact locality being discovered by the enemy. The result is scattering the guns; how, then, provide for the important matter of single control? By signalling, or better still, by telephone; and this is being done.

When howitzers began to be re-introduced, to take the place of the extremely inaccurate mortars of older times, a new feature in field warfare emerged. Indirect and high-angle fire enabled the gun that employed it to be kept completely out of sight behind artificial or natural objects; these have had a great effect on both field and permanent fortification. Trenches are narrowed; closed redoubts are either provided with efficient bomb-proofs, or are not used at all. In hasty field works, escape from the effect of the heavy, high-explosive shells is only to be obtained by invisibility, and hoping that the shell will not fall within the narrow excavations. And the chance of a shell coming down at a sharp angle falling within a depth of two or three yards is very small. Hurricanes of fire of this kind did very little harm to the Boers in their deep and narrow trenches. But against redoubts these guns are most effective; redoubts liable to such fire will not be constructed, unless there is ample time and means to make bomb-proof cover of the most ample kind.

Napoleon himself is supposed to have made the following remarks, after his fighting days were over:—"The principles of strategy may be said to be fairly fixed, but it is otherwise with tactics. There is little agreement on this matter; there are many varieties of system for disposing troops in battle; but the details are not capable of being reduced to fixed maxims. The ground, the moral of the troops and the object govern all.

"The shallow or deployed order was constantly followed by Wellington, and people have concluded that it was the best, since it triumphed over our columns. In the steep positions of Spain and Portugal, a defensive line *ought* to be deployed and to depend upon its fire. If Wellington resisted our columns at Waterloo until the arrival of the Prussians, it was because a horrible mud hindered our impulse and did not permit our guns to accompany; it was also due to our columns being over deep. The chief merit of the column is impulsion; if the nature of the ground, or the column's too great mass, prevents it from acting freely, you get all its disadvantages without its advantages.

"Jomini," says Napoleon, "proposes a very good order of battle. At bottom, it is only the attack by battalion columns, formed on the two centre companies, but it is applied to the whole line of battle. It is less deep and more mobile than the plans of his predecessors; it is less feeble than the deployed order. We often used it in the early campaigns.* But as armies increased in size, we took to columns of too large a

* It was a development of the *ordre demi-plein*, applied to several lines.

size, induced by the desire of having the greatest possible masses disposable under one's hand. Instead of being in columns of single battalions on a double-company front, we often were in columns 12 battalions deep on a battalion front; which is an enormous difference. It was above all at Albuera, at Moskowa and at Waterloo that we incurred this reproach, but we did not lose through this at Essling.

"It should be observed that if one wishes to form a deep column, or a line of little battalion columns, one should have on the flanks a battalion marching in file. The worst thing that can happen to a column is to be forced to halt because the enemy is about to charge it in flank. The battalions suggested would put it under shelter from the hostile movement; marching at length, they would quickly present a front to the enemy who wished to charge the column in flank; while the column, thus covered, could continue without obstruction its offensive impulsion.

"The English employed also a system of little battalion squares, which is not without use in attack or defence. The battalion of eight companies can form a square having three companies of front and one on each flank; the latter can march in files. This order gives in fact two lines of infantry; it shows a front of one company more than Jomini's battalion column. It is true that it has less 'go' than the latter; but if one has to have recourse to fire, it has half as much again as the column, and is less damaged by artillery fire. It is better than a deployed battalion, because it is less wavering; with such a square one can attack over any kind of ground."

He sums up as follows :—

"(1). For a defensive battle in position, first line deployed, second in columns of attack by battalions. (The English method, in fact).

"(2). To attack, on the contrary, two lines of battalion columns on double-company front, the columns of the 2nd line opposite the intervals of the 1st line, to give less chance to the hostile artillery, and to facilitate the passage of the lines. The intervals between battalions to be garnished with skirmishers and cannon.

"(3). To diminish crowding on the move and to give more front to the column, put the men in two ranks. (English system again).

"(4). To give every help to the production of impulse, the command should be distributed in depth; that is to say, each brigade should form its own 1st and 2nd line, in order that each general may have his support under his own orders, and not have to expect help from another.

"(5). One can use the system of squares both defensively and offensively, on ground that is not too uneven. It has rather less solidity and impulsion than the battalion columns, but it has advantages against cavalry and artillery. The long square has the inconvenience of narrow flanks; but it has more mobility, more fire and more front than the regular square.

"(6). Deep columns of several battalions one behind the other should only be employed when one has no room for extension. *Then,* one must have a battalion marching in files on the flanks, and plenty of skirmishers to protect the march.

"(7). All these varieties in little tactics make no change whatever in the principles of great tactics."

In discussing the battle of Talavera, 28th July, 1809, Napoleon is made to say :—
"Wellington uses the defensive-offensive; he waits for his adversary in a well-studied position, he wears out the assailants by his artillery and a murderous fire of musketry;

then, when they are close up, he avoids the danger of awaiting attack by himself pushing forward with his united forces. This system is as good as another, but it depends on the locality, the troops, and the character of the enemy. I myself have known how to act in defensive-offensive style at Rivoli and at Austerlitz. This method gained the day at Talavera, because the infantry of the 1st Corps (Victor's) attacked by one division after another. Our braves ran to the assault of the enemy's position with an admirable rashness, but a rashness that ensured their ruin; they arrived out of breath and disordered, and the hostile line, thundering against them with individual fire and by volleys, had our men at their mercy when it charged with the bayonet."

The fact is that the French had, from their experience in Europe, gained the idea that if their attack got to close range the enemy would run.

A description in the next chapter of a few of the Peninsular battles will show that, with little variation, this was the expectation of the French, and the experience of Talavera their usual experience.

When the Napoleonic wars were over, the military nations of Europe set to work to plan their future tactics in the light of what had passed. We find the Austrians adhering to their columns, based now on the French system. The order of battle of a brigade becomes one of two lines of close columns of battalions at deploying intervals; the battalions of both lines are controlled by the brigadier; if there are two brigades, they go into battle side by side; if there is a 3rd brigade, it follows in rear of the centre.

Austrians and Prussians both ordered the system of combined brigades; and the regulations still left great latitude to battalion commanders. All this was French, the foundation being battalions in close column. Both these nations had had cruel experience of French success in this formation. The new French regulations, after the Napoleonic wars, give no definite formations for brigades and divisions, but lay down rules to facilitate and expedite the manœuvres of battalions.

French, Austrian and Prussian regulations came to the point of hardly differing at all. But in some of their instructions one can surely see the effect of the British successes. Thus—"When you wish to drive the enemy from a position which he occupies, attack with the bayonet; the attack can be made *in line* or in column. You will attack in line if your troops are of a superior *moral* to the enemy; the greater part of your men are thus able to take part in the fighting; in this order you will suffer much less from artillery fire than if you were in column. It is always indispensable, as you must arrive all together on the position to be captured, that the intervening space shall be of such a nature as to permit you to march in good order; and you must not have cavalry attack to fear while thus marching in line.

"You will attack in column when, beyond everything else, you wish to give to your troops a great mobility. You will be able then, while all along maintaining order in your ranks, to surmount all the difficulties of ground you may encounter, and to readily form square when you must. It is true you will be then exposed to loss from artillery fire; but it will be seldom that you will not be able, by using judgment in the direction of your march, profiting here by a fold of ground, there by a copse of trees, and again by a hedge or some other cover, to avoid the worst effects.

"You will decide, then, for attack in line or column, according to the shape of the ground, the latest information, the moral of your troops, their degree of tactical ability;

but, once you are fixed in this particular, be sure that nothing but a rapid continuous march, followed by an impetuous shock with the bayonet, will instil terror into the enemy, sustain you under his fire and snatch the victory.

"When moving thus against the foe, guard yourself under all circumstances from halting to return his fire. Stopping will inevitably make your attack fail, and produce disorder. In a critical moment, push on; your men must be kept going.

"Whether troops attack in column or deployed in line, they must be supported by a reserve always in column, at 150 or 200 paces behind. If the attacking line be a battalion, let the reserve be ⅓rd of a battalion in strength, or at the least two companies. . . .

"After every attack, whatever be its result, your troops will be in more or less of disorder; if you have won, the supports will follow on the enemy's heels and the rest will rally and re-form on the spot, in case of counter-attack; if you are repulsed, you will rally behind your reserve.

"If circumstances should compel you to await a bayonet attack, you will deploy all your men and receive the enemy with fire. After the last discharge, delivered at 50 paces, you will meet him with an impetuous bayonet charge."

In all this we see plain traces of the effect of the success of the British tactics. We see them hankering after the *line*, recommending it "*if* your troops are of a superior moral to the enemy"—*i.e.*, acknowledging its superiority if your men are of good enough quality.

The urgent recommendation, too, that the 1st line, on making a successful attack, shall not itself incontinently pursue in disorder, seems to be taken from the British tactics.

The sketch I now give shows the Prussian regulation for a battalion attacking. Both Prussians and Austrians at this time kept their companies in three ranks. The 3rd ranks of each company were skirmishers when required.

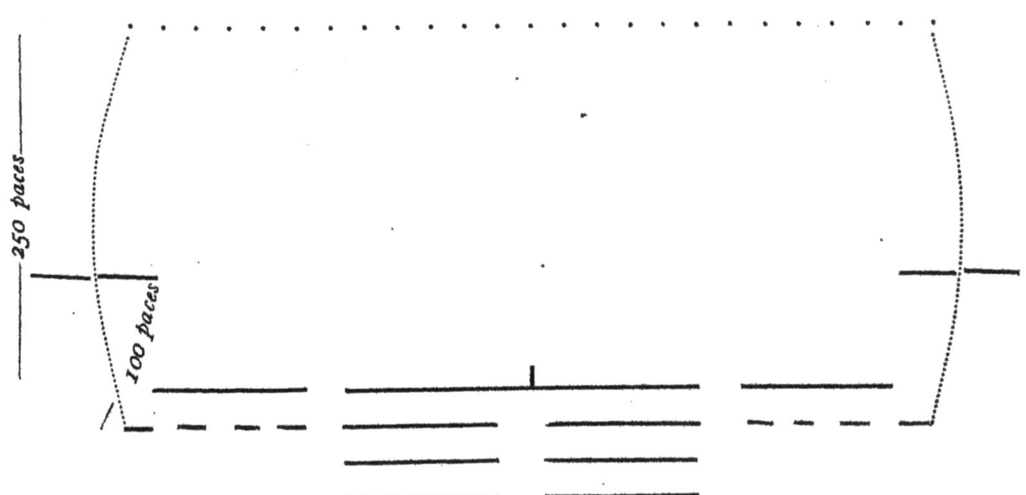

Cavalry.

As to cavalry, what we nowadays hold to be its chief use seemed at the epoch of which I am treating to be not generally recognized. It is true that Napoleon used cavalry far in advance, to find out the enemy's movements; and our own cavalry in Spain and Portugal often rendered the most effective service as scouts and screen. The French sometimes complained of British cavalry officers, singly or in pairs, dogging the march of their columns and coolly taking notes at short range of what they saw; when they sent cavalry to drive the scouts off, these would retire across country and leap obstacles which brought the French troopers to a standstill.

But in spite of this use of cavalry, it was never established as the settled *rôle* of that arm in any of the official regulations of Europe; in the Crimea we find, just before the Alma, that a mass of Russians defiled close past the left wing of the British without either side knowing of the proximity; and even in 1866 the Prussians and Austrians were capable of being within a few miles of each other in vast force without accurate knowledge of the facts.

The very great success of the German cavalry screen, and the good effects thereby produced, are apt to be held as results that would always follow such action; but the Germans themselves, since the war, acknowledge that a great part of the success was due to the French not using their cavalry in similar fashion. And they also acknowledge that they themselves were absurdly over-timid in this connection at the very outset. They say that in any future war of a similar kind, *i.e.*, directly across the frontier, those cavalry divisions that could be most rapidly mobilized would be sent boldly across the frontier at once to harass the enemy and find out what he was doing. Partly with this in view, there is now a tendency everywhere to arm cavalry with a rifle equal to the infantry weapon, so that an advance of the kind would not risk being perpetually checked by quite small bodies of infantry installed behind an obstacle. During the Franco-German War, the German advanced cavalry frequently captured by surprise magazines of French chassepôts and ammunition. The German troopers promptly discarded their carbines, and armed themselves with the captured rifles.

Frederick's cavalry under Seidlitz had on the whole a distinctly better appreciation of correct cavalry tactics than the French cavalry under Napoleon. The latter were too much inclined to manœuvring, halting and file-firing, while Seidlitz's horsemen were much more thoroughly imbued with the idea of charging home without delay.

In the matter of the amount of independence to be allowed to cavalry commanders in battle, there is an appearance of fundamental diversity in the practice of Frederick and Napoleon on the one hand, and of Wellington on the other. The two former, on joining battle, gave very general instructions to the cavalry commander, and then left him a free hand. Wellington would plant his cavalry at some point and expect it to await orders from himself before operating. But a more careful reading shows that Wellington was usually quite differently situated from the two others; when he took this line, he was on the defensive, and his cavalry force was small. Frederick and Napoleon were almost always the attackers, and they wielded great armies of horse. It is easy to imagine the position of Wellington, a supremely prudent commander, "pondering over everything the enemy can do."

Change of Tactics during a Campaign.

This is a good point at which to call attention to a peculiar feature in the development of tactics, viz., that these very often take place during the course of a campaign, under the stress of necessity. The development of Mesnil-Durand may be said to be due to the sheer logical excogitation of a method by an earnest and able soldier. But the more usual cause is the discovery in the course of a campaign that some new weapon, or some new advantage in the hands of the enemy, has rendered your previous methods obsolete. Thus, for instance, the Germans found in 1870 that supports could not reach the firing line in column, and supports began, as it were, to take the law into their own hands and extended as soon as they got within range of the enemy's fire. Again in South Africa, our men, trained to be at two or three paces apart in the firing line of the attack, very soon learned that 10 to 15 paces was not too much; and this amount of extension was actually enjoined in Army Orders by the C.-in-C. in South Africa.

CHAPTER VI.

THE BRITISH IN THE PENINSULA.

CORUNA, 16TH JANUARY, 1809.

FOR 12 days the British Army, Sir John Moore commanding, had been retreating before a superior force of French under Marshal Soult, through the mountains of Galicia, in the north-west corner of Spain. A fleet of transports and frigates put into Coruña harbour, on the night of the 14th January, 1809, and the work of embarking the sick and most of the guns and the best of the horses was set going. The difficulties of marching in Galicia at that season had delayed even Soult and his experienced troops. As they arrived, the British took up their ground for battle as shown in the map. The heights on which they stood were dominated in altitude by the French ground, but the small number of our troops made it impossible to hold the more extensive range from which the French attacked. On our left stood Hope, with ten battalions, having his outer flank secure on the muddy Mero; Baird's Division of 13 battalions extended along the rest of the height. It will be noticed that the oblique direction of the position rendered it capable of enfilade from a dominating hill, on which Soult, during the night before the battle, established a battery of 11 heavy guns. Thus Moore's naturally strongest flank was of necessity refused, while the right was "in the air." Baird kept a couple of battalions in column behind his right, and sent a detachment into Elvina village. Paget's reserve of six battalions was behind the heights, as shown; a detached battalion stood a mile to the right of Baird on the edge of the heights along which the numerous French cavalry were seen to be extending, and another battalion linked these by a chain of skirmishers across the valley.

Plate VII.

As the French had the power of manœuvring quite out of sight beyond their ridge, it was necessary to guard the St. Jago road; therefore Fraser's Division of seven battalions stood on some moderate heights above St. Cristoval village.

By the morning of January 16th, all but the fighting men were on board, and it began to look as if the embarkation would be unmolested; but at 2 p.m. a general advance of the French began, and 14,500 British infantry took their posts.

It will be seen that the ground was all in favour of the enemy; it is worth noting that Coruña itself was sufficiently fortified to be safe against a rush, but that the harbour would be at the mercy of an assailant who drove the defender into the works. An advantage we had was that the stores in Coruña had supplied many thousands of new English muskets and fresh ammunition.

The French came on, 20,000 strong. Distributing their light guns along the front, Soult opened fire from his heavy battery and descended in three columns, covered by clouds of skirmishers; advancing with their usual impetuosity, they quickly drove in the British picquets, and Elvina was carried by the left column (Mermet's). This column then divided, and a part sought to get round Baird's right flank. Merle, from the French centre, directed himself at Hope's right, and Delaborde advanced by Palavia Abaxo. The few British six-pounders were soon outmatched.

Moore, watching the fight from Baird's right, saw that the whole of the French infantry were in these three columns, or supporting them; so Fraser could now be considered a reserve, and Paget available for early action. This officer's force was therefore directed to beyond the portion of the column that was trying to outflank Baird, who now, by a wheel of his right flank, was anticipating the threat. Fraser was brought forward a little to support Paget. The danger on the flank being thus provided for, Mermet's frontal attack had to be attended to. The 50th and 42nd left Baird's defensive line, and after a sharp struggle cleared the village, the Guards' battalion filling up the gap thus left. The 50th advanced through the village, but the 42nd withdrew under some misunderstanding. Mermet, seeing this, advanced again in force; but a few animating words from Moore sent the 42nd in again.

By this time the fight was raging along the whole line. Paget's brisk advance in the valley encouraged the skirmishers and the detached battalion, and all the French there were pushed back, including Lahoussaye's Dragoons, who had dismounted to bar the way to the great battery. At this moment Moore was struck down and carried off the field; Baird having been wounded earlier, Hope became C.-in-C. The British were now making headway all along the line, and it looked as if the pushing forward of Fraser's fresh men would have brought disaster to the enemy, who had a single bridge in rear across which to retreat. But night was coming on, and Hope felt bound to proceed with embarkation; it could hardly be known for certain, also, that Ney's Corps was not closing up behind Soult.

In this battle the French right was under Delaborde, who had shown such skill and vigour at Rorica, and still it effected little. This was because its attack had to be solely frontal, and was not assisted by the raking fire of the great battery. The feature of Moore's tactics was the bold and early use of Paget; and of the fighting, the gallant action of the 50th and 42nd at Elvina. We lost 800 men, and the French more than 2,000.

Talavera de La Reyna, July 27th and 28th, 1809.

Coruña had been a defensive-offensive battle on the part of the British, with the offensive portion well begun but not pushed to extremity. Talavera was a defensive battle, successfully fought by Wellington, with 20,000 British and 34,000 Spanish troops under Cuesta, against 50,000 French with 80 guns. Wellington had 30 guns, and his Allies about 70.

The French attack on the allied position began on the evening of the 27th, Mackenzie's Division having, earlier in the day, been driven back from the Alberche with the loss of 400 men. Wellington had assumed command of both forces, and had placed the Spanish in two lines, the right resting in the front of the town, which was on the Tagus bank. The ground in front of the Spanish was covered with olive trees and much broken with banks and ditches. All the approaches of the town were defended by batteries and ditches; mud walls and log breastworks secured the whole front.

The left of the Spanish rose into an eminence, where a redoubt was being made which was garnished with Spanish guns. Beyond that the ground was generally open, and on it extended the British line, one mile long, covered in front by a small ravine—Campbell's Division first, 3,000 men, supported by Cotton's Cavalry Brigade; Sherbrooke's Division stood next, 6,000 men including the Guards' Brigade, with Mackenzie's Division, 3,000 strong, in support; then came the German Legion, and the left of the infantry line was completed by Hill's Division, 4,000 strong. This part of the line faced nearly north, looking into a valley that ran east and west. Beyond the valley, half a mile wide, rose the mountain ridge dividing the Alberche from the Tietar. The British cavalry, 3,000 sabres, was most of it in rear of the left, and there was a strong body of Spanish horse. The French came on with Milhaud's Dragoons on their extreme left; these horsemen rode close up to the Spanish front, to make it display its strength. A terrific, but ill-aimed, discharge of musketry came from the ditches, and then 10,000 Spaniards broke and fled, carrying with them a lot of their gunners. Fortunately the French horsemen were not able to cross the entangled ground, and their infantry were too far off to profit by the panic. Wellington has been blamed for not having, towards the end of the fight on the following day, ordered forward the Spaniards to take in flank the French masses, which confined their attacks to the British half of the line. The episode just related seems a sufficient reason, but it is only fair to say that some of the Spanish battalions did respectable work next day, also some of their squadrons, and that, on Wellington's own word, the Spanish artillery was excellent.

Sebastiani's Corps (4th) was following up on the French left, with King Joseph's Reserve behind. These did nothing on the 27th, but on the French right Victor, seeing the key of the British position, Donkin's Hill, feebly occupied, started to take it, using his whole corps, though the sun was already setting. Ruffin's Division made the attack, quickly and vigorously, Lapisse demonstrating against the Germans, and Villatte supporting both. Donkin repulsed the frontal part of the attack, but round his left the French gained the summit. Hill came forward to help, and after a fierce struggle in semi-darkness, with a good deal of bayonet work, the French were cleared off the hill and back to their own side of the ravine. The British had lost 800, the French 1,000 men.

During the night Victor persuaded Joseph to try again. The bulk of the French artillery was now massed on a height within effective range of the angle of the British left, and could both sweep the valley to its right and the whole front of the British position, as far as the redoubt on the centre.

Ruffin again led the attack on the right, striving to envelope the enemy's hill, advancing with two regiments abreast and the third in support in battalion columns. The French guns poured a hail of shell on Hill, until Ruffin's men began to mask their fire, which was then turned sharp on Sherbrooke in the British centre. Ruffin's formations were somewhat broken by the difficulties of the ground, and a series of small, but most virulent, struggles ensued all along the hillside. Sections here and there reached the summit, but the steady 2nd line of the British always succeeded in driving them down again. Men fell fast on both sides, and Hill was wounded. After 40 minutes of real hard work Ruffin's Brigade broke and fled in disorder, having lost 1,500 men; the massed artillery covered the flight and prevented pursuit. Our guns were no match for them either in size or number, and Cuesta, being asked for some of his, which were idle, sent *two*.

Wellington, seeing that further attempts on this side would probably be made up the valley, now massed most of his horsemen beyond Hill, and facing down the valley. There was now for three hours till noon a complete cessation of fighting, but the forecast was correct, for Ruffin was now sent towards the far side of the valley, with the object of getting him beyond our left, Villatte menacing the hill by a direct attack and preventing interference with Ruffin's march. Victor's 3rd Division, Lapisse, was to help Villatte against the hill, and also, supported by cavalry and the King's Reserve, to pass the little ravine and attack Sherbrooke. Sebastiani was to move against the British right in full strength.

Wellington was on a commanding point, from which he saw Sebastiani hurl his masses against Campbell with infinite fury. This division, with the help of Mackenzie's men and the two nearest Spanish battalions, withstood the attack, advanced with shouts against the columns when the range became short, sweeping aside the hostile skirmishers as a gale sweeps the leaves, lapped round the flanks of the masses and drove them back with huge loss, capturing 10 guns that accompanied the attack. Then, with admirable discipline, the pursuit was stayed, and line re-formed. The Frenchmen, than whom in those days no troops more quickly recovered from a repulse, re-formed in a moment, and were at it again. But again the British fire, aided this time by a flank charge of a Spanish cavalry regiment, broke them up, and the fight was over at this place.

Meanwhile Ruffin and Villatte were moving as above stated, and the wisdom of having cavalry in the valley was justified. But the ground towards the enemy was not properly reconnoitred, and when Anson's Brigade, the 23rd Light Dragoons and the German Hussars, ordered to attack, were well into their gallop, they came unexpectedly upon a small ravine. The Germans pulled up, but the dragoons scrambled through, and appeared on the other side in twos and threes, within range of the French infantry, who had rapidly formed squares. General Anson got his men together, passed through Villatte's men, and engaged with the French chasseurs in rear. Victor, who was watching the episode, had already dispatched light cavalry to the spot, and the 23rd, outmatched, only got out of the difficulty with the loss of half their numbers. Wellington wrote in his despatch :—" Although the 23rd Dragoons suffered considerable loss, the charge had the effect of preventing the execution of that part of the enemy's plan." It is hardly to be thought that the mere charge of a single regiment was the sole cause, but that the French saw several more lines of cavalry waiting for them. Joseph should have had the mass of his cavalry actually accompanying Ruffin and Villatte.

All this time, the hill at the corner was being assailed by Lapisse from the front, aided greatly by the short range afforded to the powerful French artillery. The attacks were pressed home, but always without success, and it was the very success of the British here that produced a dangerous crisis at this important point. The Guards' Brigade, in the excitement of winning, pressed forward imprudently, were assailed in flank by reserves and cavalry and battered by the enemy's guns. Falling back at the same moment that the Germans on their left rear fell into confusion, a dangerous gap was made in the British centre, towards which Lapisse with ready eye directed his rallied troops. But there was a ready eye on the other side. The moment Wellington perceived that the Guards' advance was pushing forward recklessly, he brought up the

48th from behind Donkin. This stubborn old regiment marched forward in line with the coolness of parade, let the fugitives pass through by forming temporary gaps, wheeled again into line and poured steady volleys on the advancing enemy. Cotton's Dragoons came up and charged, the Guards and Germans re-formed under cover of the 48th, and the battle was restored. For the hostile columns gradually lost their impulse and no further support arrived in time to push them on. Wellington had been ready for the crisis, and he wins who recognizes it. Ruffin and Villatte had halted after our cavalry charge, and the "go" of the French attack was over. Joseph was too timid to send in his Guards and Reserve, and the French withdrew to the hill from which they had started their attack.

By 6 p.m. all fighting ceased; the British, reduced to 14,500 men, and exhausted from want of food, could not pursue, and to use the Spaniards for any such purpose would have been madness. Two of our generals and 31 other officers had been killed; 3 generals, 192 officers wounded. Of the rank and file there were 800 killed and 3,700 wounded—about 25 per cent. of the total force. The enemy lost 7,400 in officers and men, more than half that number being of Victor's Corps.

On the British side, the chief matters to note in the "grand Tactics" are the securing of the left flank by the cavalry, and the timely sending forward of the 48th Regiment; in the fighting tactics, the fine discipline that restrained the advance of Campbell's Division, and the premature rush of the Guards. In a battle begun defensively, small portions of the line must not be allowed to make reckless advances, but when the enemy's impulse appears to be spent, the time for an all-round advance has arrived.

On the French side, as soon as they planned to attack the British only, the cavalry should have been in great force in the valley, or leaving the cavalry to "contain" the Spaniards and the British right, a much stronger force of infantry than one division should have attacked from the valley. The general plan of attacking the British only was good; had it been even temporarily successful, the Spanish would have given the French no trouble, whereas, from the mere physical difficulties of the Spanish position, time would have been spent which Wellington would surely have utilized to push forward, and then on to the French right flank.

If the French, on the contrary, had during the previous night placed their whole army beyond the British left—as some of them wished to do—Wellington had planned to keep his left wing and cavalry where he had them and to put his right wing on the heights where the French guns were in the actual fight. Success in battle would then have cut off the whole or the bulk of the enemy from Madrid.

British in the Peninsula.

A French writer says:—"The English always took care to choose a good position to offer battle in; they did not crown the crests of the heights on which they placed themselves; their infantry deployed in line, formed up from 50 to 100 paces in rear of the edge, in such a manner as not to be seen even if the slope had little steepness. The slope itself was garnished with skirmishers." (The flank companies of each battalion skirmished, and were picked men). "The sound of firing and the gradual return of the skirmishers warned them of the hostile approach. At the moment when the enemy appeared, a steady volley was fired whose effect was terrible at so short a range; then the line charged. If it was successful, it contented itself with sending the skirmishers in pursuit, and took up its former position."

He goes on to acknowledge that these tactics were very reasonable; after their fire they charged the heads of the columns and the wings lapped round their flanks. Then he contrasts the wild French method of pursuit when they were successful, causing loss of ranks and cohesion and discipline.

There is no doubt the British method was rational, for troops of their quality. First, there is the skilful choice of position, safe from infantry fire till the last moment and steady volleying on the head of a mass that has just painfully climbed a long slope. The retired position, too, would often shelter them from view of the enemy's gunners. Then the steady volley, confusing at the least the head of the column, which would be in two minds whether to advance as it was, or to deploy. Then during the hesitation, or the attempt at deployment, the charge with the bayonet. Finally, the perfect coolness and discipline following the charge.

In the attack the British also adhered to the simple line formation, with a supporting line in rear, and a reserve in the hands of the commander.

The same French writer that I have been quoting says that "in spite of all their precautions, the British ought to have been beaten every time, and the columns would certainly have done it if the technical formations of the French had been as pure as in the first wars of the Empire, and if other causes in addition had not rendered useless the courage of those admirable troops that composed the armies of the Empire."

These are simply excuses. The British tactics were on a higher level altogether than those of the French, just as Mesnil-Durand's system was on a higher level than that of Frederick. And the reasons in these cases are two. In the British case—(1), more and steadier fire; (2), a reason which was the *cause* of the power to use the higher tactics, viz., an infantry of superior quality. In the French case—(1), superior mobility, enabling units to be handled more freely and intelligently in the varying phases of the fight; (2), as in the British case, an infantry of superior quality to their Continental enemies. For it should always be remembered that there are two main factors in producing change or improvement in tactics. One is an improvement in the weapon, the other an improvement in the quality of the troops—sometimes an improvement in the individuals, and sometimes an improvement in the mass, due to training and drill. Let us hark back over the stages of development with this in view. In what was Frederick's system, with its heavy cumbrousness, superior to his predecessors and his adversaries? To talk of his system as in itself producing mobility would raise a smile; but he trained his men to such an unheard-of condition of perfection in drill that his army was able to make the stiff and complicated movements like an unerring machine, if no one interfered. The infantry of his opponents, far less trained and drilled, were afraid to make voluntary movements in the presence of a formidable enemy. Frederick had, in fact, produced men in the mass of a higher quality than his enemies, and he was able to manœuvre for a definite object, all cumbrous as his way was, in a manner forbidden to his opponents. Thus he achieved, over them, a comparative excess of mobility.

Change in Weapons and in Quality of Men both required for Changes in Tactics.

Then comes the French era. In this, and the new system it produced, a comparative excess of mobility was achieved, not by improvement of men in mass—*i.e.*, not by an immense grind of drill and training—but by an improvement of the individual soldier. The better educated and socially superior men who poured into the ranks of the Revolutionary armies made it possible to dispense with the years and years of barrack-square drill, and to adopt a tactical method suited to rapidly trained men of great individual intelligence. More freedom of movement for small units became possible because the men were individually better ; but the lack of the discipline of the barrack square sometimes made itself felt.

Then the British infantry come on the scene, and you get men of an extraordinary coolness and self-confidence. A thin line of them is not appalled by the close approach of a solid column ; they are of that quality that they wish to put it to the touch of the volley and the bayonet to prove whether the solid mass is so formidable as it supposes itself to be. Hence arises the phrase that the British quality is "not to know when you are beaten." Europe was accustomed to hold, if a French column succeeded in advancing to within a few score yards of the hostile front, that the hostile front must give ground or be broken through. But a superior individual existed in the British infantry ; he could be trusted to so form up in apparently flimsy ranks that every man could fire, and every man could use his bayonet.

Thus, not so much in mobility as in the full use of every weapon, were the British tactics superior to the French, and these tactics were possible because the individual was a better fighting man by nature, and had also received all the requisite drill and training to ensure discipline. Wellington, with the capacity of the truly great soldier, made his *début*, as an independent commander against French troops, in the campaign of Vimeiro, but he had pondered long and deeply over the causes that led to the almost continuous tactical successes of the French ; or rather, knowing the cause to lie in their intelligent and plentiful use of skirmishers, he set himself the problem of counteracting their effect. As skirmishers, he knew the French soldier was probably for the moment superior to the British, so the sending out of an equal number of these would not solve the difficulty, and more could not be spared from the line of resistance or the supports or reserves. He accordingly fell back on the method of getting more fire from the line of resistance. Ignoring the *letter* of the Drill-Book of the day, he ordered the infantry to stand in two ranks, instead of the customary three. The Drill-Book said that three-deep was to be the normal formation, and remarked that the 3rd rank's use was to fill gaps in the other two. Wellington used it for extra fire from the outset, but he made no such change in the traditional and perfectly sound principle of keeping three distinct lines of battle, what Cæsar calls *triplex acies*, for defence and attack—the 2nd line to be sent forward *before* the 1st is broken, but as soon as the 1st has done all it can, and the 3rd to be used in the same manner.

Every one has noticed that Wellington, though he fought many defensive battles, did no entrenching to speak of, except at Torres Vedras. At Talavera, the Spanish half of the position was prepared for defence, simply because this part of the force was not fit for offensive operations. Wellington wished to encourage his battalions to active work at all points of the line ; men armed with the musket had to stand up to load ; a high breastwork was required to cover a man under these circumstances, and there was seldom time for the work, and hardly ever sufficient tools. Moreover, the tendency would be to encourage a purely passive defence, and Wellington was too good a tactician to allow that. He found it better, with the weapons of those days to face, to make the best use of the ground as it was, withdrawing for instance the infantry main line 100 yards or so back from the edge of the crest, so that they might be out of sight of the hostile artillery, only sending the skirmishing companies to the edge and down the forward slope.

The scouting operations of the British cavalry officers deserve a word of comment and commendation. These, singly or in pairs, well-mounted and good cross-country riders, annoyed the French columns by hovering about just out of musket range and taking note of the strength and direction of marches. Any attempt to catch these hunting men usually failed. Thus, in the Busaco Campaign, Masséna's concentration towards the head waters of the Mondego river left it uncertain at first whether his advance would be by Belmonte in the Zezere valley, by the left bank of the Mondego, or by its right bank on Coimbra. An officer watching the movements from points right within the French Army's territory was able to report such facts as made it certain that the third route was being taken, the one that led over the Busaco ridge.

At an earlier stage, in December, 1808, when Moore was making his strategic counter-attack from Salamanca against Soult, a British cavalry officer, scouting singly, witnessed in a village an altercation going on between a French officer and a knot of villagers. The former was set upon and killed, and the peasants found a despatch on his person. The British officer, asking for the document, was refused, till he tempted them by offering all the money he had in his pockets, 20 dollars. The despatch turned out to be one from Berthier, Napoleon's Chief of the Staff, to Soult; and it gave the important information that a French corps was on the Tagus and nearer to Lisbon than Moore himself, that the French thought Moore was retreating into Portugal, and that Soult was to push on west towards Galicia. This finally determined Moore, who had begun preparations for retreat, to make his march forward against Soult, an operation which drew Napoleon from Madrid, and arrested all his plans for the rapid subjugation of the Peninsula.

CHAPTER VII.

THE WAR OF SECESSION.

FROM the middle of 1861 to April, 1865, the Southern Confederation in the United States of America kept up the unequal struggle—unequal, because the Federals greatly outnumbered them in population, were much superior in wealth and had from an early stage complete command of the sea—an advantage which enabled them not only to land at suitable points on the southern seaboard, and to obtain munitions of war in any quantity and of the latest patterns from Europe, but to prevent their enemies from thus supplying themselves.

From the free and independent nature of the American people and of their institutions, we should expect correspondingly novel developments in methods of fighting. They were very free in their use of skirmishers; their cavalry, especially that of the Confederates, acted in the boldest and most efficient manner; and if their artillery was sometimes rather timidly used, that was the universal custom in European armies of the immediate past. There were, of course, great exceptions on occasion, as when Pelham at Fredericksburg, on the extreme right of the Confederates, boldly brought his two guns out into the open, and took in flank the attack that Franklin was making on Jackson. But, when marching to meet the enemy, there was still the tendency to keep the guns far back in the line of march, and to hold back a reserve of artillery during the fight.

In the numerous forests of the Southern States, it was perhaps natural enough that great use should be made of log breastworks. Against the rifles of those days, it was easy with logs to keep out the bullets; and both sides made intelligent use of this contrivance. Entrenching was very freely indulged in by both sides; and such was the character of the combatants that on most occasions mere strategic threats did not suffice to frighten the enemy out of his position. Both parties usually wished to prove who was the best man before yielding ground; consequently the battles were often most sanguinary.

After a few remarks on the weapons in use, and a short description of some of the cavalry work, I shall describe one of their battles. Europe was at that time apt to look upon the belligerents as so much of amateurs that their tactics were hardly worth studying; but the strategy often displayed, especially on the side of the South, was fully appreciated.

At the beginning of the war, we read, with reference to the South, "arms were far scarcer than men. The limited supply of rifles in the arsenals was soon exhausted. Flint-lock muskets, converted to percussion action, were then supplied; but no inconsiderable proportion of fowling-pieces were to be seen among the infantry, while the cavalry, in default of sabres, carried rude lances fabricated by village blacksmiths."*

* The shot-guns were often most effective in the close-quarters fighting in the woods.

Comparison of Firearms.*

The rifles (muzzle-loaders) used throughout the war by both sides compare as follows with later weapons :—

	Sighted to	Effective at
American	1,000 yds.	250 yds.
Needle-gun (1866 and 1870) ...	660 "	250 "
Chassepôt (1870)	1,320 "	350 "
Martini-Henry	2,100 "	400 "
Magazine small-bores	3,200 "	600 "

By effective range is meant the distance where, under ordinary conditions, the enemy's losses are sufficient to stop his advance. The effective range of "Brown Bess" was 60 yards. The American rifled artillery was effective, in clear weather, at 2,000 yards, the 12-pounder smooth-bore at 1,600, the 6-pounder at 1,200.

In 1866 and in 1870 the rifled artillery may be taken as about equal to the above.

In 1862 the South is reported as having its batteries of eight, four and three guns, but mostly smooth-bores of the smaller calibre. The battle I shall describe took place in this year. The infantry were by now fairly armed and equipped.

At the same stage we read of Bank's Army of the North, "the artillery, armed with a proportion of rifled guns, was more efficient than that of the Confederates, and in cavalry alone were the Federals overmatched."

American Artillery.

In many fights the artillery was most effectively used. Thus, at the first battle of Bull Run in 1861, we read "a long line of guns (Federal), crossing the fields at a gallop, came into action on the opposite slope. In vain Imboden's gunners (Confederate), with their pieces well placed behind a swell of ground, strove to divert their attention from the retreating infantry, now climbing the slopes of the Henry Hill. The Federal batteries, however, powerful in numbers, in discipline and in *matériel*, plied their fire fast. The shells fell in quick succession amongst the disordered ranks of the southern regiments, and not all the efforts of their officers could stay their flight. The day seemed lost."

Night attacks, and attacks at dawn, were made on several occasions, at a time when Europe generally condemned them as too risky.

In the earlier half of the war the Confederates were specially distinguished over their opponents in their bold use of cavalry.

A very able and observant British cavalry officer (Colonel Trench) thus wrote in 1884 about the American cavalry in this war :—

Cavalry acting Dismounted.

"Hence, though the necessity of imparting to him instruction in the art of fighting on foot was long ago *theoretically* recognized, the practical importance of the subject, both in our own and Continental armies, was up to the time of the Franco-German War resolutely ignored. No stronger proof of the truth of this assertion need be adduced than the fact that the instructive and valuable experiences of the American Civil War were never taken to heart or utilized by the great armies of Europe. For what were the facts and the lessons which might have been learnt from the exploits and successes of the American cavalry on both sides? The shrewd, practical common sense of the American people, and the utter absence of tradition, prejudice and red tape, led them to adopt a system of tactics somewhat new, and peculiar to themselves, and

* Taken from a footnote in Colonel Henderson's "Stonewall Jackson."

not at all in accordance with European notions, practices or ideas." "Who that has read the accounts of what mounted troops—call them cavalry or call them mounted rifles, as you will—achieved in America can deny their great usefulness and efficiency? Though these successes of the American horsemen were patent to all the world, the cavalry of Europe steadfastly shut its eyes to obvious facts, and failed to profit by the cheapest experience of all—viz., that obtained at the expense of others. It was after 1870-71, in which both the French and German cavalry (each drilled, equipped and manœuvred in the old-fashioned style) failed to effect anything on any field of battle at all proportionate to the cavalry forces employed or to the sacrifices incurred, that the experiences of the American War were taken to heart and began to bear some practical fruit. Even on scouting and reconnaissance duties, as soon as the "Franc-tireurs" were organized, the training, equipment and armament of the much-vaunted Prussian Uhlans rendered them quite incapable of coping with foes who, though brave, must, from want of training and organization, have been contemptible in themselves."

I have already stated incidentally that the Uhlans, as time went on, took to substituting captured chassepôts for their own short-range carbines, and gradually learned dismounted tactics for themselves.

Battle of Gettysburg, July 1st, 2nd and 3rd, 1863.

This battle has been called the Battle of Lost Opportunities. It was an "encounter battle," in that both armies were on the move, the Confederate Army under Lee converging upon the field from west and north, the Federals under Meade moving in one direction, north-west. The battle is generally held to have been the turning point of the war.

Lee had 73,500 men and 190 guns, besides Stuart's Cavalry (10,000 and 16 guns). The bulk of this cavalry force had indulged in a prolonged raid round the rear of the enemy, and was not to the fore when most required.

The mass of the army was in three corps, of which the 2nd (Ewell's) was coming down on Gettysburg from the north, while the 1st (Longstreet's) and the 3rd (Hill's) were marching from the west. Lee's line of retreat was on the west side of the South Mountains, thence across the Potomac into the Shenandoah Valley.

The Federals were covering Washington and Baltimore. They had 82,000 men and 300 guns, divided into seven Army Corps (1st, 2nd, 3rd, 5th, 6th, 11th, 12th). The small size of these corps should be kept in mind. The Federal cavalry (under Pleasonton) was about 10,000 strong, with 27 guns.

Both armies had a mixture of smooth-bore and rifled artillery, the Federals being much the better supplied of the two with the new weapons.

Buford's Cavalry Division (Federal) was forward in Gettysburg before the battle began, and the nearest corps on the morning of July 1st were the 1st and 11th. The cavalry discovering an advance in force of Hill's Corps on the Cashtown and Fairfield and Mummasberg roads, these two corps were ordered to hold the Seminary Ridge and cover the town; the 1st Corps began to arrive about 10 a.m., and was at once posted along a broader ridge some 500 yards in front of the Seminary. The corps was at once engaged in a desperate encounter with superior forces, Archer's Brigade of Heth's Division of Hill's Corps being already in the woods and pushing for the summit of the

Fig. 1, Plate VIII.

Fig. 2, Plate VIII.

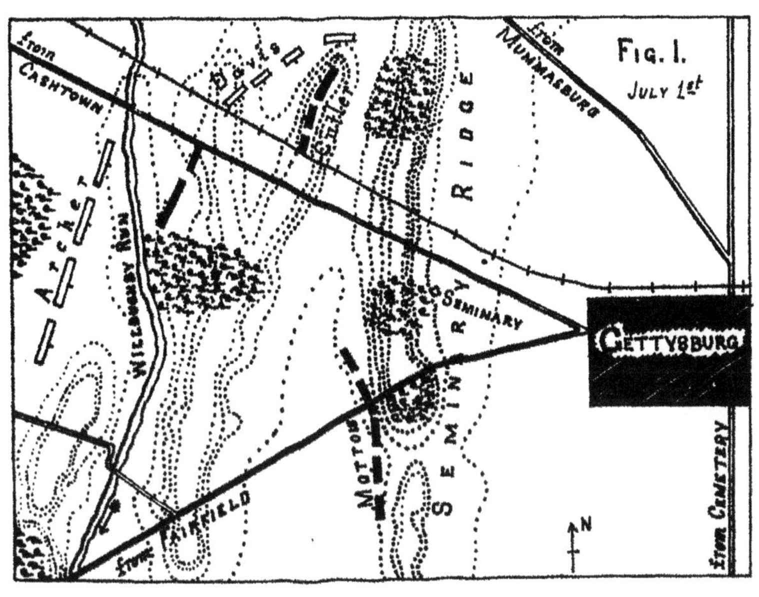

Fig. 1.
July 1st
from Cashtown
from Mummasburg
Davis
Archer
Willoughby Run
Ridge
Seminary
Gettysburg
Seminary Ridge
from Fairfield
from Cemetery
N

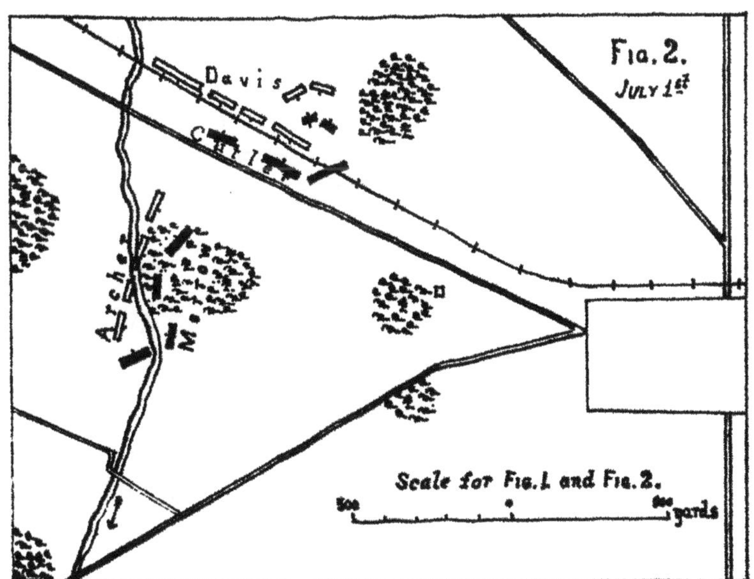

Fig. 2.
July 1st
Davis
Archer
Scale for Fig. 1 and Fig. 2.
yards

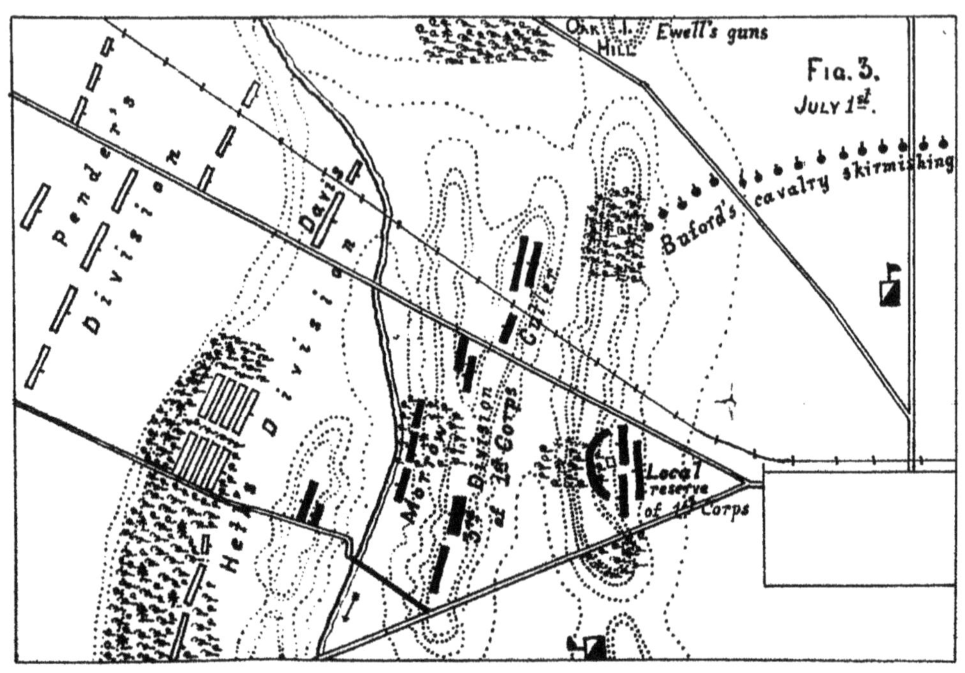

OAK HILL
Ewell's guns
FIG. 3.
JULY 1st.
Buford's cavalry skirmishing
Pender's Division
Davis
Heth's Division
Cutler
3rd Division of 1st Corps
Local reserve of 1st Corps

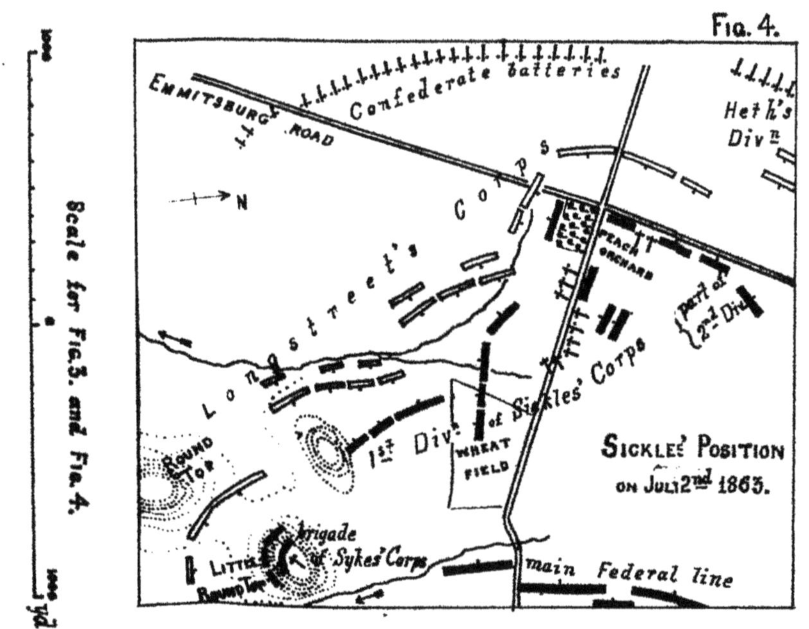

FIG. 4.
Confederate batteries
EMMITSBURG ROAD
Heth's Divn
N
Longstreet's Corps
PEACH ORCHARD
part of 2nd Divn
Sickles' Corps
1st Divn of Sickles' Corps
WHEAT FIELD
ROUND TOP
LITTLE ROUND TOP
Brigade of Sykes' Corps
main Federal line
SICKLES' POSITION ON JULY 2nd 1863.
Scale for Fig. 3. and Fig. 4.
1000 yds

PL. VIII.

Fig. 5.
July 3rd 1863.
Ewell's Corps
Rock Creek
Gettysburg
Pender
Cemetery Hill
Heth
11th Corps
Culp's Hill
Johnson's expulsion
12th Corps
1st Corps
2nd Corps Gibbon
Pickett
Doubleday Div. of 1st Corps
Artillery Park
Wilcox
3rd Corps Sickles
Reserve
Peach Orchard
6th Corps
2 divisions of Longstreet's
Sykes 5th Corps
Little Round Top
Kilpatrick
Round Top
W. & A. K. Johnston Limited Edinburgh & London
Scale for Fig. 5.
0
1000
2000
3000 yards

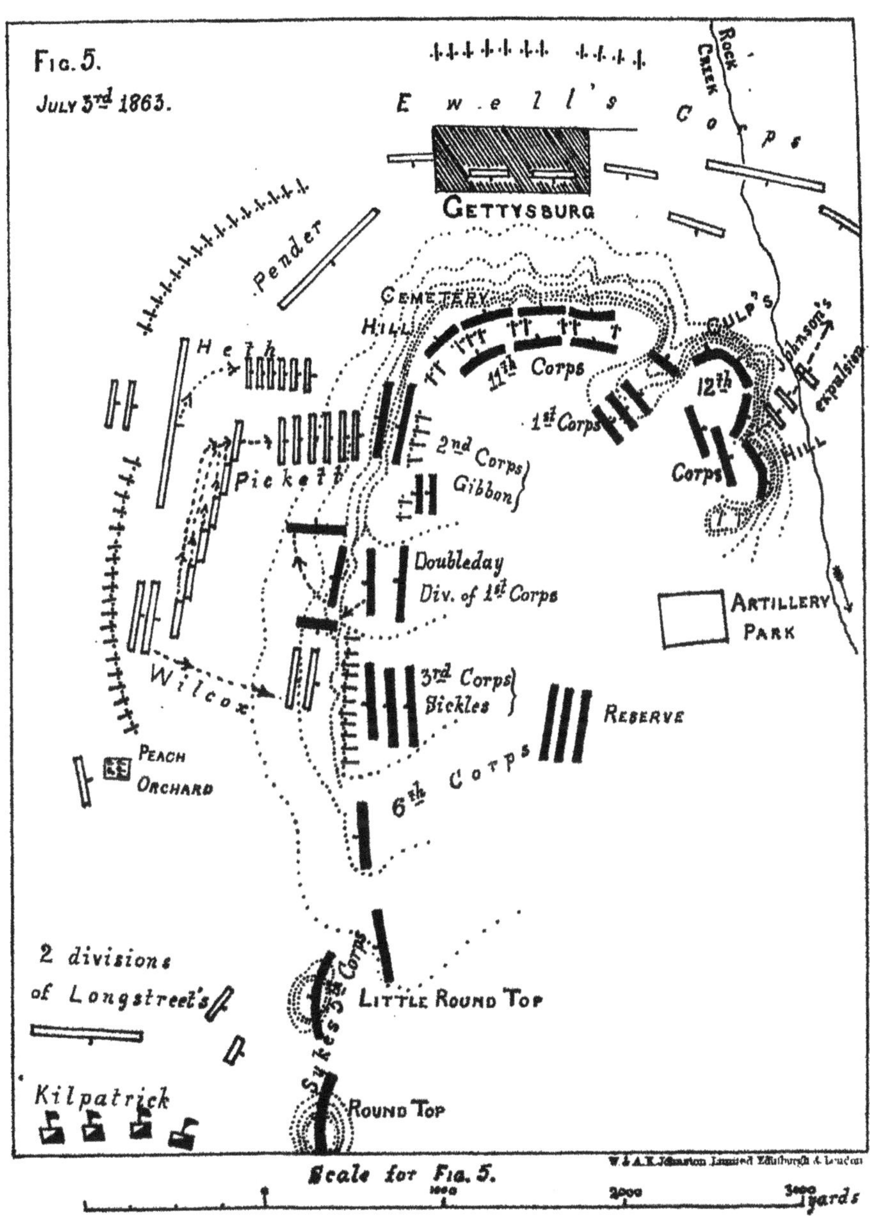

ERRATUM.

Page 49, line 15, *for* "Federals" *read* "Confederates."

ridge. The first Federal brigade to arrive deployed at the double, "charged with utmost steadiness and fury, hurled the enemy back into the run (burn), captured, after a sharp and desperate conflict, nearly 1,000 prisoners and Archer himself . . . and re-formed their lines on the high ground beyond the ravine."

This kind of fighting was going to be characteristic of the whole three days' battle; General Doubleday, at present commanding the 1st Corps, says in his report, "Final success in this war can only be attained by desperate fighting, and the infliction of heavy loss upon the enemy."

About this time Reynolds, who was in general command at the front, was killed.

At the same time that the temporary success was being obtained against Archer, Cutler's Brigade, drawn up across the north-west road and railway, was being assailed in front and flank. The division commander, Wadsworth, ordered retreat to the Seminary Ridge, but one regiment and a battery did not receive the order. Seeing these two units in great jeopardy, some battalions that had returned from the attack on Archer formed to their right, took the ~~Federals~~ Confederates about the railway in flank and restored the combat by a vigorous attack. *Fig.* 2.

But two divisions of Hill's Corps were now up in full strength, and Buford's Cavalry Brigade on the north of the town was beginning to feel the pressure of Ewell's advance. Portions of the 11th Corps began to arrive between 1 and 2 p.m., and were sent to the right to face Ewell; about half remained at the Cemetery as a reserve. The part posted out went far forward and left a gap between its left and the right of the 1st Corps. General Doubleday says he would now have retreated, but he had no orders. At 1.30 Ewell's leading battery began to fire from Oak Hill, and Cutler had to withdraw to the Seminary, being enfiladed. *Fig.* 3.

The 1st Corps continued to defend the position with dogged courage, but it was rapidly becoming untenable, the 11th Corps having been pretty rapidly driven in by Ewell. Doubleday says, "It would of course have been impossible to hold the line if Hill attacked on the west and Ewell assailed me *at the same time* from the north; but I occupied the central position, and their converging columns did not strike *together* until the grand final advance at the close of the day, and therefore I was able to resist several of their isolated attacks before the last crash came."

This dislocated action was going to be a characteristic of the Confederates throughout the whole three days' battle. When "the crash came," and the two Federal corps retreated by order to the Cemetery on the south of the town, the 1st Corps had been reduced from 8,200 to under 3,000 men. Gettysburg was occupied by Ewell. The latter was *recommended* by Lee to pursue, but he had lost 3,000 men, and did not care about it. It seemed as if an opportunity was here being lost, for the 3rd Corps (Sickles') marching from Emmitsburg, and the 12th (Slocum's) from Two Taverns, were still at a distance.

Before night, however, these two were posted on the Cemetery Ridge.

General Meade now made up his mind to accept battle in this position; his original intention had been to fight behind Pipe Creek, 18 miles to the south-west, where considerable preparations had been made. Lee, on the other hand, was already further east than he cared about; he had hoped to draw Meade much further west before engaging in the decisive struggle, but the absence of Stuart had caused him to lose touch.

July 2nd.

Before the morning of July 2nd, Lee had his whole force up—Longstreet on the right, Hill in the centre, Ewell on the left.

The Cemetery Ridge was to prove an excellent defensive position. The slopes to the front were gentle, there were plenty of stone walls and boulders towards the top to cover infantry, and both flanks had rugged and commanding eminences. Culp's Hill on the right descended steeply to Rock Creek; westwards the ground sank away before rising again to Cemetery Hill. The summit then ran pretty continuously, but of gradually less height, almost due south for three miles; across a burn from the extremity there stood two commanding knolls, Round Top and Little Round Top, which were presently recognized as the key to this part of the line.

Besides the four Corps already mentioned as being now on the spot,* the 2nd Corps (Gibbon) had camped on the night of July 1st four miles from the ridge; the 5th (Sykes') had reached Bonaughton, six miles by road from the Cemetery; the 6th (Sedgwick's) was making a forced march of 36 miles; and cavalry divisions were coming on to cover both flanks.

The troops set to work at hasty entrenching.

Meade was in two minds whether to await attack, or mass to his right and attack Ewell. It would have been a mistake, for his full strength was not up, and the men were exhausted by their rapid marches. Lee, on his side, had good reason to expect to be able on the morning of July 2nd to destroy the four corps. From want of supplies he could not stand on the defensive; he must either attack or manœuvre; several authorities say he should have done this beyond the Federal left, alleging that he might thus have cut off Meade from the 2nd and the 6th Corps. But, by an early attack in full strength, he hoped to be able to get to grips with the enemy before Meade should have the help of *three* of his corps, 2nd, 5th and 6th, which seems a surer way than running the risks of manœuvring for the sake of separating off *two* corps.

But the "early attack in full strength" was exactly what did not come off, and Longstreet is usually blamed for it. Lee at first wished to make his chief effort against the enemy's right; being dissuaded on the ground of its strength, and the consideration of the gentler slopes and fewer troops on the hostile left, it was arranged that Longstreet should attack there "as early as possible" in the morning, that the sound of his guns should be the signal for Ewell to advance, and that Hill should demonstrate against the centre.

Lee's relations with corps commanders had always been peculiar; he seldom gave them a downright order; this was an occasion when he had reason to regret not having done so. Longstreet argued half the night with Lee in favour of manœuvring round Meade's left; when a subordinate commander does that, and his advice is not taken, the order as to what he is to do should be most definite.

The bulk of Longstreet's Corps was encamped within four miles of the field; but, on one pretext or another, he did not get into motion till after noon; some marching and counter-marching was then performed, and it was 4 p.m. before two divisions were in position for attack.

Meanwhile, the enemy was hourly increasing in strength. The 5th Corps was complete on the field by noon, while the 6th arrived from its long march about 4 p.m., and was posted in reserve.

* 1st (now commanded by Newton), 11th (Howard's), 3rd (Sickles'), 12th (now Williams').

v. Fig. 4, *Plate* VIII.

Sickles had placed his corps (3rd) in a very remarkable position, tempted probably by the intermediate rise of ground. The idea was that a salient like this, projecting to the Peach Orchard on the Emmitsburg Road, would provide a jumping-off point for counter-attack, when the time should come. But such a venture as this position should be held in special strength and be powerfully entrenched; in itself it was bound to be a weak point, both of its faces being liable to easy enfilade. It could only be effective, if the assailant should endeavour to push his attack past it without dealing with it first.

Longstreet's attack took place from south-west, and it soon showed the faultiness of Sickles' salient, which was rapidly disposed of, in spite of reinforcements sent down from the ridge. A desperate attempt at diversion was made by a division of the 2nd Corps advancing almost to the Confederate batteries, but it was taken in flank by the men who were now in possession of the salient and had to retreat with heavy loss.

This kind of counter-attack, isolated as it is, can effect little; in the present state of the battle, it could not be supported.

Meantime Longstreet's right made for Round Top, not yet occupied by Sykes, and proceeded thence towards Little Round Top, half a mile to the north of it. This hill was also not yet occupied; but General Warren, Chief Engineer of the Federals, seeing what was imminent, detached a brigade of Sickles' to the hill just in time. The fighting here became desperate, and the Federals were on the point of giving up the hill, when another brigade of the 5th Corps arrived, dragging with them a gun.

Little Round Top was now fairly safe; the Confederates had made a great mistake in not utilizing their unopposed capture of Round Top to post a battery or two on it. Little Round Top would then have been untenable by the enemy, and his position at this end of the ridge would have been seriously threatened.

Sickles' salient had now disappeared, and a pursuing Confederate brigade of Hill's Corps gained the summit of the main position, and turned a captured gun there on the enemy; but not being supported, it was driven out with great loss, and the line was re-formed.

Longstreet's left division was now in the wheat field close under the enemy's main line, but it was driven back out of it by Sickles, and by the appearance of two brigades of Sedgwick's Corps (6th), which, having rested for two hours, was now taking a hand in the game. Longstreet arrived at this juncture; darkness was coming on, and he gave up the attack and withdrew his men, apparently evacuating Round Top at this time as well.

Hill's operations had been feeble throughout the day; Longstreet had been far too late in getting to work, but had fought well when he did begin. Ewell's proceedings were very curious.

v. Fig. 5.

Having been ready for business, he was to wait till Longstreet's attack began; and after all, Ewell's attack did not begin till Longstreet's was nearly over. The former's plan was to send Johnson's Division against the east side of Culp's Hill, Early's against Culp's Hill and Cemetery Hill, while Rodes's was to issue from the town on Early's right. The two last effected nothing; but Johnson established himself at nightfall in some works that had been evacuated earlier in the day, when Meade, anxious for his left, had taken away part of the 12th Corps from this end of the position.

Ewell's batteries had been able to do little during this day's fight. Why he so delayed his attack is not very clear; the difficulty of getting effective gun positions, and the formidable appearance of things after long inspection, may have induced him to wait on the chance of Longstreet's and Hill's attacks causing the enemy to weaken their right.

Culp's and Cemetery Hills being once taken, the Federals would have been beaten, it is true; but a key of this kind is sometimes too dear to buy. Success at this end, with failure at the other end, would let the Federals at least escape from the field. When Lee gave up his idea of attacking the north end alone, he wished to bring Ewell round to the Seminary, and to edge Hill southward, so that Longstreet's attack might be better supported; but Ewell and Hill were so confident that they could not only hold their own, but take the ground in front of them, if Longstreet prevented reinforcement by the enemy from left to right, that Lee gave in, and thus took on a contract too great for his force.

July 3rd.

Fig. 5, *Plate* VIII.

Lee's confidence that he could still force the Federal position is shown by his official report. "The result of this day's operations induced the belief that with proper concert of action, and with the increased support that the positions gained on the right would enable the artillery to render the assaulting columns, we should ultimately succeed, and it was accordingly determined to continue the attack."

Lee's overnight plan had been to continue the attacks on the extremities, but a reconnaissance at daybreak showed him the 5th and 6th *entrenched* on the Tops; he therefore decided to attack the centre, while Ewell was to extend his hold on Culp's Hill. But the enemy were quite aware of the danger of allowing Johnson's lodgment to remain, as it was sure to be backed up in the morning by strong forces. During the night guns were posted to bear on Johnson, and part of the 12th Corps brought back; at the first streak of light Johnson was assailed by the artillery. He could make no reply, not having been able to bring any guns up the steep slope. Hoping to disembarrass himself of the shell-fire, and no doubt in expectation of immediate support from Ewell, he charged the guns and was beaten back with frightful loss. Meade, through the inactivity of the rest of the Federals, was able to bring 6 to 1 against Johnson, who was driven off the hill by 11 a.m. Ewell's Corps did nothing more during the day.

The utility of "interior lines" has seldom been more clearly manifested. The reinforcements that had helped to the rapid expulsion of Johnson were able to be back in their positions to meet the great attack that Lee was still bent upon making against the centre.

v. Fig. 5, *Plate* VIII.

This attack was to be made by two of Longstreet's Divisions, Heth's and Pickett's; the former had suffered severely already, the latter was fresh. Heth's, on the left, was to have Pender in *échelon* on his left; Pickett's, on the right, was to have Wilcox's Brigade to cover his flank, and Wright's in support in his rear. The two latter were the corps that had reached the crest on the previous day. Far away to the right Longstreet's other two divisions were going to attempt the turning of the Round Tops, but they were skilfully kept in play there by Kilpatrick's Cavalry Division. Hill was to respond to any demands for support on the part of Longstreet.

Guns to the number of 140, in two great batteries, opened the attack at 1 p.m. Meade was thus being granted double time for the charge that was coming.

Major-General Hunt, Meade's Chief of Artillery, says in his report:—"At 10 a.m. I made an inspection of the whole line, ascertaining that all the batteries—only those of our right serving with the 12th Corps being engaged at the time*—were in good condition and well supplied with ammunition. As the enemy was evidently increasing his artillery force in front of our left, I gave instructions not to fire at small bodies, nor to allow our fire to be drawn without promise of adequate results; to watch the enemy closely, and when he opened to concentrate fire on one battery at a time until it was silenced; under all circumstances to fire deliberately, and to husband ammunition as much as possible. . . . We could not, from our restricted position, bring more than 80 to reply effectively." Note here one of the disadvantages of the "interior" position.

The report continues:—"As soon as the nature of the enemy's attack was made clear, and I could form an opinion . . . for which my position (Round Top) afforded great facility, I went to the Artillery Park . . . and ordered all the batteries to be ready to move at a moment's notice . . . to replace such batteries as should become disabled."

Here is evidently a man who knows his business.

The artillery duel was of the most violent and destructive character; neither side gained any marked predominance.

"About 2.30 p.m., finding our ammunition running low and that it was very unsafe to bring up loads of it, a number of caissons and limbers having been exploded, I directed that the fire should be gradually stopped, . . . and the enemy soon slackened his fire also." Lee came perilously near to expending all his ammunition. Hunt was keeping a good stock of canister for the attack which was now imminent.

Later in the report General Hunt writes:—"The destruction of *matériel* was large. . . . The whole slope behind our crest, though concealed from the enemy, was swept by his shot, and offered no protection to horses or carriages. The enemy's superiority in the number of guns was fully matched by the superior accuracy of ours. . . . The attacks on the part of the enemy were not well managed. Their artillery fire was too much dispersed, and failed to produce the intended effect."

The casualties in the artillery for the three days were 40 officers, 697 men and 881 horses.

About 3 p.m. Lee made his last throw of the dice. The great column, with its flankers, approached 20,000 in numbers. From the point where it formed up under cover it had more than a mile to advance under fire. It began to suffer at once from the solid shot the Federals greeted it with at this range; intermediately, shell was fired, and finally canister.

The direction of the column was first towards Doubleday's Division, but when about 500 yards distant Pickett wheeled it 45 degrees to his left, which brought it against Gibbon.† Wilcox, on the right flank, did not conform to this movement, and soon there was a large gap, leaving the right flank of the main column naked.

All observers say that Pickett's men came on in magnificent style. The shot and shell tore gaps, but the lines closed up and moved without halt; only the left half of the column, Heth's old division, being particularly exposed from the lie of the ground,

* *i.e.*, against Johnson.

† This is how the movement appeared to Doubleday; but other reports seem to show that the main attack started in two lines, the rear line in *échelon* on the left of the front line; that each line, when it had advanced about half-way, formed battalion column on the march in rear of its *left* battalion.

showed signs of wavering here and there. A battery on Little Round Top did very effective work on the mass.

As the column approached the crest, the separation of Wilcox let it in for a flank attack. Two regiments in Doubleday's front line wheeled to their right, poured in some volleys, and advanced to close quarters; the nearest part of the column crowded in towards the centre, causing confusion there, and parts retreated, but the head pressed on and got in among the Federal guns. Here it was subjected to several bayonet charges, and after desperate fighting was driven back with great carnage.

Wilcox, meantime, came up against the 3rd Corps, and was so frightfully exposed to the artillery fire from there that he halted and began shooting. Taken in flank by a wheel of two regiments on his left, in the same fashion as Doubleday's wheel against the great column, Wilcox was soon disposed of. Longstreet's outer divisions effected nothing against the Tops. Kilpatrick manœuvred and fought his two cavalry brigades and two batteries with such spirit that it seemed at one time as if he was going to reach the Confederate train in rear of their artillery line.

On the other flank, Stuart's cavalry were engaged in an attempt to get at the Federal rear; a deal of hand-to-hand fighting took place with the Federal horse, and some dismounted work was indulged in. Stuart on the whole had the best of it, and he states that he was in a situation to cut off the enemy's retreat, if Pickett's charge had succeeded.

Comments.

Lee's last throw had failed him. Whether he was right or wrong in attacking at all, we are not here concerned to enquire, but only as to the methods employed.

The success of the first day was achieved long before dark, and there was ample time for Ewell and Hill to have driven the 1st and 11th Corps from Culp's Hill and the Cemetery, and to have occupied these. There would then have been no general battle of Gettysburg; perhaps Lee would have then stood on the defensive and awaited Meade's attack, but Meade would more probably have fully prepared the Pipe Creek position and have waited for Lee to attack or manœuvre or retreat.

On the second day there was a conspicuous lack of co-ordination in the work of Lee's Corps Commanders. The length and shape of the attacking line may have been responsible for this to some extent. The circuit of the Federal line, not including the cavalry, measured about four and a-half miles, giving nearly 18,000 men to a mile when the whole force was up, and counting in all reserves. This amounts to over ten men to the yard; and the internal lateral communications were good. From the position, for instance, of the 6th Corps, when it was posted in reserve on arrival, any part of the front could be reached in half an hour.

The Confederate line, at one and a-half miles from the enemy, had a circuit of seven miles, giving a little over 10,000 men to the mile. Hill's Corps, whose function was to "demonstrate," had the longest section. But the second day was lost through the notable lack of combination in the attacks of Longstreet and Ewell. Johnson's success came so late in the day that the footing in the Federal position could not be improved or enlarged. Johnson would hardly have gained that footing but for the removal of part of the 12th Corps to face Longstreet, so it may be allowed that Ewell was right in waiting till Longstreet's attack should produce some effect. This only shows up in stronger light the fault of the latter in delaying his advance so much.

On the whole, it seems that "proper concert of action" would have given Lee the victory on July 2nd.

On the 3rd Johnson was not properly backed, and his early expulsion enabled Meade to turn his whole attention to meeting the great assault. The quiescence of the whole of the rest of the Confederate front till noon made it possible for Meade to effect this rapid expulsion of Johnson by bringing a huge preponderance of force to that point. Johnson's position was only tenable and improvable by strong backing, and by the engagement of the enemy's attention seriously at other points. If the preparations of the great assault could not be effected before one o'clock, Johnson's position was bound to become very precarious, and vigorous feints should have been made at points distant from Culp's Hill. This might have been done by Hill; and Stuart was on the wrong flank. A powerful and active mounted force working beyond Longstreet would have enabled the latter to seriously menace the Round Tops before noon, and have kept away the reinforcements that ousted Johnson. It seems as if the point for the assault was not well chosen, and it was a mistake to have to cease the artillery bombardment just as the column got under way. This was partly due to the direction of the column masking the artilley fire, but it seems that when the column was close under the Federal's ridge, the Confederate guns were able to fire for a few minutes over the heads of the men in the column. At the crisis of the fight, you should have every weapon at work as far as possible. Had the column advanced from the direction of the Seminary against the north-west corner of Cemetery Hill, the Federal artillery would have had still greater difficulty in bringing adequate gun-fire to bear upon it. In the effort, the bulk of their batteries would have had to slew round to the right, and would have exposed their left flanks to the Confederate guns, which would have continued firing throughout the assault. As the thing was actually done, the critical period of the contest found the column smitten in front and flank by both artillery and infantry fire, and the Confederates were using one arm at a time—first the artillery, then the infantry. In the suggested attack, the column's advance would have been "covered" by artillery fire, directed to taking off the pressure from it of the Federal artillery fire; and the direction of the attack against the north-west angle would have freed it from all serious flanking fire.

On the other hand, this angle of the position was strongly entrenched.

It might have been expected that Meade would at once have followed up the repulse of the column by a vigorous counter-attack in the direction of the enemy's gun positions, but nothing of the kind was attempted. During the evening and night, the Confederates entrenched their line, in reality for the purpose of covering retreat; but Lee's intentions remained doubtful in the mind of Meade, until, on the 5th, there was no room left for doubt that the Confederates had abandoned the contest.

The telegraphic and visual signalling arrangements of the Federals are deserving of all praise. The department for this work was well organized and most efficient.

The casualties of the three days amounted, on the Federal side, to 23,186 *hors de combat;* on the Confederate side, to 22,728. These figures give 28 per cent. and 31 per cent. respectively.

CHAPTER VIII.

CAVALRY DEVELOPMENTS IN THE WAR OF SECESSION.

ARE "Cavalry raids" a branch of Tactics? I am tempted to include them, and to show, from the events of the American War of Secession, their very considerable influence on a campaign.

The details of these great raids may be read in many books; they form the most romantic volumes in the long History of War.*

In many cases the chief gain of them was information, as when Stuart, raiding behind the Federals in Northern Virginia, was able to assure Lee that Burnside, the Northern C.-in.-C., was shifting his base to Acquia Creek on the Potomac, with the evident intention, therefore, of crossing the Rapahannock at or near Fredericksburg. Lee was thus able to ignore Burnside's feint southwards from Warrenton.

Sometimes the material damage done to the enemy was serious, and often the moral effect was great. In some cases the raid was merely a great reconnaissance, but, even when this was all that was originally intended, opportunities occurred for substantial destruction of stores and means of communication.

The final great raid, Sheridan's, comes under the category of a strategical operation carried out by an invasion by an army of cavalry; it had a principal share in the final overthrow of the Confederates.

"General J. H. Morgan, a Kentuckian, a man who had no professional training as a soldier, is generally credited by the Americans with being the first to realize that a long-range weapon gave the dragoon a great advantage which he never possessed with the old-fashioned carbine."

With this idea he organized a force of cavalry which could move rapidly and fight, either on foot or mounted, as occasion might require. He had an able fellow-worker in General Forrest, and to these two Southern generals, "both unprofessional men, is to be attributed the credit of having originated and adopted a system of cavalry tactics, we may even say strategy, that was new in many of its features, and a most successful adaptation of the modern improved firearm to the use of horsemen."

v. Plate IX.†

Morgan's first important raid was made into Kentucky in 1862. "He started from Knoxville, Tennessee, on July 4th, and moved through Sparta and Glasgow to Lebanon,

* Denison's "History of Cavalry"; "Campaigns of Forrest"; Duke's "History of Morgan's Cavalry"; Brackett's "History of U.S. Cavalry," etc.

† The Maps given of three of these raids show roughly the positions of the chief bodies of the armies of both sides. An examination of the positions of the troops hostile to the raiders, and the two considerations that the former were usually on the alert and that telegraphs were in working order, give a vivid appreciation of the boldness of the expeditions and of the skill with which they must have been carried through. The Maps are modelled upon those in Trench's "Cavalry," with slight alterations in accordance with the U.S. Records' Atlas. The quotations are mostly from Trench's "Cavalry."

Pl. IX.

Morgan's Kentucky Raid, 1862.
Federals
Confederates
Ohio River
Louisville
Lawrenceburg
Midway
Lexington
Cythiana
Harrodsburg
Lebanon
Glasgow
Cumberland R.
Livingstone
Nashville
Sparta
Knoxville
Chattanooga
0 10 20 30 40 50 60 70 80 90 100 miles
Stuart's Pennsylvania Raid, Oct. 1862.
Federals
Confederates
Cumberland
Hancock
Winchester
Shenandoah R.
Mercersburg
Harper's Ferry
Chambersburg
Leesburg
Poolesville
Cashtown
Gettysburg
Monrovia
Washington
Baltimore
0 10 20 30 40 50 miles
Grierson's Raid, 1863.
Memphis
Corinth
La Grange
Federals
Confederates
Mississippi R.
Pontotoc
Vicksburg
Newtown
Jackson
Hazlehurst
Baton Rouge
Mobile
N. Orleans
Gulf of Mexico
100 50 0 100 miles

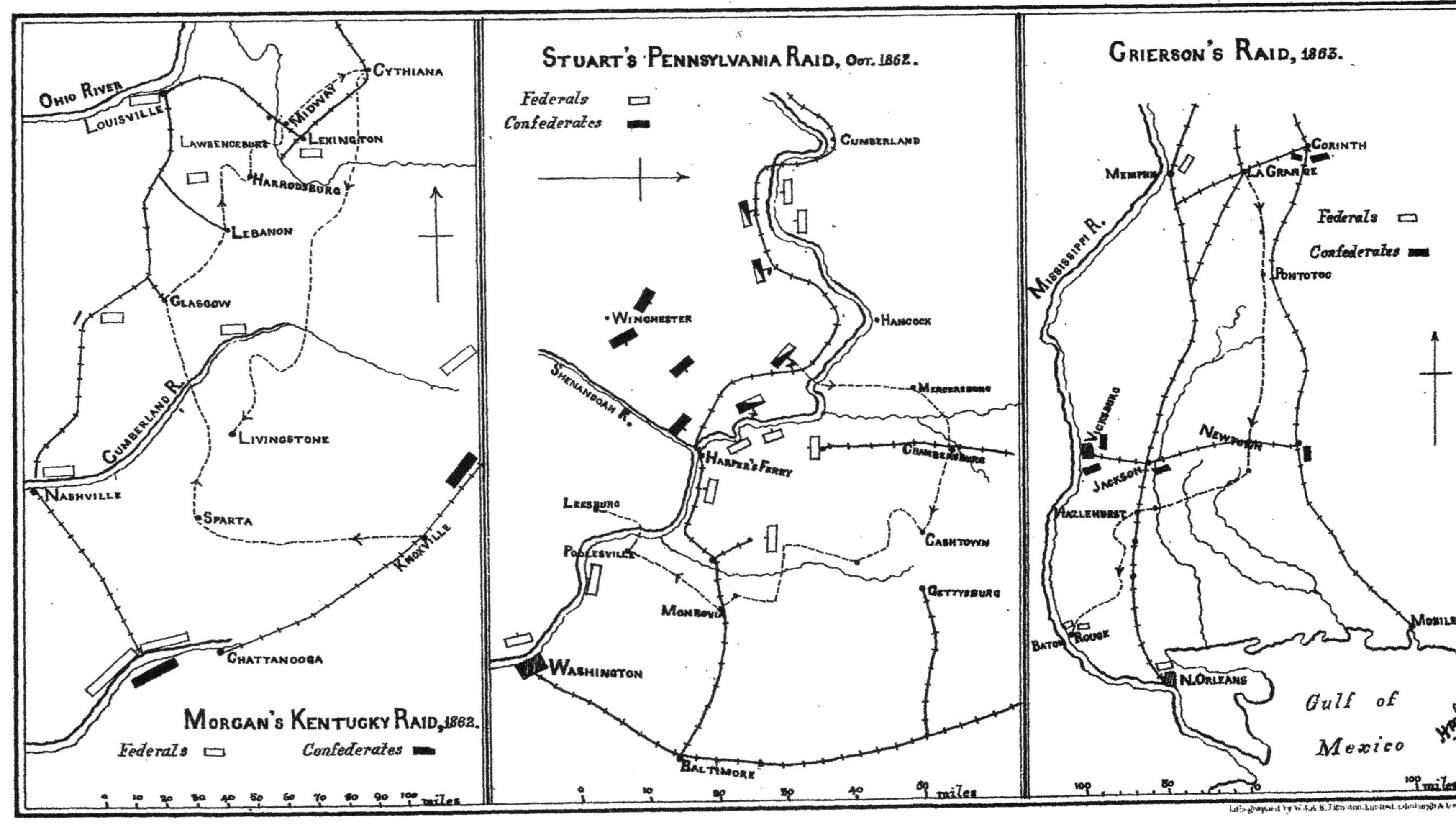

where large supplies of stores of every kind were captured. From Lebanon, Morgan marched to Harrodsburg, from there on to Laurenceburg, and then to Midway, a station on the railway between Frankfort and Lexington. The latter place was the headquarters of the Federal forces in the region, and both at that point and at Frankfort were large bodies of Federal troops much superior to Morgan's force.

"By skilful marches, by scattering his forces and threatening several points at once," he completely bewildered the Federal officers. His extreme mobility rendered it difficult for the enemy to be sure of his real whereabouts at any moment. Such was his marching capacity that, on reaching Midway his force had covered 300 miles in eight days, and the men were still fresh and in high spirits.

Morgan's own report says :—"I left Knoxville on the 4th, with about 900 men, and returned to Livingstone on the 28th with nearly 1,200, having been absent just 24 days, during which time I travelled over 1,000 miles, captured 17 towns, destroyed all the Government stores and arms in them, dispersed nearly 1,500 Home Guards (local Militia), and paroled nearly 1,200 regular troops. I lost in killed, wounded and missing about 90."

Another raid in August by the same general effected the complete severance of General Buell's communications by the occupation of a portion of the Louisville and Nashville Railway, and compelled the latter general to abandon his projects and retire to Louisville.

General J. E. B. Stuart of the Confederates has been already mentioned. "Under this brilliant commander the Confederate horsemen on two occasions made raids round the whole position of the enemy, going round by one flank and returning by the other. The first was that made round McLellan's Army in front of Richmond in 1862. It was in reality a reconnaissance on a large scale, but partook somewhat of the character of a raid, as a great deal of damage was effected upon the stores of the Federals, their ammunition and provisions."

Stuart's force on this occasion was 2,500 cavalry and 2 Horse Artillery guns. Starting from Taylorsville, the column forced its way through the Federal lines, brushing aside all opposition. Transports on the Pamunkey were taken and destroyed, great convoys seized and burnt. Having traversed the rear of the whole Federal Army and destroyed the railway, Stuart got his men safely across the Chickahominy beyond the hostile left flank, having thoroughly acquainted himself with the enemy's situation.

"This was partly in the nature of an armed reconnaissance and partly secret, as they tried to avoid fighting and to conceal their movements as much as possible, while they were quite ready to attack whatever appeared to bar their progress. The information gained by this raid was exceedingly valuable, and enabled Lee to plan the splendid operations called the 'seven days' battles,' in which Stonewall Jackson, a few days after, fell upon the flank and rear of McLellan with perfect confidence and terrible effect."

On 9th October of the same year Stuart made his greatest raid. Setting out with 1,800 horses and 4 guns he crossed the Potomac between Hancock and Harper's Ferry, reached Mercersburg at noon and occupied Chambersburg at dark. He cut the telegraph lines in every direction, obstructed the railways and captured large numbers of horses.

v. Plate IX.

He was now in rear of the whole Federal Army, 60 miles from his own main body at Winchester, and in a hostile country. Finding it too dangerous to return by the same route, he moved rapidly, as shown in the Map, and reached a point on the river opposite Leesburg. Here he met a body of the enemy sent to guard the fords, and had to fight in earnest. His horsemen distinguished themselves now by their dismounted fighting. Pushing back the enemy in front and making a dash to the left for a particular ford, he found 200 infantry well posted and guarding it. These he kept at bay till his force was across. The first part of this great ride was the raid proper, the second part was the escape. From Chambersburg the ride to Leesburg was 90 miles, and it had been traversed in 36 hours. The loss to Stuart's force was insignificant.

General Forrest's work was no less remarkable. His first action was of a unique kind. A Federal gunboat had been sent to Canton, on the Cumberland River, to destroy a Confederate magazine there. Forrest took his regiment a 30-mile night march and anticipated the gunboat. Placing his men under log cover, he engaged the gunboat at short range, his men firing with such effect that the boat had to close its ports and retreat.

In August, 1864, he made a dash at the large city of Memphis, took a large number of prisoners, and got away almost without loss.

But the northern authorities soon woke up to the usefulness of this new development of warfare. During the winter of 1862–63, they worked hard at the organization and equipment of similar bodies. In April, 1863, General Grierson, of the north, had a brigade, 2,000 strong, trained in the new methods. Starting from La Grange, Tennessee, on the 17th, he rode through the whole of the Mississippi State, ransacking the country, destroying supplies, cutting railways and telegraphs, burning bridges and stores. Travelling as shown in the Map, he reached Newtown on the 29th, and destroyed cars, engines and bridges on the railway. At Hazelhurst he cut another line of railway, and some miles to the south destroyed a depôt of cars. On May 2nd he came to Bâton Rouge, having traversed 300 miles of hostile country, done an immense amount of damage and suffered no loss to his own force.

v. Plate IX.

The Southerners, then, were the pioneers in this style of warfare; but the North, with far greater resources in men and money, eventually beat their opponents in it by superior equipment and numbers. Thus General Wilson, moving through Alabama towards the close of the war in 1865, had 12,000 horse; and Sheridan had 10,000 in the operations by which he brought about Lee's surrender. To the Southerners, shut in to ever-diminishing territories and cut off from all extraneous sources of supply, every destruction of their stores was a step towards defeat, curtailing their already limited resources. To the Northerners, the loss of magazines was rather a mere annoyance, but sometimes one that caused delay.

But a mere statement, that such-and-such things were done, is not of much value; and it is necessary to enquire into the system of working and organizing that produced these results.

Morgan's Organization and Tactics.

Morgan's Cavalry comprised ten regiments of 500 men each, and the men were for the most part educated, intelligent men, good horsemen and good shots. Morgan's preparations for a raid were made with the greatest care. The very best of his men were formed into companies of scouts, a number of whom were sent, two or three weeks before the raid began, into the heart of the district to be raided. These men disguised themselves—became spies, in fact—and made themselves acquainted with everything that might come in useful—the enemy's forces and posts, his magazines, his projected movements, the best fords and safest routes for secret marches. The raiding force rode light, carrying no change of clothes, no provisions, nothing but arms, saddle, bridle and blanket. The arms were—carbine with bayonet, 100 cartridges, revolvers. Only two companies had sabres. They thus literally had to live on the country; they could not even replenish their ammunition during a raid unless they had the good fortune to capture some that fitted their weapons. In 1863, in the raid that terminated in Morgan's capture, his men were obliged to re-arm themselves completely with captured weapons, because the ammunition they could find only fitted these.

The usual pace was the walk, at four miles an hour. On some days this would be kept up for 15 hours out of the 24. On one occasion, in 35 hours, Morgan's Command covered 90 miles. By keeping his main body thus at the walk, time was given for pretty thorough foraging along both sides of the route, and the safety of the column was more easily provided for; time was permitted to requisition horses to replace those that were giving out. There never seems to have been any difficulty in getting horses for this purpose, and this probably is the chief explanation of how such long marches were achieved.

In general neither men nor horses entered a house or a stable. The men got what shelter they could at the roadside, and slept with their bridles in their hands. Except for a long halt, no girth was unloosened. When the horses were fed they were unbridled, and what forage the foragers had been able to collect *en route* was put before them. On some occasions the horses were nearly 24 hours without food or water. Such work was naturally of a most exhausting nature to man and beast. Morgan's main idea in fighting tactics was to obtain the advantage of surprise. At the beginning of a raid he attacked energetically, to produce moral effect; when the chief object was accomplished, he was chary of fighting, and avoided it if he could.

His dismounted fighting was in this wise :—

"On arrival at the place of combat one of the troops on the flanks went forward, and, deploying in skirmishing order, covered the front. Meanwhile, the main body dismounted, and each horse-holder had to hold four horses or even eight when a large number of combatants was required.* These latter formed a single rank, with intervals of two or three yards between the men, in a line which was slightly concave, the extremities of the line being advanced towards the enemy. When once deployed, this line advanced at the double, while the skirmishers cleared the front. When the ground offered sufficient shelter, the men generally deployed only one line to make the attack.

"When, however, the ground was open, two or even three lines were formed, each exactly in rear of the line in front of it. The distances of these lines from each other was regulated by circumstances. The 1st line opened fire and lay down; thereupon the 2nd line immediately advanced, and, passing through the intervals of the front,

* At this epoch, and for some years later, the custom in our service was for each horse-holder to have only one horse besides his own, so that only half the force could fight on foot. Later on, about 1880, each holder was given two of his comrades' horses to hold, and this was also the rule on the Continent of Europe. The Boers went one better than the Americans; each man had trained his horse to stand alone, so that only a fraction of the force watched the mounts.

went forward a certain distance, opened fire in its turn and lay down." This alternate action was continued, both in frontal and flank attacks. When close quarters was reached, Morgan's men used their revolvers.

There is no word of a reserve. These troops were not to fight pitched battles, but to brush aside opposition and push on.

"With Morgan, however, it was a favourite manœuvre to hold the enemy and engage his attention by feints on his front, while with a considerable portion of his force he made a wide turning, dismounted his men on the enemy's rear and fought on foot, in order to strike an effective blow. With regard to artillery, Morgan generally took with him four howitzers and two Parrott guns. These latter long-range pieces were seldom used except during retreats." This was quite sound; a surprise was Morgan's aim, and a surprise at two miles' range would be of little effect.

Forrest's Organization and Tactics.

General Forrest had his corps in three divisions. The 6,000 horse was in nine brigades, or 18 regiments of 300 to 350 men each; 1,500 to 2,000, or one division, was the force he usually employed at one time. These were accompanied by some light vehicles, carrying provisions, tools, etc., for rapid demolitions. He also sent out scouts many days in advance. His men had a sabre attached to the saddle, a revolver and a carbine. He generally marched all day at the walk or the trot, halted two hours at noon, and all night. The force bivouacked in the simplest manner. Near the enemy each man slept with his bridle over his arm. The column generally made from 40 to 45 miles a day, marching with the ordinary precautions of advanced guard and flankers and scouts a day's march ahead. The special scout companies were of picked men who personally knew the country to be traversed. His men constantly changed horses *en route;* it was the exception for a horse to last more than three or four days. Forrest, on coming up with the enemy, would leave his advanced guard to skirmish with them, post a regiment, dismounted and concealed, some distance in rear, make a detour with the bulk of his force, out of sight. The advanced guard would gradually fall back, tempt the enemy, who so far had seen nothing but the advanced guard, out of his position, lure him on to within short range of the posted regiment; when the latter opened fire, the turning force would make its appearance. He took two light guns for each brigade. He was never afraid of risking his guns—"if they were taken, his men had to retake them." They travelled well forward in the line of march. Forrest's men used the following method for railway demolition. They heaped brushwood along the line and fired it; the heat bent the rails and damaged the fastenings, and the line was readily rendered quite useless. In half an hour 1,000 men disposed of a mile of railway.

Stuart's Organization.

Stuart had been an officer of the regular forces, and his cavalry were organized and equipped after the European pattern, and acted usually in the manner of regular cavalry. Morgan and Forrest were men of no military education whatever, and their troops were really mounted infantry. Stuart had not the independent command these two had; his force was an integral part of Lee's Army. His troops did not rough it, as Morgan's and Forrest's did; they had tents, were well provisioned and accompanied by transport. They were armed with sabres and revolvers—very few of them with carbines. When he made the circuit of McLellan's Army he had 2,500 men and two light guns, but in the following year he had a corps of no less than 12,000 horse and 24 guns.

On the Federal side the noted cavalry leaders were Grierson, Stoneman, Wilson and Sheridan.

Wilson's Organization.

In the last year of the war Wilson had a large and most carefully organized corps, fit to act as an independent invading army, capable of keeping the field for 60 days. His Alabama raid was made with 12,000 men. These took no change of clothing, but each man had on him five days' provisions, 24 lbs. of corn, 100 cartridges and a set of spare shoes. A transport of 250 carts carried the remaining supplies required for two months' work.

As in the other cases, a swarm of scouts or spies preceded the advance by some days. His force usually advanced on four or five parallel roads, using the common precautions of advanced guard and flankers. An average march was from 20 to 30 miles.

Sheridan's Organization and Tactics.

Sheridan had a special section of 60 spies, who were paid according to the information they brought in. Most of the Northerners' raids had to be made in a country where the population was bitterly hostile, and the greatest precautions had to be taken. His 10,000 horsemen carried the sabre attached to the saddle, the Spencer magazine rifle and the revolver. They only fought on horseback against cavalry, and then charged sword in hand. The enemy, however, seldom remained mounted to meet the charge, but took to the ground and musketry fire; when they had no time allowed them for this manœuvre they used their revolvers, and the Northern horsemen often found that the sabre was no match for a good revolver.

When Sheridan had to deal with infantry, his men always dismounted; and their repeating rifles and their revolvers made a most formidable foe. For attack, he had three bodies—skirmishers, a 2nd line in columns, then a substantial reserve of a whole division. In marching he preferred to have his whole command under his hand on one road, though 10,000 horse, four abreast, would occupy nearly eight miles of space. Three regiments formed the advanced guard; one on the main road, the other two spreading across country, or using parallel roads. Scouts were far out in front as usual. Except for some special purpose, a day's march was usually restricted to 16 miles. Extreme precautions had to be observed, on account of the known boldness and dash of the Southern horse.

The transport carts carried chiefly ammunition; each horse carried four days' rations, two days' oats, and the man's tent, cloak and blanket.

Sheridan at first used to take four batteries with him, but he gradually reduced his artillery to one battery.

Colonel Trench, from whose book a great part of this chapter has been taken, thus sums up :—"The boldest and most rapid raids were generally made by bodies of horse not exceeding 2,000 men or so, and the cavalry which made the longest marches in the 24 hours was that of Morgan, whose general rate of march was the walk. Both the Federal and the Confederate cavalry in the course of their operations marched as concentrated as possible, in order to be at the disposal of their chief as soon as they were wanted. Horses were never changed systematically during a raid, but at any time when any horse was done up his rider replaced him by a fresh one as soon as he could pick up one on the march.

"Artillery scarcely played any very prominent or definite part in the raids. All the chiefs who led these expeditions progressively diminished the number of guns which were originally deemed to be necessary, in order not to hamper the mobility of the main body. In the principal and most famous of these expeditions it was the scouts who most contributed to the successes which were obtained. These scouts, who were nearly all young men, well educated and intelligent, were often better informed about everything which concerned the enemy's forces than the generals of those forces themselves. When operating in a country which was hostile to them, their task was naturally a far more difficult one, but owing to their experience and a special aptitude which they developed, they carried out their most difficult duties with a skill to which ordinary soldiers would never have attained."

We have seen then how the belligerents in the War of Secession produced a novelty in Tactics of which the military world should have taken serious notice, but which, as we shall see, was for long completely ignored in Europe. The splendid work done by the special companies of scouts is particularly worthy of notice; this scouting on the great scale requires picked men and much training, and it is worthy of very careful consideration whether we should not have in the British Army a *corps d'élite* of scouts, specially recruited, specially trained, specially paid. To take men from the squadrons at random in the field for this work is haphazard; the ordinary trooper, even if given some practice in the work, is apt to feel tied down to his squadron. The scout proper, working in front of the cavalry screen, should be quite independent of the squadron behind him on the road, except to the extent that he should know, if possible, its route for the day and where it will probably halt for the night. But if he belongs to the squadron and therefore receives his orders from its commander, both parties are prejudiced. The squadron is apt to look on their man, or their officer's scouting patrol, as the case may be, as a sort of advanced guard, which is no part of the business of a scout. The man, or the scouting patrol, will be apt, reciprocally, to be paying too much attention to the squadron in his rear. Now the squadron is itself the advanced guard of the independent cavalry mass, and its business is to fight. It should entirely look after its own safety. And the scout, or scouting patrol, should have its C.O. travelling at the head of the mass in close touch with the chief cavalry commander, should look to that point for orders and should have no connection with the squadron beyond that of being able to use its troopers to transmit important information. With all this in view it seems proper that the scouts of an army should have the *habit of independence*, while thoroughly understanding their trade and its great importance.

There is no doubt that the remarkable successes achieved by these raids in America were due in great part to local conditions that are not often found. In a country having a prosperous, but not crowded, population, a great many riding and driving horses will usually be found, and we have seen that the celerity of the movements of the raiding columns—a celerity on which their eventual safety in great measure depended—was due to the apparent ease with which knocked-up horses were replaced *en route*. In many countries this convenience could not be expected to anything like the same extent.

Again, the fine work of the scouts and spies must have been considerably helped by two special conditions that existed, and we have seen that this work was the other great cause of the success of the raids. First, one language was common to both belligerents; second, in a country where every able-bodied man rode when engaged about his ordinary business, the appearance of a stranger on horseback would excite no remark in villages or farms. These points should be kept in view when considering the feasibility of raids in other countries and under other conditions.

In reading about these raids and the remarkable work they sometimes effected, one is apt to be led into giving them more credit than they deserve as helpers towards the successful completion of a campaign, which, after all, is the true test to apply towards operations of war. More than once it happened in America that the cavalry was far away, raiding, when the C.-in-C. really ought to have had them under his own hand. Thus, in the campaign of Gettysburg in 1863, Lee's cavalry under the renowned General Stuart, raided round the east side of the Federal Army under Hooker, while Lee's force was marching on the west side of Hooker. The lack of information that then ensued on the part of Lee—for Stuart was completely out of touch with him—brought it about that Lee was compelled to fight an offensive battle at a time and place that did not altogether suit him.

CHAPTER IX.

PRUSSIA AND AUSTRIA IN 1866.

IN the year following the termination of the American War of Secession, Prussia and Austria came to blows. Austria had at the outset the help of some of the German States, but, with the exception of Saxony, these did not succeed in uniting their forces with her. By the 29th of June, 1866, the Hanoverian Army, 20,000 strong, was threatened with superior forces and capitulated. A Prussian force, three divisions strong, under General Von Herwarth, entered Saxony on the 18th June, and took Dresden without striking a blow, the Saxon Army having retreated the day before into Bohemia.

The Prussian Army for Bohemia, under the King's command, was in three groups, at Torgau, Görlitz and Neisse, as much as 170 miles from end to end. The right wing coming from Saxony was called the Army of the Elbe; the Görlitz or centre force was the Ist Army, commanded by Prince Frederick Charles—it had three Army Corps and a cavalry division. The Neisse force, the left wing, was the IInd Army, commanded by the Crown Prince; it had three Army Corps and the Guard Corps. The total was 278,000 men and 840 guns. The Austrian force that could be brought into Bohemia was calculated by the Prussian General Staff at 244,000—a very close estimate. In Bohemia it might be possible for them to concentrate, at one of the passes leading into Prussian territory, as many as 60,000 or 80,000 men, before the Prussians were ready to invade, but it was not expected that Austria would venture upon such a scheme as invading with so small a force.

The heads of the Prussian columns, to all of which had been given the neighbourhood of Gitschin as a *rendezvous*, were met by Austrians, and had to fight several stiff contests. I shall choose for description and comment the combat at Trautenau, where the Prussians were badly handled and beaten back through the pass.

TRAUTENAU.

To the I. Prussian Army Corps was assigned the pass of Trautenau through the Riesengebirge. On the 26th June its main body was at Liebau, which is more than 10 miles distant from Trautenau, where the pass is cleared.

The Austrian 10th Corps was hastening to close the pass; its main body appears to have been within five or six miles south of Trautenau on the evening of the 26th.

The Prussian Corps started at 4 a.m. on the 27th in two columns, the right from near Liebau, the left from Schömberg. The former column detached a squadron, two battalions and two guns, which were to reach Trautenau from Altstadt. The main columns were to meet at Parchnitz and rest there two hours, the advanced guard of the right column ($7\frac{1}{2}$ squadrons, five battalions, 16 guns) occupying Trautenau. The left column reached Parchnitz at 8 a.m., but the other had been badly delayed; to the latter had been intrusted the occupation of Trautenau. When, at 10 a.m., the advanced guard approached the town, it found the bridge barricaded; but the defenders were only

v. Plate XVI.

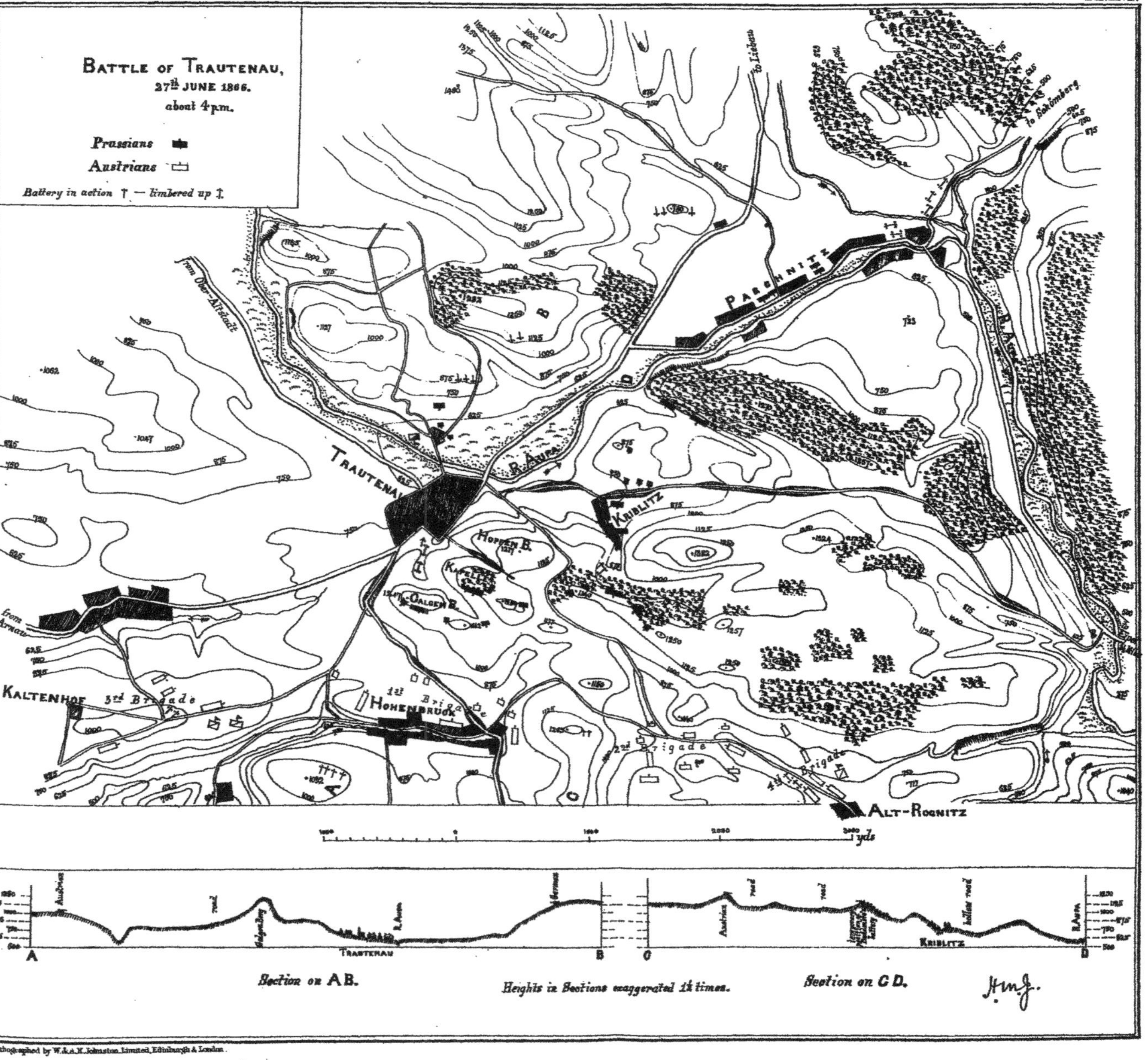
BATTLE OF TRAUTENAU,
27th JUNE 1866.
about 4 p.m.
Prussians
Austrians
Battery in action — limbered up
TRAUTENAU
PARSCHNITZ
KRIBLITZ
HOHENBRUCK
KALTENHOF
ALT-ROGNITZ
R. Aupa
Hopfen B.
Galgen B.
1st Brigade
2nd Brigade
3rd Brigade
4th Brigade
from Ober-Altstadt
from Arnau
to Liebau
to Schömberg
yds
Section on AB.
Heights in Sections exaggerated 1½ times.
Section on CD.
Lithographed by W. & A. K. Johnston, Limited, Edinburgh & London.

a handful of dismounted cavalry, who soon withdrew. Some squadrons of the advanced guard passed through the town to the south-west exit, came upon five or six squadrons of the enemy and charged them, at the same time as the cavalry of the right flank detachment joined in from the direction of Altstadt. A sharp fight ensued, but, infantry appearing on both sides, the cavalry withdrew.

A study of the map shows that the heights south-east of the town* and the plateau extending from them constitute the key of the position. The leading Austrian Brigade was already hurrying there, and a battery was opening fire on the Prussian advanced guard from the north-west of Hohenbrück. Austrian infantry were about to enter the town, when they were withdrawn for some unknown reason, and retired. But detachments reached the heights south-east of the town before the Prussian advanced guard was ready to attack in strength. A part of the latter garrisoned the houses, and presently a few companies advanced, skirmishing. Two companies pushed into the hallow road leading to Kriblitz; the 10 guns of the main body of the advanced guard opened fire from the other side of the Aupa, 300 yards from the bridge. The 4-pr. battery of the vanguard had unlimbered to the west of the town.

The skirmishing action was making no progress—only one company was moving against the Galgenberg—when the two battalions of the right flanking detachment entered the town, and immediately joined in the attack of the Kapellenberg. General von Bonin, the commander of the I. Corps, was now on the scene (11 a.m.), but his main body was still short of Parchnitz. He recognized that, if the heights were held in force, frontal attack would be very costly and difficult. He accordingly ordered three battalions, and shortly three more, to be detached from the main body towards Alt-Rognitz, to gain the plateau and thence take the enemy in flank. "The three battalions already in action were withdrawn to the hill north of Trautenau, in order there, supported by a battalion and the cavalry, to support the advanced guard in an eventual retreat, or to oppose a possible attack from the direction of Arnau."

The student of German Artillery tactics in 1870 will note the comparative feebleness of this proceeding; also that the cavalry should have been discovering whether there was to be an "attack from the direction of Arnau." The General in command of the left column ordered the 44th and 45th Regiments and a battery, under General Buddenbrock, to carry out the movement against the Austrian right. They crossed at the far end of Parchnitz, but immediately found themselves in ground where the steep banks could only be climbed in loose order, and the scrub traversed in single file; no horses could get through. The movement proved very exhausting to the troops and very slow; some attempt should have been made to examine the ground.

Before Buddenbrock could appear on the scene, the renewed attacks of the advanced guard had proved successful. The Galgenberg was first captured, and then the rest of the Austrian Brigade began to retire slowly, fighting, from the Kapellenberg and the Hopfenberg. The retreat was made towards Hohenbrück, 16 Austrian guns covering it from the height (1162).

The opponents, so far, were about equal in numbers, and the Prussians had had no help from their guns; they had, on the other hand, the advantage of the breechloader, which a man could load without standing up.

* Galgenberg, Kapellenberg and Hopfenberg.

No further advance was made until, after 1 p.m., Buddenbrock's battalions appeared. His 44th Regiment came through Kriblitz and up through the wood to the south of it. The 45th passed on a broad front between (1382) and (1324), and the battery established itself on (1140), and engaged hostile artillery beyond Hohenbruck (1032), the range being nearly 3,000 yards.

Some companies of the main body of the laggard right column were now in the firing line.

The 24 Austrian guns now in action forced Buddenbrock's battery to retire, but the Prussians at this stage greatly outnumbered their enemy in infantry. At 2.30 two Prussian batteries were brought up to the Galgenberg, attracted the hostile artillery fire, and enabled the infantry to advance. Rapidly the Austrians abandoned Hohenbruck and Alt-Rognitz and the ground between, which were occupied by the Prussians; but the troops engaged were by now too exhausted for further work. By 3 p.m. firing ceased along the whole line.

Where were the Prussian cavalry? Now was the time for them to be following up the enemy. Von Bonin assumed that the fight was over, and he allowed the Guard Division, now at Parchnitz, to continue its march to Eipel. Had his cavalry informed him of what was coming, he would have brought that division on to the plateau, as well as all his available artillery.

By 3.30 it was clear that the enemy had been reinforced; a whole brigade had, in fact, arrived to them, and it was attacking the Prussian left, the original brigade again advancing against the main front.

Bonin thereupon established three battalions in Trautenau, brought up two more batteries to the hill north of the town, and ordered the main body and the advanced guard to advance to the attack. The latter never received the order. The 45th Regiment, on the left, began to be outflanked, and between 4 and 5 p.m. was in retreat on Parchnitz. The other battalions began to follow, the two batteries on Galgenberg were forced to retire. Two battalions held on for a bit to the wood south of Kriblitz, and delayed the enemy; two more stood in the plantation of Kapelle and as far as (1257), while two battalions of the reserve infantry held the edge of the great wood north-east of Kriblitz and the slopes towards the village. Under cover of these battalions the main body gained the other bank of the Aupa without much trouble.

By this time a 3rd Brigade of Austrians was on the scene, and batteries unlimbered on the height east of Kaltenhof (range to Trautenau 2,500 yards). Forty Austrian guns were now firing at Trautenau and its heights. The fresh Austrian brigade attacked with vigour, and one of the Prussian reserve battalions lost eight officers and 238 men in an hour. Then a 4th Austrian Brigade appeared on the field, and the advance made to the east of Kriblitz began to render the position about Trautenau so precarious that at 6.30 the final retreat began. Then ensued some very active fighting. All the troops got across the river, the battalion that held the edge of the great wood opposite Parchnitz keeping its place against all attacks till 9 p.m. Another battalion prevented the enemy for the moment from using the Trautenau bridge, and later two batteries, on (780) north-west of Parchnitz, prevented his advance along the high road. This was the end of the fighting. The total available Austrian force comprised four brigades of infantry, say 22,000 men, and 60 guns, of the bulk of which they made the best use.

The Prussian total was 25 battalions say 20,000, and no less than 96 guns; of the latter no less than 54 were kept in reserve, and we have seen how little use was made of the rest.

The Prussian official account says:—"The I. Corps was, during the whole day, thus held in a disadvantageous position, because, at the commencement, Trautenau and the heights commanding it had not been immediately occupied, and the *débouché* of the main body thus secured."

But this statement is misleading to the extent that it does not go to the root of the matter. We have seen that "Trautenau and the heights commanding it" were in the hands of the Prussians quite early, and that by 3 p.m. the enemy had been pushed back south of the line Hohenbruck-Alt-Rognitz. The main faults were the neglect of following up thereupon with cavalry, and the failure to bring up all available artillery on to the plateau. The engineers too should have made bridges between Trautenau and Parchnitz to help forward the guns, and to provide for their retirement, if that should be necessary. The Prussian Commander seemed to think that the chief use of artillery is to prevent the enemy from following you.

The account goes on:—"The infantry fought almost alone . . . the Austrians, with full freedom of movement, used all their arms against the Prussian infantry, and could bring the whole superiority of their fire to advantage." I have referred already to the loss of strength in allowing the Guard Division to march off to Eipel; this was due to ignorance of the situation.

Between 1 and 3 a.m. on the 28th, the Prussian troops arrived in their last night's bivouacs, supremely exhausted.

The Prussians lost 56 officers, 1,282 men and 78 horses; the Austrians 196 officers, 5,586 men and 185 horses.

The Austrians here, as we shall find again at Sadowa, adhered to the attempt to advance in solid columns against the breech-loader. The Prussians had company columns supporting swarms of skirmishers.

Sadowa or Königgrätz.

The decisive battle of the campaign was fought six days later, on the 3rd of July. The Ist Army and the Army of the Elbe had been united under the command of Prince Frederick Charles. The Army Corps of these armies had been divided up into independent divisional commands, except the II. Corps, which was under General Von Schmidt.

The force was, on the 2nd July, posted as follows:—

The III. Corps had its divisions, 5th and 6th, and some extra brigades at Miletin and Dobes.

v. Inset of *Plate* XI.

IV. Corps—7th Division at Horitz, 8th at Gutwasser.

II. Corps—3rd Division at Aujezd, 4th at Wostromer.

A cavalry division at Baschnitz, another at Liskowitz.

These were the troops of the Ist Army.

The Army of the Elbe had advanced guard at Smidar, 14th Division at Chotetitz, 15th Division at Lhota, 16th at Hochwesely. Large bodies of artillery at Belohrad and Welhost. These forces were thus fairly concentrated.

The IInd Army had been delayed by its combats.

Its four corps were thus distributed on July 2nd :—

I. Corps, Von Bonin, at Prausnitz.

Guards, Prince of Würtemberg, at Königinhof.

V. and VI. Corps, von Steinmetz and von Mutius, at Gradlitz.

The cavalry division was at Neustadtl, miles in rear.

The Prussians were unaware that the Austrian Army was concentrated within a march of them, so inefficient had been the cavalry action.

The enemy, on 2nd July, despatched their trains to the rear across the Elbe, but even they had only a very vague knowledge of the German advance. The Prussians expected to find their enemy across the Elbe, with his flanks resting on Josephstadt and Königgrätz. Such a position would be a difficult one to attack, so the first necessity was to find out the truth. Orders were issued for 3rd July; they sound like an order for a reconnaissance in force, with the anticipation of a battle resulting.

v. Large Map.

But, before any one of these orders could be carried out, fresh news came to hand. A colonel occupying Cerekwitz with a battalion reported Austrians at Lipa. Officers' patrols were sent out. They found the heights near Dub occupied, and prisoners stated that four Austrian corps were behind the Bistritz, the 3rd at Sadowa, 10th at Langenhof, 1st behind it, the Saxons at Problus, ten regiments of cavalry and a mass of artillery about Lipa. This information was correct enough, for on 2nd July the Austrian forces were disposed as follows :—

The Saxon Corps of two divisions, under its Crown Prince, at Neu Prim.

A cavalry division at Dohalicka, on the Bistritz.

3rd Corps of two divisions at Sadowa.

10th Corps of same strength, von Gablenz, at Lipa.

6th Corps of same strength at Wsestar, with a cavalry division in rear.

1st Corps of same strength, Count Clam-Gallas, in front of Königgrätz, with a cavalry division on its left.

8th Corps of two divisions, Archduke Leopold, at Nedelist.

4th, 2nd and a cavalry division extended north-east in front of the Elbe, the last being five miles from Wsestar. Behind the 2nd Corps was a cavalry division.

Thus at present the greatest strength lay along this line.

The whereabouts of only four of these corps were known to the Prussians, these being the corps against which the invaders had fought their way.

The fresh orders issued on account of the new reports directed the Elbe Army to get forward in the morning to Nechanitz; 7th and 8th Divisions to be at Cerekwitz and Milowitz respectively at 2 a.m.; 5th and 6th Divisions to move at 1.30 a.m. south to beyond Horitz; 3rd and 4th Divisions to be at Psanek at 2 a.m. Army Reserve Artillery towards Horitz. The Crown Prince was requested to bring down two corps in front of Josephstadt, and a third to Bürglitz.

On the other side Marshal Benedek, from the information at his disposal, concluded that he was to be attacked on the 3rd, but that it would be most likely on his left wing only. He ordered as follows :—

The Saxons to occupy the heights above Popowitz, refusing their left wing and covering it with their own cavalry; about Problus another cavalry division; 10th Corps in front of Langenhof; 3rd, Lipa and Chlum; 8th Corps, as support to the

Saxons; 4th Corps, Chlum and Nedelist; 2nd Corps, from the 4th to the Trotinka; the cavalry division on the right to Nedelist; 6th Corps, on the heights in front of Wsestar; 1st to Rosnitz—these two to keep concentrated and ready to march; the other cavalry divisions behind Wsestar and Rosnitz; the Artillery Reserve near the main road behind Wsestar.

The 1st and 6th Corps, the five cavalry divisions and the artillery just mentioned were thus the reserve, and they were kept concentrated and under the general's own hand. So far, good. But why keep these masses of cavalry in such a position? No doubt some of them did good work in the moment of retreat; but would not some of them at least have been better employed by pushing out to delay the Crown Prince on the right, and to threaten in flank the Elbe Army?

A considerable amount of fieldwork was rapidly constructed—batteries, breast- *v. Plate* XI. works, rifle-pits and abattis; and the villages of Lipa and Chlum were placed in a state of defence. Problus and Prim were also dealt with. Abattis were made along the front of the wood between Problus and Charbusitz, and in the Lipa wood. The batteries constructed are shown in the plan; they were covered or flanked by breast-works of a very solid make. A glance at the plan shows the faulty siting of Nos. 1, 2 and 3 batteries, which were easily commanded.

The orders for the Elbe Army and the Ist Army arrived rather late, and few of the corps were in their appointed positions till after dawn. The Crown Prince (IInd Army) had only its I. Corps across the Elbe, and had sent his VI. Corps towards Josephstadt. It was evident that he was going to be late on the field; he was prepared to send his I. Corps at once to Bürglitz, but he held, in the present uncertainty as to the disposition of the other Austrian corps, that he should keep back the Guards and the V. Corps as a support to the VI., which he thought might become heavily engaged on its reconnoitring expedition.

But by 5 a.m. the Crown Prince knew the true state of affairs and ordered the I. Corps to Bürglitz, nine miles (there was still about four miles to reach the enemy); cavalry to follow; Guards to Jericek, seven to ten miles (still three miles to the enemy); VI. Corps to Welchow, seven miles (still four miles to the enemy); V. to follow the VI. two hours later.

There was fog and dense rain, which reduced great tracts to swamps, especially along the streams, and filled the latter to the brim. The Ist Army, in advancing, found that the enemy's outposts were being withdrawn towards Sadowa. The 8th Division moved as A.G. on the left of the Sadowa road; the intention was to get into touch with the Austrian main front, and to attack sufficiently to "contain" it until the two flank attacks should become effective. The Crown Prince could not be expected on the field till midday at earliest.

At 6 a.m. the Prussian advance was in full swing, and the enemy stood as follows:—

The Saxons had troops across the Bistritz at Nechanitz; there was a battalion in Nechanitz, and another in Hradek. The 10th Corps had outposts across the stream; a brigade of the 3rd Corps was in Sadowa; part of another brigade of the same Corps was at Cistowes; the 4th Corps had outposts in the Maslowed Wood; some battalions of the 8th Corps had remained at Horenowes, when the corps itself had been shifted over to the left to support the Saxons; a brigade and six guns at Maslowed.

At 6.30 a.m. von Herwarth's A.G. debouched from the wood in front of Kobilitz, saw the enemy at Alt-Nechanitz, deployed three battalions and brought up a battery, and quickly drove the Saxon battalion across the stream. The battalions found the bridge broken, were hotly shelled by a Saxon battery, and could not get across the stream till their own battery put the Saxon one to flight. The quality of their artillery proved itself great in single instances like this, but as a whole the arm was not sufficiently used in this battle.

It was now 8 a.m., and the cavalry of the Ist Army was at Sucha, with three batteries of horse artillery. To its left rear, on another road, was the 3rd Division. The 4th Division had a very hard march in rain-sodden ground, emerging between Stracow and Dub. A 4-pr. battery went on with some cavalry to Mzan, and was promptly shelled by four hostile batteries from behind Dohalicka at a mile range. As the 8th Division's A.G. appeared through the mist, it was received with artillery fire from the round between Sadowa and Maslo wed. The Austrians were making good use of their guns.

The 7th was very early in place at Cerekwitz, and prepared to move on Benatek, as soon as it should hear the guns of the 8th. At 8 a.m. it was advancing, saw an Austrian battalion retiring into Benatek, and was itself fired on by artillery from Horenowes and Maslowed, and from Skalka Wood. A battery here was also engaged with the 8th Division's guns on Roskos-Berg, but, being now overpowered, retired to Lipa. A few shells thrown into Benatek set the village on fire, the enemy abandoned it, and the 7th Division entered.

A heavy fire from Maslowed Wood now stopped the advance.

The King, arriving on Dub about 8 a.m., ordered an advance on that side to the Bistritz. He had the 5th and 6th Divisions in immediate support of the 8th, and the 3rd and 4th not far off to the right. Even if, as was quite likely, the Austrians might have a great preponderance of force in their immediate front, he held that the advance of the IInd Army would settle the day, and would be facilitated by a vigorous "containing" action at Sadowa.

But the I. Corps did not get its definite order to advance till nearly 8 a.m., and up to this time the Guards had no orders. The VI. Corps was on the left bank of the Elbe.

Benedek began to see that a pitched battle was imminent. Afraid of extending too much he ordered the Saxons to consider Prim and Problus as their main position, and to abandon the original intention of holding Nechanitz and Hradek Hill in force. The 10th Corps sent three battalions across the Bistritz to the Sugar Factory, and had a brigade about the Sadowa Wood in support. Several of its batteries took up a good position at Dohalicka, and it was these that fired on the first appearance of Prussians at Mzan.

The 3rd Corps, bivouacking to the west of Cistowes, was held to be too far advanced, and was ordered back to Chlum and Lipa. A brigade and two batteries were, however, to go forward into Sadowa, to act as a sort of rear guard to this retrograde movement.

It cannot help striking one that there was too much shifting of troops at the last moment. No doubt there are cases when such movement may have the effect of nullifying the enemy's previous reconnaissances, but that cannot have been the case here.

Benedek must have long ere this have made up his mind, either to a defensive attitude or the opposite. He was not expecting any reinforcements; when such are expected, but the enemy reaches you too soon, a contraction of your position and consequent shifting of units at the last moment is natural enough, but here the effect was pernicious for the reason, partly, that some of the carefully made fieldworks had to be abandoned.

On the right, also, the positions actually taken up were not those originally intended; but in this case a great advance to the front was made, and the most carefully entrenched position was left 2,000 paces in rear. The 4th Corps was to have been from Chlum to Nedelist, the 2nd on from there to the Elbe. Instead of this, these two corps occupied the line Cistowes-Maslowed-Horenowes, the hill in front of Sendrasitz. This position, however, was much superior for defence to the former one; but the 4th Corps was never informed of the earthworks that had been constructed, and made little use of them in their retreat.

The Saxons, too, had constructed a powerful battery on the height of Schloss Hradek, above Lubno, but were forced to abandon it, owing to orders at the last moment that they were not to extend their left flank beyond Problus and Nieder Prim. The Saxons only abandoned Nechanitz after a skirmish of some duration, and then the battalions retired to Lubno, and eventually to behind the wood of Popowitz. A Prussian battalion followed into Nechanitz and on to the height south of Lubno. Another battalion moved through the swampy meadows to Komarow, waded the stream and attacked Lubno, which was held by a battalion assisted by two Austrian batteries firing from the direction of Popowitz Wood. When reinforced by a second battalion the Prussians captured Lubno, but only with considerable loss. Before it was captured, at about 10.15 a.m., two Prussian companies, shortly reinforced by two battalions and a battery, saw, from the abandoned Saxon earthwork above Lubno, 12 Saxon squadrons and a battery on the march from Lubno to Nieder-Prim. "The effect of the battery's fire on the column was evident, but no disorder could be seen. In perfect order, and instantly filling up the gaps which the shells tore in their ranks, they marched to the reserve position east of Nieder-Prim."

At Alt-Nechanitz, another battalion of the Elbe Army's advance guard had turned off to the right; but the bridges were destroyed at Steiskal, and the buildings held by the enemy. Moving on, the battalion drove two detached companies out of Kuncitz, repaired the bridge there, and followed to Hradek Castle, where it was shortly joined by another battalion.

Lubno, and the whole of the Hradek position were now (11 a.m.) in the hands of the Elbe Army's advance guard—two battalions about Hradek, two and a-half battalions and two batteries along the heights towards Lubno, two squadrons and a horse battery behind the heights, two and a-half battalions in and near Lubno. These forces constituted a *tête-de-pont* behind which the 15th Division was beginning to form east of Nechanitz.

As we have seen, the 7th Division's advance guard had already gained a firm footing by the capture of Benatek. Two more battalions now arrived, and a push was made for Maslowed Wood, held at first only by Austrian outposts; but these had now been reinforced by the bulk of a brigade of the 4th Corps. The wood covered a very steep ridge, and there was a good deal of undergrowth.

At about 8.30 a.m., the four battalions prepared to open the attack, while the main body of the division was to deploy in the dell, north of Benatek. A battery was brought up to the crest east of Benatek, and another to its left rear, but both were shortly overpowered, harassed by flank fire of skirmishers issuing from Horenowes, and had to retire leaving one dismounted gun. Two other batteries came through the village, unlimbered, and assisted the infantry attack, and were later joined by one of the two that had been driven off the crest.

The enemy failed to hold the skirts of the wood, and portions of it were captured by the Prussians; but the main body of the Austrian 4th Corps was approaching. Very soon 40 Austrian guns were firing from the side of Maslowed, and the 18 Prussian guns did not suffice to draw off the fire from the infantry.

A confused struggle was going on among the trees; here and there small bodies of Prussians, with all semblance of order lost, penetrated to the south edge. Great havoc was wrought by the Austrian shells. The wood was first penetrated at the west end, and here two battalions established themselves, preparatory to attacking Cistowes. The projecting north-east part of the wood was taken just in time to forestall strong hostile reinforcements, and the furious attack of these was repulsed by steady file-firing. Fresh battalions from the 7th Division's main body then carried the opposite slope to the south, and cleared the enemy out of the east of the main wood. But a desperate struggle still raged in the interior.

Cistowes fell temporarily into the hands of two Prussian battalions that had worked through that end of the wood; but more brigades of the 4th Corps were arriving. The leading one, in two lines, advanced on Cistowes from the east, one battalion working off to the right into the wood, the other six marching on Cistowes, while the battery added itself to the 40 guns at Maslowed. This overwhelming mass drove the Prussians out of Cistowes, but part of a battalion held on doggedly to some detached houses west of the village.

By 9.30 a.m., the bulk of the Austrian 4th Corps, 28 battalions, was on the field, and 96 of their guns were firing; to these the Prussians were for the moment opposing only six battalions and 18 rifled guns.

The Austrian 2nd Corps was now getting into position from Racitz to Horenowes; several of its batteries opened fire from near the latter place. The whole of the 7th Division (Prussian) now joined in the struggle, in and about the great wood. All attempts to issue towards Maslowed village failed. The confusion of this fight is indescribable; it may be judged from the fact that, towards 11 a.m., while some Prussian companies held houses beyond the west of Cistowes, and an earth bank on the north side, great parts of the wood were again in the hands of the enemy, some of their parties in the wood having Prussian detachments both north and south of them; but the help of brigades and guns of the 2nd Corps was giving the Austrians gradual repossession.

Thus for three hours had the 7th Division, very little aided by artillery, "contained" the whole of one Austrian Army Corps and a large part of another. Its position was now critical in the extreme.

Attention must now be turned to the other divisions of the Ist Army—the 8th, 3rd, 4th, 6th and 5th, namely.

At 8 a.m. General von Horn had the 8th Division in readiness behind the Roskos-

Berg, and began his move on Sowetitz. On his left flank a battery opened fire, and three on his right flank from the hill. Prohaska's Brigade of the 3rd Austrian Corps was the immediate opponent. Its battery, firing from the north side of the high road, was soon silenced, and withdrew before 9 a.m. Another battery took its place, and was also driven off. Then came the news that Benatek was taken, and progress being made with Maslowed Wood, whereupon two light infantry bridges were constructed west of Skalka Wood.

By 9 a.m. the 4th Division was formed for action near Mzan, and the 3rd at Zawadilka; their batteries had been firing since 8 a.m. The orders now were for the 8th, 3rd, and 4th to secure the passages of the Bistritz, the 5th and 6th to close up in support of the 8th, the reserve artillery (16 batteries) to go forward to the Roskos-Berg.

Prohaska was soon overpowered, and retreated to the position of the 3rd Corps at Lipa. Four battalions of the 8th Division met with no resistance in crossing and gaining Skalka Wood; they then turned towards Sadowa Wood, and on the way were shelled from Lipa. Six battalions now entered the wood, and found few of the enemy occupying it, but on reaching the other side they were greeted with such a hail of shells from the massed guns of the 3rd and 10th Corps, stationed from Lipa towards Stresetitz, that all further advance was stopped, and the position of the troops in the wood became very unpleasant. During this halt it was that the Prussians had been driven out of Cistowes; a battalion attempted to go to their help from Sadowa Wood, but was quickly driven back by the hostile gun fire. Three rifled batteries opened fire from the spur rising from Skalka Wood towards Cistowes, but proved insufficient to draw the enemy's artillery fire.

The three rifled batteries of the 4th Division were meanwhile engaging across the Bistritz, but losing heavily, the 12-pr. smooth-bore battery had to be withdrawn. At 8.30 the 3rd Division's artillery arrived and opened fire from the right of the guns just mentioned. Two rifled batteries of the Reserve were also brought to bear from between Mzan and the high road, whence they were able almost to enfilade, though at long range, some of the hostile batteries.

The 10th Corps' Artillery (Austrian) seem to have been at first well forward towards Dohalicka; its commander requested assistance at this time from the 3rd Corps and was refused.

The Prussian gunners began to be troubled by Austrian skirmishers from the Sugar Factory and Unter-Dohalitz, so a battalion was sent against the former. The three Austrian battalions rapidly quitted the place, leaving 25 prisoners. The battalion then moved on Dohalitz, followed by the rest of the advanced guard, while the main body of the 4th Division made for Sadowa. The commander of the 10th Corps had no intention of holding these forward positions, and withdrew by degrees, and without much fighting, to the main position in front of Langenhof and Stresetitz. The leading Prussian battalion waded the stream, made its way to Ober-Dohalitz, and took part in the holding of the wood; 10½ battalions, the main body of the 4th, was across the Bistritz close behind the wood.

By this time (towards 11 a.m.) Dohalicka was in the hands of the 3rd Division. This division had formed for battle about Zawadilka, and advanced on a front from Kopanina to Johanneshof. It made little use of its artillery; one bold attempt was made by a single horse-battery to enfilade from between Johanneshof and Tresowitz, but it failed with loss. By 9.30, however, the advanced line of Austrian guns began to retire, and the 3rd Division commenced its advance. It was soon evident that the Austrians were not seriously defending the line of the Bistritz. In Mokrowous 9 officers and 60 men were captured. Dohalicka was entered just as the enemy were leaving it. Part of it was at once prepared for defence.

The great Cavalry Corps, "that had been held together at the cost of many a sacrifice," was at 10.30 still in its first position near Sucha. A request for support came from the Army of the Elbe; half of the corps was sent southwards. And thus, in the words of the Official Account, "it was now, on the decisive day, separated into two halves."

On the Prussian right, the enemy's main position had not yet been touched, and on the rest of the front it had only been approached. Only 12 Prussian batteries were across the Bistritz, and only seven of these had been able to act, while the enemy were firing from 250 well-placed guns.

IInd Army.

The IInd Army was meanwhile making for the field. The 1st Guard Division reached Choteborek at 11 a.m., its advanced guard being at the same moment at Jericek, two miles forward. On its left, the 11th Division, VI. Corps, was moving with great difficulty across country towards Racitz, having the 12th Division marching on *its* left. At 11 a.m., the advanced guard had nearly reached the Horicka-Berg near Racitz, and was there fired upon by artillery.

The 9th and 10th Divisions, V. Corps, with their artillery were approaching Choteborek at this time; but the I. Corps' advanced guard had not yet reached Bürglitz. The great cavalry division behind this corps was prevented from getting forward by the corps artillery, which would not give way.

So difficult had been the marching, that the rear of the columns of the Guards and VI. Corps were still as far back as the Elbe.

The height of Horenowes commands a perfect view of several miles to north-west, north and north-east. The approaching Prussians saw a single battery on it, but expected to find the whole Austrian right, on closer approach. But the enemy seemed to have no expectation of early attack on this side. His 4th and 2nd Corps were both occupied west of Horenowes, and his main Reserve, the 1st and 6th, were at the moment further from Horenowes than the heads of the Prussian columns.

The commander of the Guards saw at once that the heights of Horenowes were his proper objective. At 11.30 his two leading batteries opened fire from near Wrchownitz. By noon the main body of the 1st Guard Division was forming between this place and the Jericek road, along with a heavy cavalry brigade, while the advanced guard had already pushed on two battalions towards the village. A clump of trees on the height was indicated to all as the general direction.

The 11th Division had possession of Racitz by noon; and shortly afterwards the whole of the Horicka-Berg and its wood were in the hands of the 12th Division.

By this time 48 guns of the IInd Army had opened fire.

The Austrians had been making preparations, about 11 a.m., for a combined attack of 4th and 2nd Corps, for the annihilation of the 7th Division, which was holding on doggedly in the greatest peril from Benatek to the wood. But the new aspect of affairs put an end to this project, and the 2nd Corps was recalled to make a front to the north, 40 guns firing from the east slope of Horenowes to cover the movements.

The 4th Corps also received definite orders about 1 p.m., to fall back to the originally planned main position, namely, from Chlum to Nedelist. A battalion still held Maslowed, and a brigade remained in Cistowes. These movements left the artillery on both sides of Horenowes unsupported, but the guns held on till the leading battalion of the Guards had taken the village and was advancing against them. Those on the west side withdrew first, battery by battery, then those on the east. Towards 1 p.m., the Prussians had 90 guns shelling the retiring enemy, and their fire was doing great havoc. Soon after 1 p.m., the Prussian infantry was on the heights of Horenowes.

Two cavalry charges were made; five or six squadrons, trotting round from north of Horenowes, saw an Austrian battalion retreating in some confusion. An immediate charge was made, and 3 officers and 70 men taken prisoners. Proceeding against the main body of the battalion, they found it massed to receive them, and failed to break it.

Three squadrons had reached the Horenowes heights and saw a solid column of the 2nd Corps marching on the road from Maslowed to Nedelist, guarded in rear by the battalion that had been the last to leave the former village. The squadrons had cover till they topped the Maslowed spur, but the charge failed to break the battalion, which had at once formed square.

The 11th Division, moving from Racitz, had little difficulty in clearing the plateau north of Sendrasitz; the leading battalion went on to the village, and found a detachment of the 12th Division already there. They drove off a battery firing from the south-east, and captured one of its guns.

The 1st Guard Division was, by 2 p.m., established on the Maslowed spur, and in the village; the VI. held Sendrasitz, and its left wing was nearing Lochenitz, where there was an important bridge. The Austrian 2nd Corps was getting on slowly with its flank march to meet this movement; a brigade was on the high road guarding the bridge, but only one other brigade had got past Nedelist. The 4th Corps was gradually withdrawing to the entrenched position, its movement being to a great extent screened from the enemy's view; but Prussian shells reached Benedek's massed Artillery Reserve near Sweti, and forced it to shift its position. The 64 guns remaining here were now brought to the west of Nedelist, with 40 guns of the 4th Corps in the entrenchments to the left. Chlum was now the Guards' objective. More than 100 guns were ranged against them, many of them entrenched, eight battalions were among the batteries and breastworks, one in Chlum, and five and a-half battalions behind the village. There was a second line of seven or eight battalions; and stretching south-east from No. 2 Entrenchment, a brigade, mixed up with batteries from the Reserve.*

Benedek had seen what was going on here, but seemed to fear no danger; the 6th Corps, massed in reserve south of Rosberitz, was given no order to front eastwards.

The attack on Chlum was planned as follows:—While the 1st Guard Division was forming behind the Maslowed spur, eight batteries were brought forward across the Maslowed-Nedelist road. The one on the right fired on Cistowes, to prevent any interference on that flank; the left battery fired on masses of Reserve visible towards Sweti; the rest engaged the powerful artillery lately referred to. The division, advancing on Chlum from the north-west, had one battalion spread out in front, followed

* The plan of the battle exhibits the positions of all troops at this stage.

by two brigades, each of which also threw out companies of skirmishers. The advance was made quickly in spite of the severe shell-fire. Entrenchment No. 3 was soon reached; the Austrian gunners held on to the last, some of them firing grape at 50 paces. The Prussian riflemen turned their attention to the teams, and prevented the removal of 14 guns at this point. The two battalions between No. 3 and Chlum gave way quickly, pursued by three or four Guard battalions, whose advance cut off the battery in No. 4. A battalion now wheeled against the village, and got in without much difficulty; but serious resistance was met with in the interior. When it was cleared, and a second battalion occupied the south half, an attempt made to debouch from the west side was greeted by a hail of grape at 200 yards from a battery of the 3rd Corps. Collecting the two battalions for a spring, the battery was gallantly stormed as it stood, only one of its eight guns escaping. At 2-45 p.m., Chlum was taken. A battalion reached the hollow way leaving Chlum for Sweti, and dislodged a hostile battalion from it. The Austrian battalions further east offered little resistance, and the earthworks were abandoned.

Four Prussian companies moving on Rosberitz were charged unsuccessfully by several regiments of Austrian cavalry, which lost 200 men in a few minutes. The infantry, confident in its fire, had received them in single line.

By 3 p.m., Rosberitz was occupied, Nedelist cleared of the enemy by the 11th Division and several more guns captured.

Meanwhile, an Austrian brigade had been left in Cistowes; it broke out to the east, and in attempting to reach Nedelist got mixed up with the rear troops of the Prussian Guards.

Part of it made an attempt on Chlum from Lipa Wood; a battery and some cavalry that accompanied it were dispersed and destroyed. The remains of the brigade were forced back to Lipa.

Prussian batteries were now brought forward between Rosberitz and Sweti, and the 11th Division was working on between Nedelist and Lochenitz, where it came upon a mass of Austrian Reserve Artillery and captured a battery. The 12th Division came on from Trotina, had a sharp fight at a camp of huts outside Lochenitz and then attacked the village.

The Guards had captured 55 guns. Their head was on the great Königgrätz road, and was thus in rear of the enemy's main position; "but it was now in face of the mighty masses of the Austrian Reserves, the 1st and 6th Corps and a powerful artillery, within a distance of less than a mile." These were in the low ground between Sweti, Wsestar and Rosnitz, and from Langenhof there was the threat of a huge body of cavalry.

We saw the 7th Division in the utmost jeopardy at noon; only when the Guards' attack became effective was it possible to re-form and reorganize the division. It had lost 84 officers and over 2,000 men.

Shortly after noon the Ist Army had its 5th Division across the Bistritz at Dohalitz, the 6th across at Sadowa, the 4th and 8th in and behind the Sadowa Wood, the 3rd Division had crossed at Dohalicka and Mokrowous.

On this side the struggle became an unequal artillery combat. The Official Account says:—"We may assume that 200 pieces of artillery (Austrian) came by degrees into action at this spot (the heights from Lipa towards Tresowitz), under the most favourable circumstances imaginable. More than 80 Prussian guns remained beyond the

Bistritz, because no use could be made of them further forward." "The artillery of the 4th Division could find no space for action, and was withdrawn over the Sadowa bridge." "The cavalry was in a similar position; sufficient ground had not yet been gained to afford room for its formation across the stream, a circumstance which was not without influence later in the day when the fruits of victory were reaped."

By noon, only 18 Prussian guns were firing from the north side of Sadowa Wood, 24 from near Dohalicka. A little later four more batteries unlimbered alongside of the latter. Again, a little later, the 18 guns on the Cistowes spur were increased to 42.

Things were looking very bad towards 3 p.m., and the commanders began to have an eye on a possible retreat, "when suddenly the fire of the nearest batteries on the opposite side ceased." In fact, the Austrian 4th Corps had marched off to the right at 2 p.m., masking the movement to the last moment with their guns.

The artillery combat south of Sadowa Wood was going very badly for the Prussians, "a want of unity in the direction was painfully evident." Several batteries retreated across the stream. "The difficulty of keeping up the communication between the batteries and their wagons" was great. "It arose from the scarcity of passages over the Bistritz."

For more than five hours the men of the 8th Division passively held the Sadowa Wood—a most severe trial. At 1 p.m. an attempt to advance was made and failed. Then an Austrian brigade attacked the wood and Dohalitz without success.

We must now turn to the proceedings of the Elbe Army. We left it at 11 a.m., with its advanced guard in stated positions from Hradek to Lubno, while the leading Division, 15th, was beginning to form east of Nechanitz. The 14th Division was still on the other side of that place, with the 16th still further in rear.

The Saxons cleared out of Popowitz, and retired entirely to their main position—2nd Division behind Problus, 1st Division in front line from Stresetitz to Nieder-Prim, three battalions being in Problus and two in Nieder-Prim; 34 guns in action on the brow above Nieder-Prim; 8th Corps in the wood between Problus and Charbusitz, with cavalry and the reserve guns in support behind it; part of its infantry in Ober-Prim and the wood to the south of it.

The Prussians had now 24 guns on the rise above Lubno, but the range to the Austrian artillery was nearly 3,500 yards, and little effect was produced on either side. The hostile position was seen to afford the enemy an open field of fire in the centre, so the attack was planned to take place from the two flanks. The troops at Hradek were to advance by Neu-Prim and the wood beyond Jehlitz, followed by the 15th Division; the force at Lubno to go to Popowitz and through its wood towards Problus, followed by the 14th Division; 16th Division in reserve. About noon five and a-half battalions had reached the extremity of the Popowitz Wood towards Problus, and had then to halt. About the same time a battalion on the right wing had passed through Jehlitz into the wood, another debouched from Neu-Prim towards Problus, while a third drove detachments through the wood to the east.

This weak-looking, widely-extended attack induced the Saxons to make a counter-attack towards Hradek. A brigade and the garrison of Nieder-Prim advanced, drove off the two hostile battalions, inflicting heavy loss, but was brought to a standstill by the appearance on its left flank of the battalion in the wood.

This happened shortly before 1 p.m. The Saxon Crown Prince ordered up another brigade, and, about 2 p.m., he tried again; but a considerable part of the 15th Division was now on the spot, and the stroke failed. The Prussian artillery, now reinforced to 66 guns, remained too far off to effect anything of consequence, and the pursuit was checked. After a short pause, seven battalions closed round Ober-Prim and took it, and then repulsed a most determined attempt at recapture. The defeat of these two brigades exposed the flank of the main position at Problus, so three batteries were brought up to fire from the hill east of Nieder-Prim to check the assailants. Ober-Prim was soon in flames, and had to be evacuated by the Prussians. But the guns now came up from Lubno Hill to within 1,600 yards of Nieder-Prim, and by their help this place was quickly taken.

The 14th Division was ready to leave the Popowitz Wood by 2.30 p.m.

The Austrian resistance at Problus was over; the Saxon Crown Prince had seen the beginning of the retreat of the 10th Corps on his right, the approach of the IInd Army at Chlum, the threatening direction of the 15th Division's attack. About 3 p.m., he gave the order to retreat.

It remains only to describe the Austrian attempt to retake Chlum, and to imagine the confusion of the retreat that followed.

Only the 1st Guard Division and a few battalions of its 2nd Division were as yet available to hold Chlum and its hill; but before the Austrians made their counter-attack, there was time to capture and clear the Lipa Wood.

"The wood was scarcely taken, when the Austrians advanced to the assault, and attempted to regain possession of the hill to the west of Chlum. *In close order*, with skirmishers only a few paces ahead of them, the columns—apparently one brigade—crossed the high road between Rosberitz and Lipa, and ascended the face of the hill with great bravery, utterly disregarding the fire that was poured on them from Rosberitz." On the heights there were available to meet this attack only five or six companies; these reserved their fire till the enemy were within 100 paces; they delivered two volleys and continued with steady and rapid file fire, beating the enemy, who lost heavily.

Lipa was now attacked from three sides, and taken, but with great slaughter on both sides. Near this place an Austrian battery, that had remained firing till 50 paces only separated the combatants, was captured. The Austrians now made four separate attacks on the west side of Rosberitz, and then retook the village by a concentric attack of two brigades on its narrow end.

The Prussian artillery was now up, and was pouring a heavy fire on the great masses retiring about Rosnitz, and doing great execution.

Meanwhile the 11th and 12th Divisions were pushing on further to the east, and, soon after Rosberitz was retaken, these troops were within a mile of the Elbe Army. Direct pursuit by the 1st Army was slow, for the Austrian artillery, at every rise, unlimbered and covered the retreat with devoted bravery. Their cavalry, too, charged frequently, fighting hand-to-hand with the Prussian squadrons that were now pushing into the fight. The latter at one point reached a hostile battalion in square, broke it and took it prisoners.

The I. Corps had begun to reach the field, and helped the further advance of the Guards; the V. Corps did not come under fire. The 16th Division of the Elbe Army was still (4.30 p.m.) coming through Nechanitz.

The concentric advance had now brought the Prussian forces on to a very narrow front of a few miles. Any attempt to spread to the left in pursuit brought them under the fire of Königgrätz. Soon the hostile infantry were out of reach, and the exhausted Prussians gave up the pursuit. Most of them had been on the move since the very small hours of the morning, many had marched 16 miles in heavy country, and had fought for ten hours. "None of them had been able to cook their rations, the horses had not been fed and very few of the men had provisions with them."

The Prussians capture 113 guns, and the Austrian loss amounts to 40,000. The Prussians lose 10,000.

Next day, touch with the Austrians is lost.

CHAPTER X.

COMMENTS ON SADOWA AND TRAUTENAU.

IN 1866 the breech-loading rifle came first into use on an extensive scale. A considerable change in the infantry tactics of the fights was naturally expected, but the change exceeded all expectation. Previously to this, skirmishers were looked on as a mere adjunct, the line of battle being the close line or columns coming on behind ; but now the skirmishers were found to *be* the line of battle,—that is to say, men in open order were capable of bringing about a decision.

Before the campaign, there was by no means a consensus of opinion that this would be the case. Even Prussian Generals, before the war, were preaching to their officers, "throw out very few skirmishers—with their breechloaders they will do as much as twice their numbers used to do." Experience, however, very soon showed that the exact reverse was the best practice. Theoretically, the argument in favour of few skirmishers seemed sound enough ; but every man, above all things, wished to get the full benefit of his splendid weapon. It was essentially the sound, offensive spirit that had been instilled into the Prussian soldier that led the company officers to extend their men in the freest manner. The enemy had a good weapon, but a muzzle-loader. The free skirmishing of the Prussians was due, then, not so much to a desire to escape loss as to a desire to inflict as much harm as possible. They, therefore, skirmished in great numbers, and a company column followed at 150 yards or so, as support. Four years later, these same troops, armed with the same weapon, met the French with a much superior one. This time the change came from the necessity for diminishing losses, and the solid supports dissolved into a second line of skirmishers. It might well have been expected that no European troops would have ever attempted the advance in column after 1870 ; but we find the Russians trying it at Plevna, with the result of the annihilation of whole regiments.

The pervading phenomenon of the fighting in 1866, as compared with anything that preceded it, is an extraordinary appearance of extension of front on the part of the attacker, with a small degree of depth ; and above all appears an inclination to lap round the enemy's flanks, whether we look at the whole battle or the attack of a part.

In the course of the fighting, too, great mixing up of units takes place. The tactics of the company columns consists in throwing out swarms of skirmishers ; and the column in support often joins the firing line and extends almost immediately. The firing line, *i.e.*, the line of battle, has now the appearance of a confused attack of irregulars ; the longer range of weapons causes shots that pass over the heads of the 1st line to hit the 2nd line, following at old-fashioned distance ; these men must either get cover for their company column, or they must get forward to fire, for they object to being hit without having a chance to return the fire. No wonder, then, if very often it happened that nearly every man was soon in the firing line, that that line had extended enormously in consequence, and that there were no reserves left near enough to be of much use.

Thus we find a Prussian Officer writing shortly after this war :—

"The far and sure-carrying arms of precision of the present time forbid supports and reserves in closed columns, except where the ground is favourable for such formations. It is apparent that either the old secondary formations have become impossible on account of the greater range of missiles, which causes the necessary distance to become so much increased that there can finally be no relation between the supports and the engaged line; or the supports see in the loose, opened-out formation the proper means to adopt, and thus, of their own accord, they rush up into the 1st line."

May (Ouvry's Transl.).

The possession of a weapon that a man could load lying down gave an advantage that requires no insisting on. Many a time the Prussian firing line was brought to a dead stop, at this point or that; a hurricane of shells and bullets flying over them as they lay, they could still load and fire, which both kept up the men's spirits and enabled them sometimes to rise and advance again, even before any support came. You can imagine a fairly close-quarters fire-fight between a Prussian and an Austrian, both lying down—both fire and miss. The former now loads and waits for his opponent to get up and load, for the Prussian soldier, both then, before 1870, and now, is trained to hold his fire unless he sees a reasonable chance of hitting. The next time he fires the Austrian is *hors de combat.* A few more of such episodes here and there along the line, and the company is ready to be up and at them.

But this rushing to the front of all who come under fire brings in the necessity of providing a strong reserve that shall keep out of fire, at least until the crisis of the whole battle justifies the engaging of almost the last man and gun.

A most striking phenomenon of the campaign is the extent to which the battles were essentially begun, continued and ended by the infantry. The other arms played a mere secondary part, and not infrequently their influence on the result was *nil.* Particularly on the side of the victors was this the case. They knew the splendid quality of their infantry, and the superiority of its weapon; so the infantry rushed on, and when its advance was checked, in its impatience it sought a way by extending to right and left, till advance was again found impossible. It was the triumph of individualism; the improved arm and the improved soldier had come together. This combination, as I have insisted earlier, is necessary to an advance in tactics; or rather the improved man is required for the full efficacy of the improved weapon.

The Prussian artillery had good weapons as the time went, they were good technical gunners, and the bold action on occasion of individual batteries shows that they were not wanting in pluck. But the tactics of artillery had not been studied and practised like those of infantry. They usually fired at absurdly long ranges, and showed little aptitude for *finding* artillery positions. Thus no less than 80 guns of the reserve artillery on the Sadowa road did nothing.

But after the war the General Staff pondered over all these things, and by 1870 the Prussian artillery understood their business. Curiously enough, the infantry tactics, in the matter of having the supports in solid columns, were retained, until the slaughter among the Guards at St. Privat made an end of it.

There is no doubt that, as at Gravelotte so here, the Prussians had a very difficult task. The Austrian position had long smooth slopes to the front, and it is all the more wonderful that the artillery were not made to prepare the way of the infantry more effectually. The difficulties and the losses of the infantry were thereby much increased.

"It is remarkable, also, that when the Austrian retreat began and continued, out of 113 guns captured 108 fell to the infantry. Where was the cavalry?" So writes a Prussian Military Critic soon after the war. But it was the very fact of the open ground—supposed to be so favourable for cavalry action, but only really so in modern times in peace manœuvres—this fact, and the self-sacrificing devotion of the Austrian gunners, that made it impossible for formed bodies of cavalry to do much. If, on the contrary, the Austrian guns had done more bolting than firing during the retreat, then Prussian cavalry would have had a chance of making a haul, and the Prussian infantry no chance at all.

Artillery.

Till 2 p.m. the Ist Army carried on essentially an artillery action; but during that time they made no substantial impression on the Austrian artillery, by which a way would have been prepared for the further advance of the infantry. The Prussian artillery barely held their own in the prolonged duel, and 80 guns stood idle a short distance to the rear. If you wish to win, you must bring a preponderance of force at the required point.

The Austrian artillery only gave way when the IInd Army began to get in their rear. Now, it has been argued that it was the business of the Ist Army merely to hold the Austrians in their place till the IInd Army got behind; but what if Benedek had known his business better, and had added, in good time, his 50,000 reserves to his right flank? He would certainly have delayed very greatly the success of the Crown Prince's movement, if he had not even prevented it altogether. If anything decisive, then, was to be effected before daylight failed, the Ist Army would have had to attack in earnest while the Austrian artillery was still in position; and the slaughter would have been so great that the attack might well have failed.

The military world was asking itself the question after this campaign, "what element is now necessary to secure a preponderance, when all belligerents will have the breechloader? A considerable party began to think that the bayonet must come in again as the arbiter; but the difficulty here is that the man on the defensive will certainly fire till the last moment. Others thought that cunningly contrived weapons of the mitrailleuse species would do it. Prussia experimented with these after the war, but deliberately rejected them—they were not, of course, on the principle, nor had they the lightness and handiness, of the Maxim gun. The Federals in America had tried them, and thought little of them. As we know, the French alone supplied them to their armies and reposed a pathetic faith in their powers. Even now, the lineal descendant of the mitrailleuse is only a weapon for occasions.

It was not unnatural that a strong body of opinion should arise that the defensive would be so much advantaged by the breechloader that it was bound to win; that armies would manœuvre till one got the enemy into such a position that he would be forced to attack tactically, though the other was strategically the attacker. But the Prussians never adopted this view. They came to the conclusion that infantry, properly helped by a powerful artillery, would still be able to attack with success, provided the infantry were so trained that they would make intelligent use of accidents of ground, and would instinctively work round the hostile flanks.

The French, on the contrary, as soon as they found themselves in possession of the best rifle in Europe, not unnaturally, perhaps, hoped to avail themselves of its full effect by occupying a good line and shooting down the approaching foe before he could do anything with his own shorter-ranging rifle. In reality, the idea was essentially foolish, for the longer-range advantage would still be with them in attacking. But they took the narrower view, and we see it in their reluctance to attack at Spicheren and at Vionville; at both of these fights they had the greater numbers—at Spicheren during the first part of the action, at Vionville all the time.

The Prussians practically concluded, and their conclusion was no doubt due to the negative success of their guns in 1866—that "in the next war that side will obtain an unconditional tactical preponderance which best knows how to make use of its artillery—the side whose artillery has had the best tactical training."

This was a sound deduction; and we shall find also that from the very failure of their cavalry in 1866 to perform its natural duty as a "screen" and as a great scouting arm—that from this very failure they learned the lesson which, when learnt, bore such good fruit in 1870.

It is a common saying that people learn much less from the experience of others than from their own. From what perverse faculty in man this arises, I shall not stop to discuss, but it is true in many things military, as in all others. The Federals try the mitrailleuse, and find it of little use; the French pin their faith to it five years later. Both Federals and Confederates show of what advantage a bold use of cavalry may be; a year later Prussians and Austrians, in possession of great masses of well-equipped horsemen, make no proper use of them. The Prussians have learnt the lesson, and profit by it, in 1870; but the French have learnt nothing in this connection from America, Prussia or Austria. Our line beat the French columns in the Peninsula and at Waterloo, and the French give up the columns; but in 1854 Russia has not learnt the lesson, nor again at Plevna in 1877 has she learnt the lesson of Chlum and St. Privat.

The conclusion, therefore, generally arrived at after this war was that officers and men must be trained to a high standard that should make them capable of "individual" fighting. The spirit was to be of more importance than the letter. "An army which cannot venture to trust in the individual worth of its soldiers, so far as to let them fight in this manner (*i.e.*, in intelligently irregular formations) cannot reckon on the advantages to be derived from the operation of the breechloader."

A vivid illustration of the action of any particular arm in a fight can be had by comparing its losses with those of the other units engaged. The illustration is specially trustworthy, if the fight has been hotly contested and lost. This was the case at Trautenau, where a few days before Sadowa, part of the IInd Prussian Army was worsted. They had 16 batteries, *i.e.*, 96 guns, available. This artillery had seven wounded on a day when the whole Prussian loss was 1,338; 78 horses fell that day, of which only two belonged to the gunners.

As the battle of Sadowa progressed, a part of the Prussian Ist Army's artillery crossed to the ridge between Ober-Dohalitz and Trisowitz. There was room for 36 batteries, but only about 42 guns got into position there, while the enemy had 250 more or less able to oppose them. The objection that a retreat would have been most difficult and hazardous, as the Bistritz was unfordable from the rain, and the guns would therefore have had to file singly, within range of the hostile guns, through Sadowa—this objection simply leads to the question, "Where were the engineers and the pontoon trains?"

The Elbe Army's artillery, also, played a trivial part; it exchanged shots with the enemy at 4,000 to 5,000 paces.

There were still many smooth-bore guns in both armies.

In the extract from Jomini which I quoted and discussed in a former chapter, embodying the tactical ideas prevalent after the Napoleonic wars on the question of artillery, I said that these ideas allowed one-third of the available guns to deal with the hostile artillery, while the remainder were to pound his infantry and cavalry exclusively. After 1866, we find something of the same idea prevailing. Thus, the same Prussian writer whom I have quoted before says:—"To silence the defending artillery cannot be the most important object of the attacker. Even if this should be done, the advance of the infantry on a position defended by breechloaders is not possible. To silence the attacking guns is the essential object of the defender's artillery, then the infantry will know how to repulse the attack of the enemy's infantry; but artillery on the offensive should, on the contrary, make it the principal object to play upon the defender's infantry. An attack can only be thought of when this has been weakened. It has only to engage with the defending guns in so far as is absolutely necessary, always having the principal object in view."

We should hardly agree to this nowadays. The fact is that the small-arm was becoming the relatively more effective weapon. Though the gun was decently accurate, its rate of fire was slow as compared with the present day, and the fuzes of the time did not allow of accurate bursting of shells in the air. Even in 1870, fuzes for shells were in their infancy. But our quick-firing field guns, with their accurate shrapnel fire, cannot be coolly ignored as recommended after 1866. We learn, in fact, from Manchuria that the field-gun has shown its ability, if the field of fire be decently open, to completely stop an infantry advance far beyond any reasonable rifle range; this, of course, when the guns are not interfered with by hostile artillery. The attacker's artillery must evidently deal with this type of gun before the attack can hope to be successful.

Suppose for instance, the defender's infantry entrenches in the Boer manner, *i.e.*, deep, narrow trenches skilfully concealed. There is evidently quite a new problem. No amount of bombarding has any great effect, unless your own infantry advances and threatens sufficiently closely to compel the hostile infantry to man his parapets. During the early advance of your infantry you would certainly turn the whole attention of your artillery to *his* guns, in order to facilitate the advance of your infantry, which, till it has arrived within (say) 1,200 yards, has nothing much to fear from anything but the enemy's shrapnel.

After 1866, people were able to say that if your infantry advances quickly and skilfully, it has nothing to fear from the enemy's artillery; even then, however, it was of doubtful truth, as witness the inability of the Prussians to get beyond the edge of the wood of Sadowa on the Austrian side. Now, it seems to be the very opposite of the truth.

Cavalry.

The Prussians formed large masses of cavalry, independent of the Army Corps. To keep these large masses concentrated all the time was difficult, on account of their supply; and after all, very little use was made of them. After Sadowa there was a totally defeated Austrian Army retreating in haste for over 100 miles over ground that afforded few, if any, strong positions, and only once, during all that time, did the cavalry find an opportunity. Up to this period it was always an understood thing that the peculiar opportunity for cavalry came during a pursuit; and the Prussian cavalry were blamed for achieving so little. But the same thing happened in 1870, and one can only hold that the time had come when cavalry can no longer effect what it did so brilliantly of old. Even on the one occasion after Sadowa, viz., at Rokeinitz on 15th July, when the cavalry passed through the rear guard of the Austrians and took the marching column further on by surprise, we find that that rear guard was first attacked and pierced by the infantry, the cavalry then following through the gap.

But there still remained for the cavalry a most important and honourable task whose execution, in 1866, was conspicuous by its absence—and this, too, on both sides. I refer, of course, to its work as "screen" and scout. The keeping of the cavalry masses safely to the rear during the advance into Bohemia led to the result that on 2nd July, the day before the battle, the Prussians were in the dark as to the whereabouts of the enemy, who turned out to be concentrated no more than five miles beyond the outposts. It is almost incredible that on the morning of the 2nd a mass of Austrian troops was five miles north of Racitz, that it withdrew to the battlefield on the same day, and that the Prussians knew nothing of it.

Engineers.

In the 1864 Campaign against the Danes, the artillery and engineers had played a conspicuous and useful part. In the 1866 Campaign, the engineers did very little. I have referred to the neglect of bridging the Bistritz. On the Austrian side at Sadowa some well-constructed batteries and breastworks were made. Part of the IInd Army, hastening to the sound of the guns, arrived at a brook swollen by rain; there was one narrow bridge, to file across which would have taken the Army Corps several hours; the engineers were acting as escort to the transport in rear! And why could not the engineers have prepared epaulements for the Prussian guns in their unequal duel on the Bistritz? There was plenty of time, for it was well known all the time that the Ist Army must fight for hours a "containing" action merely. When the wood of Sadowa was gained, but no advance beyond it could be made, and it was still uncertain whether the IInd Army would be able to be up in time, why was not the outer edge of the wood prepared for defence? If the Crown Prince had not arrived in time, if Benedek had boldly sent out the bulk of his reserve against him, the Austrians in Lipa and Langenhof would certainly have made a great attack on the Sadowa Wood.

A brief survey, then, of the battle indicates the triumph of the infantryman with the breechloader; an insufficient use of the great power of artillery; hardly any use at all of cavalry or engineers; great extension in the lines of infantry attack, instinctive tendency to lap round flanks, as the line of least resistance; the immense importance of the company leaders, and of the training of the individual man's intelligence; the self-sacrificing action of the Austrian cavalry and artillery in the retreat; the inability of pursuing cavalry to effect much by direct pursuit, if the defender's cavalry and artillery are staunch; the increasing power of artillery to stop infantry when the artillery is not required at the same time to engage with a superior hostile artillery.

The 1st, 2nd, 3rd and 4th of the Austrian Corps each numbered from 27,000 to 28,000 men; the 6th, 22,000; the 8th, 15,000; the 10th, 18,000.

Therefore the reserve, viz., 1st and 6th, and the cavalry and artillery, amounted to 49,000 foot, 14,000 horse and 3,000 gunners. The fighting line, viz., 2nd, 3rd, 4th, 8th, 10th and Saxons, amounted to 139,000 men. Thus the reserve, counting the cavalry, was not far off one-third of the total force, which seems a large proportion. On the whole, however, it must be acknowledged that the force was well placed. The Austrians were not greatly outnumbered; the battle was going to be won by the breechloader and superior tactics.

The 140,000 men in the 1st line extended about seven miles, which gives 20,000 men to the mile, or nearly 11 to the yard; a dense formation, but it must be remembered that the Austrians had muzzle-loaders. If the whole army be included, there were no less than 16 men to the yard.

On the Prussian side, the Elbe Army's three divisions amounted to 39,000 men and the Ist Army 85,000. These two thus upheld the first three hours of fighting with 124,000 men. The IInd Army amounted to 97,000, thus giving the Prussians eventually 221,000 on the field of battle.

It will be noticed that the 97,000 of the IInd Army were met by the Austrian 4th and 2nd, amounting together to only 55,000. Faulty information is here seen.

Comments on Trautenau.

A few remarks on Trautenau will bring us to the end of our discussion of the 1866 stage in the evolution of tactics.

"At the very beginning of the action a misunderstanding arose." The aim of the Ist Corps was to reach Arnau and occupy it. As a first step, the 1st Division was to secure Trautenau and the heights south-east of it with its advanced guard; but the 2nd Division, arriving at Parchnitz two hours before the other, merely halted and waited. However, the Austrians having only one brigade as yet, the seven battalions of the advanced guard and the eight sent forward by the 2nd Division pretty easily bore the enemy back beyond the line Hohenbrück-Alt Rognitz. Eight battalions, without cavalry or artillery, were at 3 p.m. on this line, themselves forming a thin line a mile long, at a distance of a mile from Trautenau; the ten battalions that had not yet been engaged, and practically all the cavalry and artillery, were east of Trautenau in the Aupa Valley. Thus, in view of a march to Arnau, the eight battalions were a flank guard—but a guard without guns or horses, and this in a country where no long view was to be had.

We have seen how they were overpowered, by the Austrian artillery chiefly; their retreat, under the circumstances, was commendably deliberate.

The further retreat of the whole corps through the pass seems quite unnecessary. It was late at night, and long after any action was perceptible on the enemy's part, before the last bodies filed into the wearied column.

CHAPTER XI.

FROM 1866 TO 1870.

In the last chapter I discussed the campaign of 1866 in Bohemia, described the strategical movements that led up to the battle of Sadowa, gave a brief outline of the battle, and then enumerated the following points as embodying the chief lessons to be learnt from the fighting:—The triumph of the infantryman armed with the breech-loader; an insufficient use of the great power of artillery; hardly any use of cavalry or engineers; great extension in the lines of infantry attack, instinctive tendency to lap round flanks, as the line of least resistance; the immense importance of the company leaders, and of the training of the individual man's intelligence; the self-sacrificing action of the Austrian cavalry and artillery in the retreat; the inability of pursuing cavalry to effect much by direct pursuit, if the defender's cavalry and artillery act thus; the increasing power of artillery to stop infantry when the artillery is not required at the same time to engage with a superior hostile artillery.

We saw, in fact, to what an extent the victorious Prussians owed their victory to their infantry. But to owe your victory throughout a campaign to one arm only is to fail in utilizing properly all your strength. The Prussian General Staff did not fail to note the loss of power thus accruing, and four years later, in 1870, we find they have learnt the lesson. Their infantry are as good as ever; in the action of their cavalry and artillery and engineers, there is a change for the better amounting to a revolution.

The cavalry now "cover and discover" in the most effective manner; the artillery, the long-range arm, practically opens each battle, posts itself with the most skilful appreciation of the tactical situation, comes boldly forward to the most effective range, prepares the way for the infantry attack by concentration of fire on the chosen objective, or, in that part of the field where attack is not likely to be successful, helps the infantry to hold the enemy in his place, while the main attack is being delivered elsewhere. For this enemy, though strategically always, and tactically often, outnumbered, is no mean antagonist. He has a very powerful infantry arm; and his work at Fröschwiller, on the Spicheren plateau, on the centre and left at Gravelotte, on the right at St. Privat, in the struggle for Bazeilles and in many another bloody scene, shows that there is high spirit, contempt of death, and a good standard of military virtue.

The engineers are now well to the front; they expedite marches by throwing bridges before the bulk of the column arrives; they lay field telegraphs in all directions; they make lines of investment in a few days into a band of steel; they demolish the enemy's railways; the first German across the Moselle is an engineer officer, bent on the gentle art of demolition. The bridge-making about Sedan is worthy of note. The 1st Bavarian Corps, on the evening of the day before the battle, have their bridge-train well forward, and throw two pontoon bridges across the Meuse near Remilly, but leave it undecked for the present. When their commander determines to cross and attack Bazeilles before daybreak, the decking is completed most skilfully in the dark in dead silence. The XI. and V. Corps having orders to cross about Donchery, the former, arriving first, throws a bridge, and the whole corps is across before daylight. The V. Corps, coming up in the small hours of the morning, has had the foresight to send ahead a staff officer to choose a site for its bridge, and to have its engineers well forward. These construct a bridge of eight spans, of which only the two centre piers are floating, the other piers being trestles. It was reported that this bridge was ready for use in 40 minutes from the time the first trestle was launched. Opposite the bridge was a high railway embankment; a ramp was made up this, while the bridge was under construction, for the guns to get over the obstruction. As the XI. Corps had crossed earlier, both bridges were at the disposal of the V. and the 4th Cavalry Division, and these reached the north bank with great celerity.

Von Boguslawski thus sums up the chief points to be noted in the tactics of 1870–71 :—

"On the German side—

"(1). The attack is directed on the enemy's flank, an assault on the centre following this, sooner or later.

"(2). In most cases, very powerful artillery fire to prepare the way.

"(3). Extensive employment of skirmishers.

"(4). Cavalry action in battle restricted.

"On the defensive, the Germans show generally skilful choice of ground, concentration of artillery and a proper system of firing.

"On the French side—

"(1). A strict defensive maintained, even against flank attacks.

"(2). Isolated counter-attacks without sufficient results.

"(3). Likewise very strong swarms of skirmishers.

"(4). Want of combination and of superior direction in the employment of artillery.

"(5). The cavalry behaves very well when it comes into play, but acts as if there were no such thing as a breechloader.

"On the offensive, in the first period, gallant, impetuous advances of great swarms of skirmishers, who shoot too much, and thus retard their own movements, often opening fire at absurd distances. In the second period of the war, bad officers and inability to manœuvre ; hence attacks unskilfully made and soon checked."

Cavalry in 1870.

In 1866, as we saw, great corps of cavalry were formed by the Prussians, laboriously kept in masses, supplied and moved with some difficulty, and after all effecting almost nothing. The question then arose what the modification for the next war was to be, and a considerable body of opinion was in favour of attaching more cavalry to the Army Corps, and having the remaining independent bodies no larger than brigades of two regiments.

But the authorities took a middle course, which worked to perfection. We may suppose that, before 1866, they had still the idea, lasting on from Napoleonic times, that charges by huge masses of cavalry might still complete a victory. This idea was

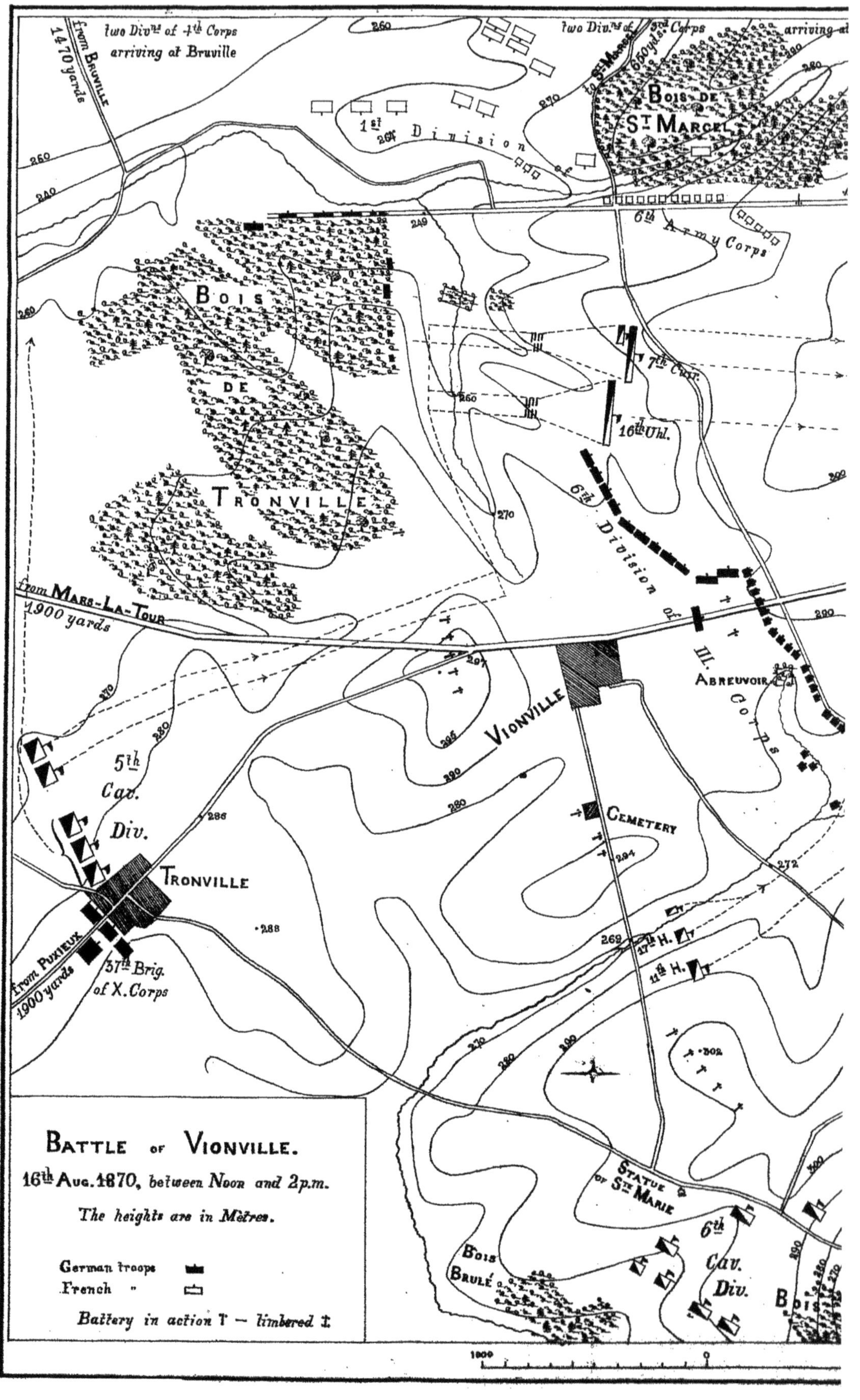

two Divns of 4th Corps arriving at Bruville
from Bruville
1470 yards
two Divns of 3rd Corps arriving at
Bois de St. Marcel
1st Division
6th Army Corps
Bois de Tronville
7th Cuir.
16th Uhl.
6th Division of III. Corps
from Mars-La-Tour
1900 yards
Vionville
Abreuvoir
5th Cav. Div.
Cemetery
Tronville
from Puxieux
1900 yards
37th Brig. of X. Corps
17th H.
11th H.
Statue of St. Marie
6th Cav. Div.
Bois Brulé
Bois
Battle of Vionville.
16th Aug. 1870, between Noon and 2 p.m.
The heights are in Mètres.
German troops
French
Battery in action — limbered

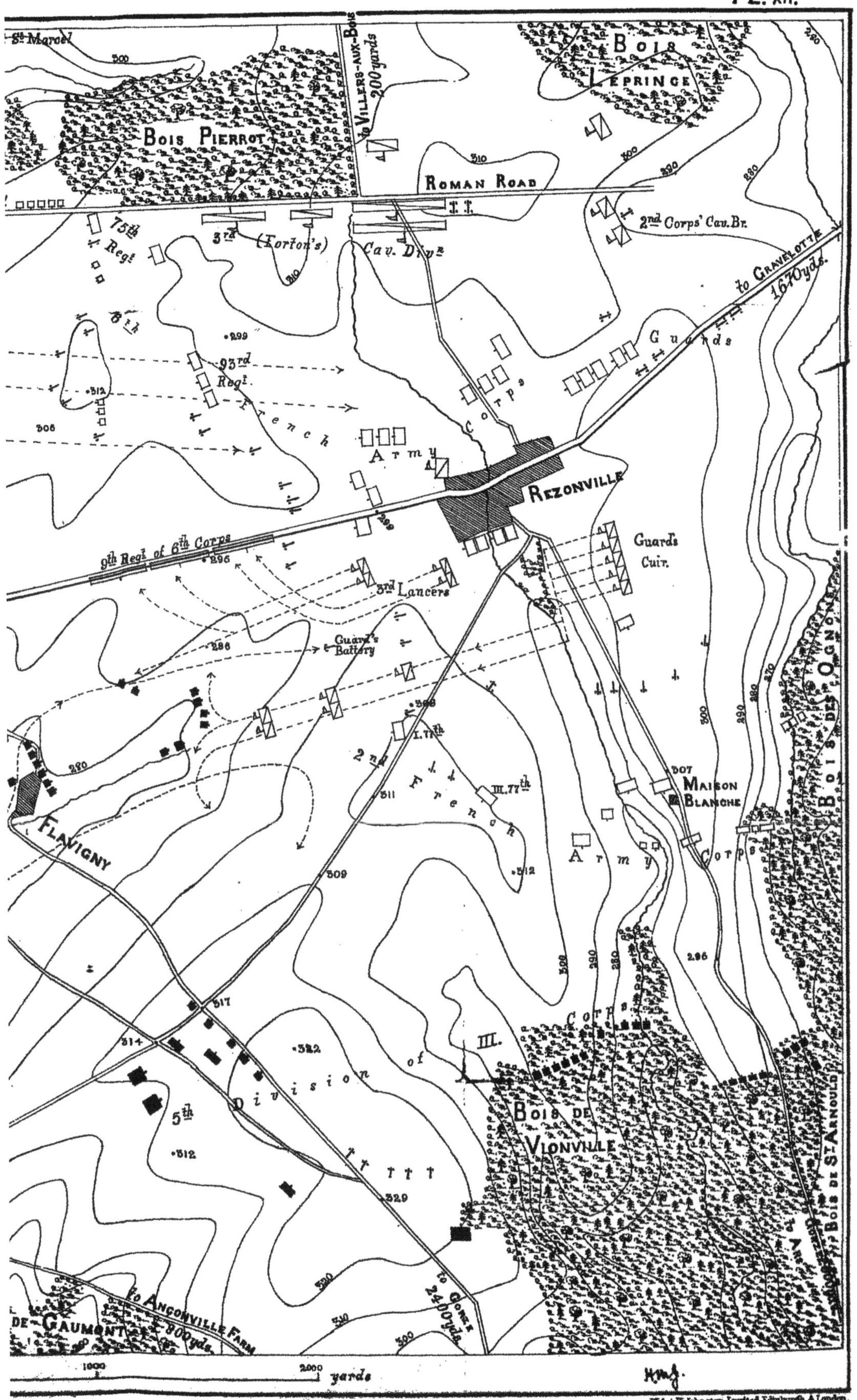
St Marcel
Bois Pierrot
to Villers-aux-Bois 200 yards
Bois Leprince
Roman Road
75th Regt
3rd (Forton's) Cav. Divn
2nd Corps' Cav. Br.
to Gravelotte 1670yds.
6th
93rd Regt
French Army Corps
Guards
Rezonville
9th Regt of 6th Corps
Guard's Cuir.
3rd Lancers
Guard's Battery
2nd French Army Corps
I. 77th
III. 77th
Maison Blanche
Flavigny
Bois des Ognons
III. Corps
5th Division of
Bois de Vionville
Bois de St Arnould
to Gorze 2400yds.
to Anconville Farm 800yds.
de Gaumont
1000
2000
yards
W. & A. K. Johnston Limited, Edinburgh & London.

now abandoned, and the chief use of cavalry was now to be to act as a screen, and as a scout. It was at once seen that, to do this effectively, a very large force in advance would be required, and that it should be directly under the control of the C.-in-C. of the army to which it belonged.*

The arrangement made accordingly was to form moderately strong cavalry divisions, which should act as independently as possible. These, moving in front of the army, would incidentally produce security for the masses in rear by giving early information of the enemy's position and movements; but their chief business was not to act as a sort of extra advanced guard to the infantry columns. The advanced cavalry were to have two great duties, the first in relation to the enemy, the second in relation to their own army in rear. The former duty was to keep touch with the enemy and find out as much as possible about his movements; the latter was to screen their own army from view, as it were, and thus prevent the enemy from becoming aware of the directions of the columns in rear.

The cavalry divisions were to be of two brigades, *i.e.*, 4 regiments, *i.e.*, 16 squadrons, or about 2,400 horses; but some were eventually made up to 20 and even more squadrons.

The supply of these divisions was independent of that of the individual Army Corps— a very important point, as the cavalry were not thus tied down to the slower-moving units.

In pitched battle the cavalry did not effect much. Successful charges against infantry were few; in the first period of the war, one important case; in the second period, three or four of such, but the latter were against half-trained infantry, who even, in some cases, were firing their chassepôts for the first time.

In 1866 there had been several combats between cavalry and cavalry; at Custozza in 1859 the Austrian cavalry attacked two infantry divisions with the greatest bravery, but did not break a square; at Balaclava, British cavalry successfully charged guns and infantry, but the sacrifice was useless. But, at Vionville, on 16th August, 1870, Bredow's charge arrested a most threatening advance of the French, and the heavy loss incurred was more than justified.

A short description of this battle, up to the time of Bredow's charge, will illustrate the point.

On the night of August 15th and the morning of the 16th, Bazaine had the French forces stationed as follows:— *Plate* XII.

2nd Corps of two divisions, between Rezonville and Flavigny, with the 3rd Cavalry Division (Forton's) and the Corps Cavalry Division (Valabrègue's) at, and to the west of, Vionville.

Guards Corps of two divisions, and its cavalry division, at Gravelotte.

6th Corps of four divisions (Canrobert's) between St. Marcel and Rezonville; but the 2nd Division had only one regiment on the field.

3rd and 4th Corps, of four and three divisions respectively, several miles to the north and east.

* It is perhaps worthy of note that recent critics of German cavalry work in 1870 have found fault with the fact that the Supreme Command kept no cavalry for its own use, and have argued that the work would have been better done if Von Moltke had had most of the independent cavalry in his own hands.

This army, of five corps, was starting upon its march to Verdun, where it hoped to be joined by MacMahon with four corps.

The object, then, of the Germans was to prevent any such junction, and they were pushing their forces across the Moselle south of Metz. Their III., X., and Guards Corps were across, and were able, in whole or in part, to join in the battle of the 16th. The III. was nearest, and was preceded by the 5th and 6th Cavalry Divisions. The commander of this corps (von Alvensleben) arrived on the scene with his advance guard, and was for some time under the impression that the bulk of the French had escaped and that he had their rear guard to deal with. His 5th Division marched from Gorze by the Vionville road, while the 6th, far behind, had been ordered towards Mar-La-Tour. Alvensleben determined to attack vigorously.

Unfortunately, as Alvensleben himself allowed, the cavalry divisions, in their haste to engage, had warned the enemy and afforded them time to prepare. The French account says:—"The untimely cannonade of the two cavalry divisions had obtained no other result than to alarm the French camp. . . . As for the French infantry, . . . it had taken arms and deployed before the two columns of the Prussians were in a position to profit by the initial surprise. This surprise turned, then, to the detriment of its authors."

The French had been keeping no decent look-out to the south, but the precipitate action of the enemy just referred to had saved them for the time. At the later battle of Beaumont, near Sedan, very much the same thing happened.

The 5th Prussian Division, marching from the Moselle in a narrow ravine, was in the following formation:—

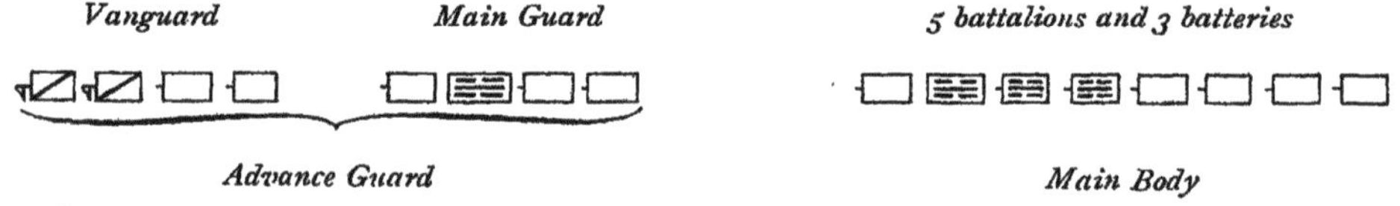

The head of this division, as it debouched near the Ançonville Farm, might have been crushed, for the French counter-attacks against the leading Prussian troops, after the first surprise was over, had driven off the cavalry batteries; but here, as in many other cases, French counter-attacks, successful at first, were not pushed home.

I shall not follow the details of the infantry and artillery struggle that took place; a most clear story of the swaying backwards and forwards of the fight is given in the French account. It shows that the German success in debouching and bringing their whole strength to bear was due in great measure to the artillery, while the French exposed to that fire, on open ground, great masses of infantry, which made these isolated, fractional counter-attacks of which mention has been made.

At 10.15 a.m. the French were still holding Vionville with two battalions, Flavigny with five battalions, and the ground from the cross-roads (314) to the Bois de Vionville with 13 or 14 battalions, and at the two former places had forced the Prussian cavalry batteries to withdraw. Before noon, the 6th Prussian Division had deployed and captured Vionville and the cemetery hill, and was extended in a long line from (272) in front of Flavigny to (260) in the valley on the East of the Bois de Trouville. The 5th

Division had taken entire possession of the plateau (317) (329) and of the woods to the right, forcing back the enemy to (311) (312) and Maison Blanche.

Alvensleben had now the whole of his men in front line, and no reserves except cavalry.

The map shows the situation at 12.30,* and gives a clear idea of the dangerous situation of the Prussian left wing, if the enemy should develop any vigour in attack. The Prussians, on the other hand, had no less than 132 guns working. The resolute attitude of Alvensleben had deceived the enemy, and led them to believe he must have another corps at his back.

Plate XII.

The map shows that the bulk of the French Guard cavalry had come forward to Rezonville, while the Prussian cavalry, except the eight squadrons† in the valley below the cemetery, was far off.

French Charge.

The French 3rd Lancers first received the order to charge. Starting in two lines from the south of Rezonville, without any reconnoitring of the ground, they were thrown into such confusion by the hedges and other obstacles that abounded that they had to be recalled and rallied behind the village. Starting again on a better line, they crossed the crest (299); but they had been given a vague order "to charge the Prussian batteries." They no doubt saw the guns on the horizon at Vionville; but when they reached (286) they found close in front of them a hurriedly-formed square of two Prussian battalions. They were aiming for it in fine style, when someone in the leading squadron cried "*À droite!*" and only part of the 2nd line went near the square. It was a blow in the air; 3 officers, 34 men and 46 horses were lost.

The Guard Cuirassiers had been ordered to support the charge. They also had some difficulty with the hedges and enclosures, and were too late to help the Lancers. Getting clear, they went off in three lines (two squadrons, two squadrons, one squadron). An officer who rode with them says:—"Nobody had thought of pointing out an objective, or examining the ground, or supplying a reserve. . . . We climbed to the plateau (308—311—312) without seeing anything to attack; but from the crest we saw, about 1,500 *mètres* away, Flavigny in flames and several lines of infantry marching towards us. At our appearance the whole lot halted; the skirmishers, who were much scattered, rejoined in haste the masses forming in rear. Some of them, too far off to their right, dipped into the bed of the rivulet."

Now was the moment when the horse-battery of the regiment should have been putting in some work on the groups that were forming; but the battery only opened fire when the charge had failed.

The rejoining of the skirmishers masked the Prussian fire for a few minutes, and the intervening ground had no obstacle, says the same officer, though previous accounts have stated that the charge failed mainly on account of the litter left by the 2nd Corps, on this their camping ground. The left squadron of the 1st line missed the Prussians altogether; the right squadron split upon the point of the wedge-shaped mass that they had formed. The Prussians had fired their first volley with admirable coolness at very short range; continuing with rapid individual fire, only 20 men on horseback remained to the right squadron when the affair was over.

* The troops on the north of the high-road are shown as they were an hour later, when Bredow charged.

† 11th and 17th Hussars, of 13th Brigade, 5th Cavalry Division, and 1 Dragoon Guard Squadron.

The 2nd line had travelled more slowly and arrived late; a very few of them reached the Prussian bayonets, the line of dead horses hindering them. The single squadron in 3rd line was broken up by the return of riderless horses, and went back to the starting point. Of the five squadrons, 200 men only returned to Rezonville.

The sacrifice was of no avail, nor was it called for. Bazaine had still whole Army Corps and masses of artillery unused.

Prussian Charge.

Of the eight Prussian squadrons before-mentioned, four were now sent off in first line towards Flavigny, followed by the other four, the latter moving as a right échelon. They were too late to meet the French cavalry, but did some execution among the men who had been dismounted in the charge. The front échelon passed Flavigny on the north; the second was delayed on marshy ground, swerved more and more to the right, and finally in no way supported the first. Fired on by the French 77th (1st Battalion) on the road near (311), it returned without effecting anything.

The front échelon passed through intervals in their own infantry beyond Flavigny, and then came under fire of a French battalion, and of a battery north of the high-road. At this moment appeared, too late, a horse battery of the French Guard (*v.* map). The right half of the battery arrived first, and was barely able to fire three rounds of grape when a troop of the Prussian Hussars was among them. The limbers, closely pursued, dashed into a battery in rear and dragged it into the flight.

At this moment Bazaine himself, with staff and cavalry escort of two squadrons, was about (299). The escort charged the scattered Prussian horsemen, and the escort of General Frossard, commanding 2nd Corps, completed their rout, and the Prussian Hussars disappeared off the field.

The *object* of this charge on the part of the Prussians had been simply to pursue and slay the French Cuirassiers, and it had started too late.

The Prussian artillery very skilfully used the opportunity afforded by the partial cessation of fire during the cavalry actions to advance to better positions 15 of the 21 batteries they now had in action.

2nd Prussian Charge.

The mass of Prussian cavalry (23 squadrons) behind Ste. Marie was now (close on 1 p.m.) put in motion; but the result was a veritable *coup manqué*. Moving in masses towards the slopes between (317) and Flavigny they got mixed up with the batteries that had just advanced from (302). Moreover, they had omitted to deploy under cover, and tried to do it on the crest, just when their appearance attracted a hail of shell and bullets from the direction of Rezonville. Swerving to the left towards Flavigny, they were further disordered by the routed Hussars, and all the "go" was taken out of them. They got no further, and retired.

The critical period of the fight for the Prussians was now approaching. Both official accounts indicate this. "Alvensleben's Corps was now practically *immobilized* on the positions it had conquered, and in no condition, not only to attack, but even to undertake a retreating action with any chance of success." The artillery had lost so many horses that it could only move with difficulty and slowly; and "it could no longer count on the infantry's support if pressure came." But "thanks to the deplorable measures adopted by the French commander, . . . the hope of von Moltke was realized, but under conditions such that the German advanced corps would have been infallibly crushed by any other than Marshal Bazaine," says the French account.

Before 2 p.m. all that part of the infantry of 2nd Corps that had been facing west and south-west had retired from the field, and their place taken by the Guards. The leading division of the X. Corps was not expected to arrive till 3 p.m. at Trouville, but some of its guns had been in action already.

There was now every appearance that the French were about to take the offensive against the Prussian left wing, and cavalry alone were available as a reserve. The bulk of the 5th Cavalry Division had been sent, a little earlier, towards Bruville, in response to information of the approach there of the 4th French Corps. There remained, to the south of the wood, only two regiments of Bredow's Brigade, 7th Cuirassiers and 16th Uhlans. These marched between the wood and Vionville, dipped into the valley (270), went down the valley to about (260), and formed to the right. The Prussian infantry here had been forced back into the wood.

Bredow's Charge.

The objective of the charge was to be the French guns about (312). The lie of the ground sheltered the two regiments from infantry fire almost to the last, for the French 93rd Regiment had shortly before this fallen back behind the crest. Another advantage gained was due to the shifting of some of the French batteries just at this moment. The cavalry, with the Cuirassiers on the left, were soon among the guns; galloping on, they came under fire of the 93rd, but for so short a time that the infantry were hardly able to fire more than one round. But Forton's Cavalry Division, lined up, backs to Bois Pierrot, at only 400 yards distance, launched its light brigade (Murat's) on the now disordered Prussians, and carried on a running fight eastwards. The Prussians were now charged in front by the 2nd Brigade of the 2nd Corps' Cavalry Division, and from Rezonville by a regiment of the 1st Brigade. Wheeling round to the right-about, the remains of the two regiments started for home, and were charged again in flank by another regiment from Bois Pierrot. The masses of French horse, con verging, hindered their own speed, and the Prussians won clear, but were fired on at sh ort range by the infantry whom they had to repass. Those who collected behind Flavigny a few minutes later were only able to form one respectable squadron out of the six that had charged.

"As for the infantry and artillery which were fighting previously (*i.e.*, before the charge) on (312), they had been singularly cut up, mixed or dislocated. The 12 batteries between Rezonville and the Roman Road all quitted their fighting positions. Two batteries only, which the charge had not actually touched, came back into position. The ten others retired on their reserves—'to replenish' (the account says sarcastically)—but took no further part in the fight. The 75th Infantry, as well as most of the 93rd, backed off to the Bois Pierrot, and did no more shooting."

"This heroic charge had then really attained the object aimed at by Alvensleben. The ground over which it had swept was for the time abandoned by the bulk of the hostile infantry and artillery"; and the 6th Prussian Division, relieved of the terrible pressure it had been experiencing, took heart of grace, and began to advance again.

The French account has some very wise remarks on the subject of the apparent inefficacy of the French artillery. "If then we make a mental picture of the general situation of the artillery of both sides towards 2 p.m., we find that 39 French batteries were then in action (while the enemy only had 21, all of which had been engaged from the start), and that, at a very short distance in rear, that is to say, about Rezonville, Villers-aux-Bois, St. Marcel and Bruville, 18 more, absolutely intact, were available." In a footnote, the account adds:—"And seven other batteries, besides, which had fought and were now near Rezonville," some of which had not suffered much. "Nevertheless the French artillery did not obtain, over the whole field, the result one would have had a right to expect" from such a superiority of numbers. "One cannot conclude that the fire of the numerous batteries had been really ineffective from the point of view of material damage, since the German batteries were badly damaged and many of them were hardly capable of changing position, while, in general and without exception, the *personnel* and the horses of the opponents (French) had not been highly tried." "It can be made clear that the supremacy achieved by the Prussian artillery at Rezonville was far from being solely due to its destructive effects on the French batteries, which effects were far less—battery by battery—than those which the French guns inflicted on them. A superior direction, energetic and always directed towards a distinct object; a tactical and technical command, suitably organized by groups; an employment rationally conceived and correctly executed, of men, horses and ammunition; finally, the sentiment of faith in the power of their weapons . . . things which, on our side, were all exactly reversed."

In reading the full account, one certainly finds an extraordinary number of useless movements of French batteries; and their regulations, or their previous drill, do not seem to have provided any means of replenishing ammunition but the method of withdrawing the whole battery out of range.

Artillery.

If the employment of the German cavalry in 1870 shows a vast improvement on that of 1866, this is no less the case with the artillery. The very change of the name "Reserve" artillery into "Corps" artillery indicates the change in its tactics. Every advanced guard of a corps had now artillery with it; and the main body had usually the whole corps artillery practically at its very head, preceded often by no more than a single battalion of infantry. In the interval between 1866 and 1870, any spectator of Prussian peace-manœuvres would notice the increased activity of the field artillery, and their efforts at finding good positions early at effective range. Also the extent to which the batteries massed and were directed by one commander. These practices bore good fruit in 1870.

Boguslawski lays down the following as the new developments :—

"(1). The batteries approach to within easy distance of the enemy, and do not blaze away at enormous ranges.

"(2). They engage in sufficient force in advanced guard actions.

"(3). They concentrate into masses and cannonade the enemy's position, preparing the way for the infantry which follows.

"(4). They have got rid of the prejudice that the loss of guns must be avoided at any price."

The Germans, in fact, had learnt the *tactics* of artillery, while the French had not.

As early as the battle of Wörth, the workmanlike use of the German artillery was manifest. At the very outset, the advanced guard batteries of the V. and XI. Corps were at work, firing from good positions opposite Wörth and above Gunstett. The cannonade brings forward rapidly all the corps artillery of both these corps, and the guns are ranged almost continuously along the heights from Görsdorf to Gunstett at an early stage of the fight. French batteries, though they are on the defensive, and have had plenty of time to plan their work, come into action in a desultory and piecemeal fashion. The commander of the V. Corps Artillery of the Germans is able in a few minutes to ensure the concentration of all, or any desired portion of, his fire, now upon this corner of Fröschwiller, now upon that edge of Elsasshausen, and again upon a fresh brigade of guns he sees arriving on the scene. The commander of the XI. Artillery sees his infantry recoiling from the south-east corner of the Niederwald, and in a minute 60 guns are pouring shells into that corner. He sees French cavalry deploying rapidly for a desperate charge, and two or three batteries are on them before they are ready to move. He sees a mitrailleuse battery unlimber, bent on sweeping the slope crowded with German skirmishers. In a few minutes his shells have found the mitrailleuse ammunition wagon and blown it up, causing such havoc that the battery with difficulty withdraws from action.

At Gravelotte the Prussian artillery of the IX. Corps at Verneville had the choice of long range and comparative safety, or a dangerous proximity to the enemy and short range. In 1866 they would have chosen the former position; in 1870 they advanced without hesitation to within 1,500 paces, compelling the infantry to come up in support, and maintain themselves there throughout the day opposite a point where the enemy were splendidly posted, where they had entrenched in three tiers, and had a great force of guns and mitrailleuses. Nearly 300 guns were eventually deployed at effective range against St. Privat, battery after battery pushing boldly forward as opportunity offered.

Great skill and resolution was also shown on many occasions in rapidly occupying with guns any point carried by the infantry, as at Spicheren.

Sedan was the great artillery battle. On the east side, along the ridge above the Givonne rivulet, all the artillery of the Saxons and the Guards deployed into line and shattered the French. On the north end of Bazeilles and the approaches from Sedan the Bavarian Corps Artillery poured an endless stream of projectiles from the other side of the Meuse, beginning operations on the day before the battle, and only ceasing when the infantry penetrated. On the north front the V. and XI. Corps marched their whole artillery boldly through the defile round the loop of the Meuse, immediately after the advanced guards. In a very short time they had the control of the situation, and their accurate fire on the Garenne Wood, assisted by that of the Guards from the east, had much to do in spreading discouragement in the French ranks. At various points the French tried to advance their infantry against these batteries, but were brought to a standstill again and again at 2,000 paces. But of course in these cases the French had no choice but to attempt frontal attack.

The Germans ended the war with a whole-hearted contempt for the mitrailleuse; some of them jeered at it as the "half gun." The weapon itself was, of course, a poor affair compared to a Maxim, and the mistake was made of brigading into batteries, and exposing these to the fire of field guns.

Each discharge of 25 bullets sent them all in the same direction, there being no sufficient cone of dispersion. They were in this particular inferior to the fire of a few riflemen; but they aspired to working at long range, and in this respect they were inferior to field guns, not only in range, but in the impossibility of observing the effect of a discharge that did not reach the mark. It was usually impossible to tell whether the shot went over the heads of the enemy or was too short. The field gunner, on the other hand, was able to judge by the point at which his shell exploded.

Similar to this is one of the weak points of long-range rifle fire. You cannot *see* that you are hitting; and you can very seldom indeed see the splash of the bullets in the dust.

On the defensive, the German artillery turned its attention to the advancing hostile infantry, and they seem to have made up their minds that this is its proper *rôle;* but they were only able to do so to the extent they did, on account of the inferior handling of the French artillery. Of course, when an infantry attack is getting dangerously close, your guns must attend to it regardless of the enemy's artillery; but, in the case of foes anything like evenly matched, the attackers' artillery will at this stage be attending chiefly to your infantry trenches, all his efforts being now turned towards breaking your infantry firing-line.

During the second period of the war, German guns were sometimes temporarily captured by the French, and retaken by the German infantry.

From the experience of the War of Secession in America, of 1866 and of 1870, it began to emerge that the attacker must expect to lose much more heavily than the defender up to the moment of success, and often through the whole affair. When the pursuit was close, and the artillery could quickly take up good positions to enfilade the line of retreat, then the balance of loss was often more than redressed. It is plain, therefore, that mere killing did not cause defeat; it was the gradual inculcation into the troops of one side that they were outmatched, and that nothing they could do would stop the enemy. It was a moral effect.

At Sedan there was naturally no pursuit, in the ordinary sense of the word. As soon as the French began to give ground they were retreating into an *inferno*, across every part of which German shells were bursting. In this battle the French lost much more heavily than the Germans, partly on account of the impossibility of getting out of range, and partly because it was only at two or three points that the German infantry were allowed to attack seriously. At these points—at Bazeilles, for instance, and at Floing—the loss of the attackers was very great.

In discussing the German infantry tactics, I shall touch upon the question of whether they did not on many occasions incur unnecessary loss.

Combined Arms.

A very remarkable feature of this campaign was the development of the combined action of the three arms, and especially of the infantry and artillery. These two supported one another in a manner they had quite failed to do in 1866. This is quite clear from the few remarks I have made on the action of the German artillery. In the most exemplary manner it risked a great deal in order to prepare the way for the infantry—and this not blunderingly, but with astonishing skill. That skill was the outcome of thought and practice in time of peace. Only thus can the best tactics of artillery or of any other arm be perfected. The French had no such practice, for their peace-manœuvres used to be carefully arranged according to programme—were, in fact, nothing but glorified ceremonial field-days. Their enemy had the experience of 1866 to work forward from, and that kind of sound sense that makes a man capable of benefiting from his own past errors. Never since the fall of Napoleon I. had there been such a display of the power accruing to the army that can make a combined use of its varied strength.

In my opinion, however, the Germans in 1870 lost men unnecessarily. At Wörth, for instance, their losses amounted to 9,000, and this was incurred in beating four French divisions only. Some regiments lost 40 per cent. of their strength. The 10th Division, V. Corps, left 3,500 on the field. These great losses were due to frontal attacks being pushed home before the flanking movements had time to develop.

We find the same kind of thing on a great scale at the battle of Gravelotte.

Having now discussed the cavalry and artillery work of the German armies in 1870, and referred also to the splendid co-operation of the artillery with the infantry in the battles, I come to the work of the infantry themselves; but, before enlarging upon that, I shall say a few words on the more general subject of attack.

Unnecessary Attacks by the Germans.

We have always understood that one of the chief advantages of the attack over the defence, both strategically and tactically, lies in the fact that the defender must, speaking generally, hold his own at all points, while the attacker has usually only to win at one or two points in order to be on the fair way to victory.

We can see, in the 1870 campaign,—as in most others,—the truth of this exemplified in the field of strategy. On the 6th August, the French front is from Saarbrücken on the left to Strasburg on the right. The breaking of this front at two points simultaneously—viz., at Saarbrücken and Wörth—caused retirement of the French all along the line. The 5th Corps, in the intermediate position at Bitsch, could not dare to hold its ground. Spicheren alone would have caused the whole front to collapse and retreat, for MacMahon would not, after the retreat of Bazaine, have dared to hold on at Wörth, even if he had not been attacked.

It is not here implied that therefore the Crown Prince of Prussia should not have risked loss by fighting; for war cannot be very successfully waged without bloodshed—by mere manœuvring. And in this case, at the very beginning of a war against a brave and capable nation, it was necessary to take every opportunity of discouraging and disheartening the enemy's troops by inflicting all the loss that can be heaped on to him by a pitched battle.

But the position may be different in the battle itself, as a tactical matter, when you know that by beating the enemy you will have him compelled to surrender, or compelled to submit to investment. If the battle is in the open, so to speak, and the enemy when beaten has a clear field for escape, you may well think it right to attack vigorously everywhere, for what damage you do must be done *there*, on the field. But it might be argued that if he cannot escape, and is acting on the defensive, you might save yourself great loss by attacking less universally.

Prudent Avoidance of Loss at Weissemburg.

As an instance, take the prudent action of the Crown Prince at Weissemburg. That town was in the possession of the French, and was garrisoned by a couple of battalions and 16 guns. It had a wall and ditch. The 2nd Bavarian Corps came on it first, from the north-east. There is no doubt that it could have stormed it then and there, but the loss would probably have been great. So the Crown Prince halted the corps, and waited till the V. Corps had crossed the Lauter on the south side of the town, and sent some guns and battalions against it from the south-west. Even then, the chief German losses for that day were incurred in the taking of the town.

Next, take the battle of Gravelotte.

Gravelotte—Unnecessary Loss.

On the morning of the battle, the French were posted as follows:—On the left, the 2nd Corps held the plateau about Point du Jour and Rozerieulles, and the St. Hubert Farm, with Lapasset's Brigade of the 5th Corps guarding its left flank and rear on the lower ground from Rozerieulles to Moulins. Next on the right came the 3rd Corps, which carried on the line northwards to beyond La Folie Farm, and had skirmishers across the valley in the Bois des Genivaux. Then came the 4th Corps, holding the front by Montigny la Grange and Amanvilliers, and having detachments to the front in farmhouses and copses towards Verneville and Habonville. Finally, the line was completed by the 6th Corps which carried it on to St. Privat and Roncourt. The Guards and a large artillery were in reserve about Fort Plappeville. Now, what was the strategical situation? Bazaine had intended to escape west through Metz to Verdun and Châlons, on the 14th, when the German corps were partly facing him on the French Nied, and partly marching towards crossing points on the Moselle south of Metz; but none of them were yet across the river. The German Ist Army, von Steinmetz, had perceived the beginning of the retreat on the afternoon of the 14th, and had promptly attacked. Two French corps, the 3rd and 4th, had halted and counter-marched, and the battle of Borny had resulted. This battle, beginning so late in the afternoon, was in a sense unfinished and indecisive, but it had a great effect in delaying Bazaine. The other three French corps, viz., Guards, 2nd and 6th, got through Metz on the 15th, and were on the road between Metz and Gravelotte. The 3rd and 4th took the whole of the 15th to refit after their fighting, and on the 16th the march to Verdun was to begin, the 2nd, 6th, Guards and two cavalry divisions taking the Mars La Tour road, the 3rd and 4th the more north road by Conflans.

The object of the Germans was to stop this march, and drive the French into the fortress; by doing this, a junction of MacMahon and Bazaine would become practically impossible. Had Bazaine any clear notion of the danger of allowing himself to be fore-stalled and then invested? His dilatory movements up to the 16th, and his swinging back of his line, after the indecisive battle of the 16th at Vionville, until it faced towards Paris, would seem to show that he had no proper appreciation of the danger. At the same time his report to the Emperor, that he would be ready in two days to march by Briey, seems to imply that he knew he must try to escape. Whatever his intentions and knowledge may have been, he is found on the 18th in the strong defensive position I have described.

Now, if he meant to get away from Metz, what was he doing with his sole, but powerful, reserve behind his left, among the Metz forts? And if the German intention was to keep him from getting away, why should they trouble to do anything more than "contain" the French left and centre?

The VII. and VIII. Corps of the Ist Army were given the task of dealing with this French left wing, and von Steinmetz was pretty clearly told that his troops, with the II. Corps in reserve, as well as the IX. opposite Montigny and Amanvilliers, were chiefly intended to keep the enemy in place, until the Guards and the Saxons should complete their march to the north and lap round the French right at St. Privat and Roncourt. But we find von Steinmetz attacking from the very first with the utmost fury, and keeping it up till after nightfall, thereby incurring terrible losses. He inflicted little damage on the enemy, and by night the troops of the French left wing were not far wrong in holding that they had won. Much the same was the result at the centre, at Montigny and Amanvilliers. Good tactics aim at striking with superiority of force at the decisive point. The French left and centre were not strategically decisive points; and so strong were the positions along this portion of the line that superiority of force would have required twice the number of attacking troops and willingness to lose in huge numbers. The German commanders cannot be excused the incurring of such unnecessary losses. Was their action due to the steady inculcation of the offensive in their training? If the French right had held out, and the French left had attacked vigorously at dawn, von Steinmetz would have had cause to regret the useless waste of yesterday's fighting. To us it seems as if he would have been wiser, and would have done all that was needed, if he had driven in the French advanced posts, brought up all his artillery to bombard Frossard and keep him occupied, while entrenching himself along the east edge of the Gravelotte plateau. Then, if a 2nd day should be needed to complete the turning and rolling up of the French right, the decisive point, the VII. and VIII., thus entrenched, could have been safely left to hold their ground, while their reserve, the II., marched north to help at the other flank.

Was their offensive action due to their training ? If so, it is quite possible to have too much of that branch of training. Or had they begun to think that every German attack must succeed ? At the next great battle, Sedan, they were much more chary of close attack everywhere at once, and they achieved the victory with less than the former ratio of loss compared with the loss they inflicted.

So little had they succeeded against the French left and centre, that the German Corps opposite these points lay down at night as they fought, and rested literally with their arms in their hands. On the other side, at these points, there was satisfaction, almost jubilation—men expecting to receive at dawn the order to advance—infinite disgust and indignation when they were told to retreat to Woippy.

Infantry.

A German officer (von B.), commenting on the work of their infantry, says :—"The German infantry moved under artillery fire often in double columns of half-companies with full intervals, often only with 30 paces between battalions. It could do this without suffering great loss, because of the very moderate accuracy of the enemy's artillery fire. This formation had undoubtedly considerable advantages in great battles, such as Wörth, Mars La Tour, Gravelotte, Sedan, etc., for it enables a commander to keep his masses in hand, and to move them with ease in any direction. The German infantry bore artillery fire uncommonly well. We are not going too far when we assert that the cases were rare indeed when the advance of our infantry was sensibly delayed by artillery fire. With the French the very opposite was the case. Both old and new troops stood our artillery fire very badly. In the first place, it is not compatible with the French temperament to endure without being able to act, and in the next place perfect discipline is necessary to make troops stand this test.

"When a shell burst in the middle of a German battalion it closed its ranks again, and every soldier advanced instinctively, obeying the voice of his leader; but a French battalion would disperse in the same case, and some time was required to get it together again. On broken ground, or amongst woods, company-columns with skirmishers were thrown out some way to the front to guard against surprises; this was always done as soon as it was intended to engage the infantry. Half-battalions were rarely made use of in front line. Occasionally it happened that company-columns were formed at too great a distance from the enemy. This always resulted in less precision of movement, and increased the difficulties of command. When the attack was commenced in earnest, the 1st line of a brigade was almost always formed of company-columns side by side, rarely of two companies thrown forward and followed by the other half-battalion. The French, who extended very strong swarms of skirmishers, and threw out supports of from two to four companies at a considerable distance from them, opened their musketry fire at very long ranges, from 1,000 to 1,400 paces. It is true that even at this distance we had men killed and wounded, and that this surprised our people unpleasantly. It would be a mistake, however, to draw any positive conclusions from this. If you look into the matter closely, you will not find any case in which our troops were really shaken by such fire. Our infantry generally extended at least one "züg" per company at once. This was, however, rarely sufficient when we came within effectual range of the chassepôt, about 500 paces. When at about 400 paces from the French skirmishers, our men were obliged to seek cover, or if it was level ground, to lie down and answer the fire, for which purpose the skirmishers were usually reinforced by another "züg" per company, if this had not already been done."

It should be understood that the Prussian company, standing on parade, was told off into two "züge," and the whole was formed up in three ranks. Before coming into action, the whole of the 3rd rank men were formed into a separate "züg," two deep. Thus in action there were three "züge" per company. The one last formed was the first to skirmish. It seems rather a clumsy method; it was an obvious relic of older times.

It may be supposed that, although German troops have not fought since 1871 against a civilized army, they have nevertheless learnt, or will have to learn on the field, that times have changed, that they will have in future to be more careful of their men, especially in the matter of never exposing any sort of mass, be it only a company-column, to the fire even of distant artillery.

In fact, during the campaign they often found the fire of the chassepôt so murderous that it was quite impossible for supports to remain in close order, and at the same time sufficiently near the firing line to act *as* supports, unless the ground was very favourable for cover. Moreover, the firing line at these times had such gaps made in it, either by the enemy's bullets or by the instinctive swerving of parts of it to gain cover, that the supports no longer reinforced as a whole, but rushed forward in driblets as required; and the remainder, seeking cover, thus automatically extended the supports. Note most distinctly that the losses suffered by the supports were not due so much to the long range of the chassepôt as to its comparative flat trajectory—*i.e.*, the supports were in most cases hit, not because the French infantry aimed at them, but because the flat trajectory caused bullets, passing just over the heads of the firing line, to catch men much farther to the rear than would have been the case with (say) the needle-gun.

To meet the situation there were the alternatives of keeping the supports further back than the then normal of 150 paces, or of extending them. It was quite natural that the latter method was most often resorted to; and the result often was that the whole of a regiment was extended by the time the crisis approached, and that a second regiment became the support. A few minutes later there was a confused intermingling of regiments,—a distinct danger if a critical situation should arise by the sudden appearance of fresh hostile reserves. But it is very easy to exaggerate the danger of this commingling. No doubt the wooden troops of such times as Frederick the Great's would be for the time a helpless mob if they were all mixed up; the risk of it was much feared before 1870, and in many quarters even after 1870. I can remember having it impressed on me in my military youth that everything was to be done to avoid it. But it has ceased to be a bugbear of anything like the dimensions it was, and this development is undoubtedly due to the spread of individual intelligence.

The German skirmishers advanced by a succession of short rushes, alternated with steady firing. The Austrian idea of 1866 that it was possible to take a position by advance without firing was gone for ever. The crisis of the infantry fight took place at from 400 to 100 yards.

Neither French nor Germans ever succeeded in bringing troops in close order into front line, or in pushing battalions or companies forward to fire volleys. The Germans had practised these things on the parade-ground, but found them impossible. Once or twice, when the French were surprised, a few volleys were fired by the Germans, and most effectively at Villersexel by moonlight. Night attacks were conspicuous by their absence. Bayonets were never once crossed in the open field. In most cases the time for the final rush came when the French infantry fire perceptibly slackened—a condition that was very often due to the flank attack that was so often carried out simultaneously, and sometimes to failure of ammunition. This failure was due to the French habit of commencing fire at long ranges, sometimes even to the extent of firing high into the air far beyond the sighting of the rifles. The Turks did this freely in 1877. As men found their ammunition failing, they slipped off to the rear, and, by the time the Germans arrived, all were gone. A counter-attack often at this stage drove back the Germans, who, however, usually rallied quickly; and throughout the Germans showed extraordinary tenacity in clinging on to any ground they had won. At the battle of Gravelotte, in particular, the swaying back and forward of the firing lines was very remarkable, and proved the gallant, warlike qualities of the antagonists.

The infantry beat cavalry in this campaign with great ease, even when the former were scattered as skirmishers. Cavalry rode *through* them sometimes, but never drove them off the field, or inflicted any great loss on them. German infantry often took guns when the infantry escort had been first driven off. At battles like Wörth many guns were thus captured, but at Gravelotte only seven. In this fight the Germans were not everywhere successful, as we know; hence there was no *débacle* of the enemy, and his guns were saved.

On the defensive, the Germans reserved their fire till the French were within 300 or 400 paces; the French opened fire as soon as the enemy was visible. There is no doubt of the superiority of the German system in this matter. Of course, if they had had a weapon equal to the chassepôt, they would have begun sooner than they did; but still would have adhered to really effective ranges.

The war produced a great many instances of fights for localities, *i.e.*, separate struggles going on for the possession of a village, a farm, a copse. In this sort of fighting, especially in the case of house-to-house fighting in a village, the French showed great aptitude; but even here, the superior discipline and training of the German told. On both sides, there was often visible a great disinclination to leave cover, whether for advance or retreat; in the latter case, this was a frequent cause of prisoner-making.

A very remarkable thing was the similarity of the tactics of all the German troops—just as if they had all been trained by one man; and in fact, the cause was the identity of the few great principles on which all were trained. Keep moving forward; avoid the enemy's front as far as possible and attack his flanks; fire only when there is a reasonable chance of hitting; hold on doggedly to any ground you have won; bring every gun to bear on the points at which you are aiming; your risk of death is greater when you turn your back on the foe than when you hold on or advance; you cannot expect orders in the firing line, so be quite sure you know what is wanted of you before you come under fire; when you are coming under his effective fire, look ahead for cover to rush to at each stage, but do not stay there; when in support, keep your eye on the firing line, and fill gaps in it as they occur.

CHAPTER XII.

EPISODES OF GRAVELOTTE.

(1). GERMAN RIGHT WING.

THE VII. and VIII. Corps of the German Ist Army under General von Steinmetz were to attend to the French left wing in front of Gravelotte. At about noon the commander received the following instructions :—"The separate action in front of Verneville (against French centre) now audible does not necessitate the general attack of the Ist Army. The latter should not show any large force ; in case of necessity, merely the artillery for the preparation of its subsequent attack."

We shall see how curiously von Steinmetz interpreted this order—an order that was both strategically and tactically sound. It is to be observed that it states "in case of necessity," even in the matter of artillery attack, thus putting far into the background all idea of a strong infantry attack.

Between 12 and 1 o'clock a division of the VIII.* with guns emerged on to the plateau between Malmaison and Gravelotte, the guns opening fire from the west of the road, and being assailed by long range fire from the neighbourhood of Moscou Farm. Finding the range too long, they advanced after 2 p.m. across the road, and there were then seven batteries of the VIII. to the left front of Mogador, two to the right front, and three in front of Gravelotte, on the other side of the Metz road. About 1.30 p.m. the VII. Corps had seven batteries at work from the south of Gravelotte, of which five were still to the west of the Ars road and just south of the village.

v. Plate XIII.

French Position.

"On the summit of the ridge the farms of Moscou and Point du Jour had been arranged for defence by the French, and connected by a carefully laid system of shelter trenches, which stretched to the north and south through these points. The roads leading from the high road to the Bois des Genivaux had been utilized for the formation of long lines of fire under cover of their high banks. As an advanced bulwark lay, half-way up the slope, the farm of St. Hubert, likewise fortified, the quarries and gravel-pits formed excellent points of support for the defence. The wood lying in the bottom of the deeply-sunk Mance Valley was so far disadvantageous to the defender that it to a certain extent concealed the preparations for attack.

"The great road from Gravelotte, the only passage of the valley suitable for all arms, leads down into it from both directions as a deeply-sunken hollow way, and then crosses it in the form of a high masonry embankment. Eastward of the valley the road is edged with deep quarries, so that deployment off it can only take place close in front of the St. Hubert farm. The defile is therefore over 1,200 yards in length, and in consequence of the dam-like elevation in the bottom can be taken under fire from St. Hubert in nearly all its extent as far as Gravelotte."

Plainly, St. Hubert must be the first objective. Its distance from Gravelotte was one mile, and from the bottom of the valley, 800 yards.

The 3rd French Corps held Moscou and the ground northwards. The batteries of

* In *Plate* XIII., the numerals VII. and VIII. as applied to the Corps Artillery should be interchanged.

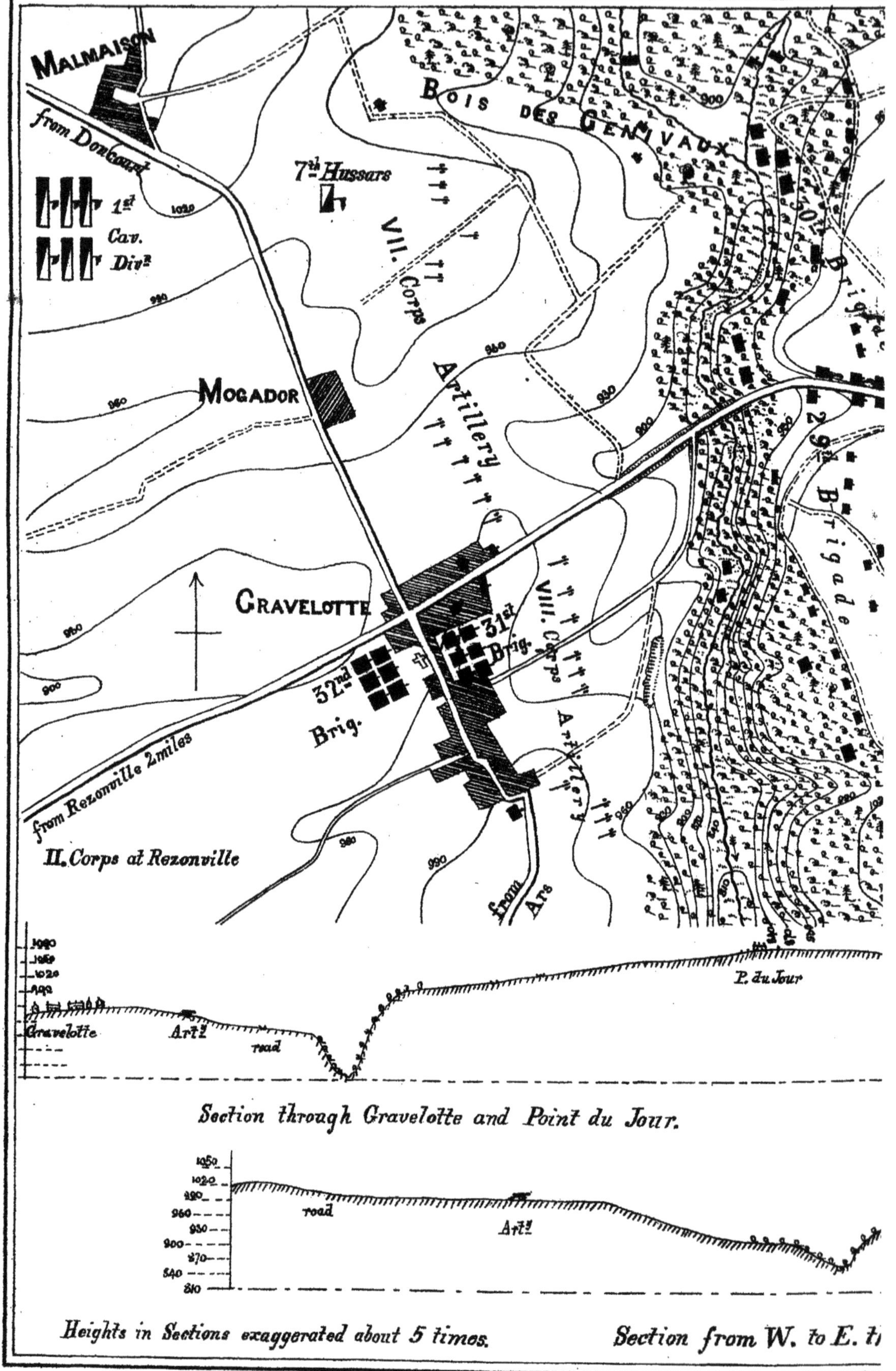

MALMAISON
from Doncourt
7th Hussars
1st Cav. Divn
Bois des Genivaux
VII. Corps
Artillery
MOGADOR
GRAVELOTTE
31st Brig.
32nd Brig.
VIII. Corps
Artillery
from Rezonville 2 miles
II. Corps at Rezonville
from Ars
30th Brigade
29th Brigade
P. du Jour
Gravelotte
Artl
road
Section through Gravelotte and Point du Jour.
road
Artl
Heights in Sections exaggerated about 5 times.
Section from W. to E.

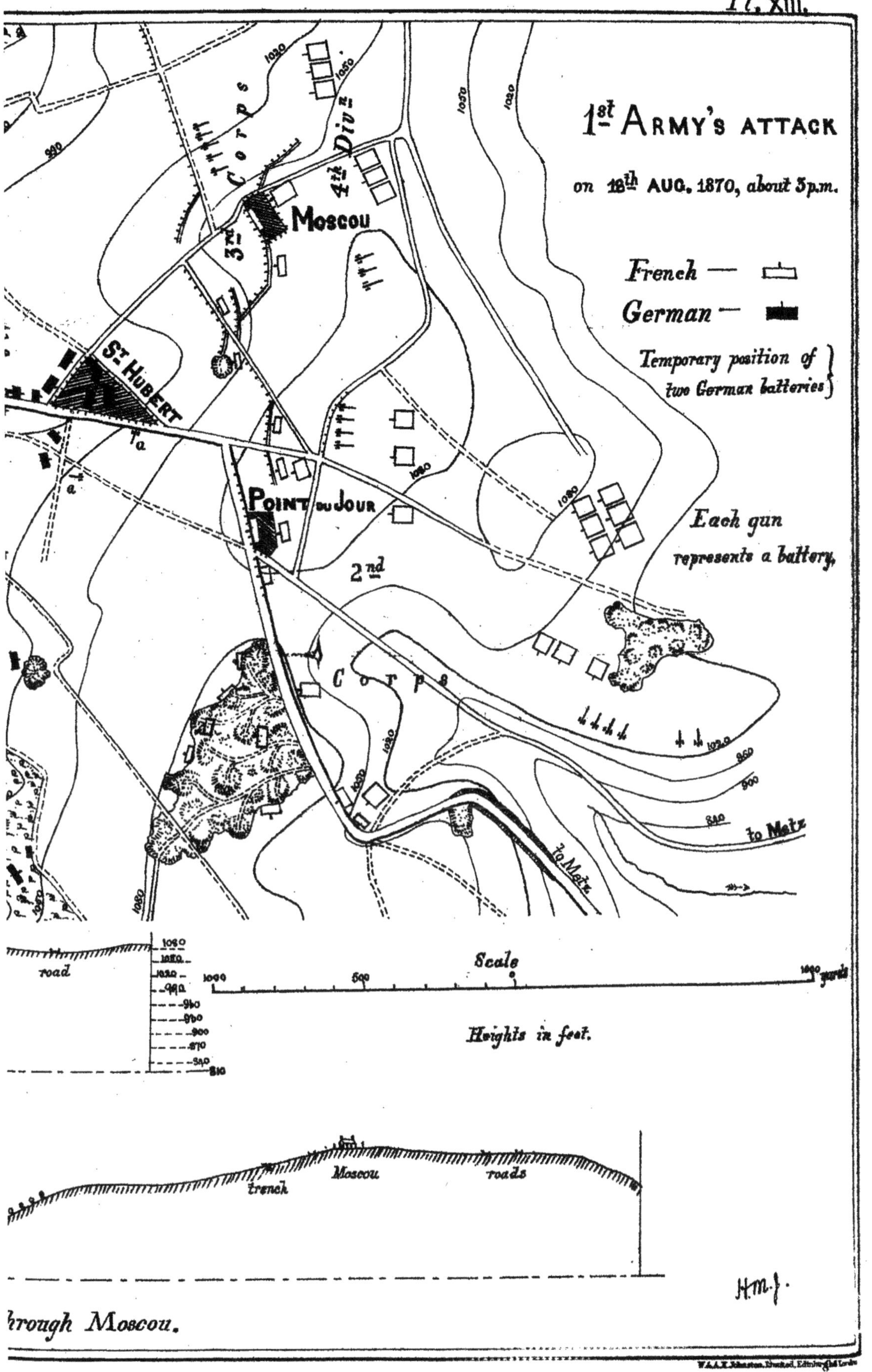

hrough Moscou.

W. & A. K. Johnston, Edinburgh & London

two of its divisions stood to the north and south of this farm; a mitrailleuse battery was most effectively placed in the prolongation of the Gravelotte high road. A regiment occupied St. Hubert.

The 2nd French Corps held from the bend of the high road southwards. A rifle battalion held the prepared buildings of Point du Jour homestead, whose buildings were connected by an earth rampart and strengthened with breastworks beyond the flanks. A brigade filled the road ditches and the great quarries adjacent, and a detachment was posted in the gravel-pit to the front. The batteries behind these troops were in covered positions. There were 50 French battalions and over 100 guns opposing von Steinmetz.

By 3 p.m. the two brigades of the 15th Division of the VIII. Corps were at close quarters.

The 29th had gained the advanced gravel-pit to the west of Point du Jour, and was extended north to the high road; the 30th was surrounding St. Hubert on three sides, and stretched to the left as far as the bifurcation in the Mance Valley. All attempts to advance across the open towards Moscou had been rudely checked. Seventy-two guns were working in front of the Gravelotte-Malmaison road. The 16th Division was close in rear of Gravelotte—31st and 32nd Brigades, namely. Gravelotte was being prepared for defence. In front of the Ars-Gravelotte road, ten batteries of the VII. were firing. The fire of the 132 German guns was proving too severe for the French, who began to withdraw their forward batteries, and showed generally a slackening of fire. The buildings of Point du Jour and of Moscou were on fire.

At this moment St. Hubert was captured, 14 battalions were disposed in and about it, and preparations made for further advance. Von Steinmetz erroneously concluded from what he saw that the enemy were beaten. He ordered up the 1st Cavalry Division, with the intention of sending it across the valley, while the commander of the VII. Corps ordered his artillery to cross, with the 29th Regiment at its head. At 4 p.m. the latter was in the defile and the guns were beginning to follow; eight battalions were descending into the wood to the right of the road.

Four batteries had pushed in before the 29th Regiment, and these alone got across; for the cavalry division appeared with its H.A. battery and blocked the road. The French had, indeed, retired from their original advanced positions, but their main line was still intact, and their fire brisked up again when they saw what was going on. Owing to the movement in progress the German artillery fire had greatly slackened.

Two light, one heavy, and one horse-battery trotted on; two of the commanders were badly wounded, and the whole team of the leading gun was destroyed.

The leading battery and the heavy battery gained the open ground south of the road, but suffered so severely from the fire converging from every point from Moscou to the great quarries that they retired almost without firing. The other light battery, taking post close to the road-wall (*v. Plate* XIII.), fired at a line of French guns west of Moscou; this battery was fairly sheltered from its immediate opponent, but was exposed in flank and rear.

The horse-battery likewise unlimbered and joined in the unequal struggle, losing ultimately 37 men and 75 horses. The idea at this time was to break into the enemy's front at Moscou, and thus get on the flank of the formidable position at Point du Jour, all attempts to reach which from the west and the south had been easily frustrated by the enemy.

The leading cavalry regiment now arrived, moved to the right at a gallop, was very soon brought to a standstill, and began to lose men so rapidly that it retreated out of sight into the valley. By 4.30 p.m. the whole of the cavalry division was back again in reserve.

These retreats naturally affected the confidence of the infantry. The skirmishers who had been holding the open ground towards Point du Jour retired into the wood, the French advancing menacingly at several points.

The 16th Division, VIII. Corps, was now swallowed up in the struggle, and several strenuous efforts were made to get at Moscou, both from the Bois des Genivaux and from the quarries near St. Hubert; but the only result was further slaughter.

During all this time the two batteries had continued firing from near St. Hubert. After a two-hours' struggle, however, the strength of the completely-exposed horse-battery was now exhausted, as the surviving gunners were but sufficient for the service of a single gun. Spare teams were brought up, and the battery removed, leaving the single battery behind the wall to continue the firing.

The VII. Corps had done no such desperate attacking, but had practically awaited events. After a pause the Germans were at it again, but when night fell the French still held their main position as shown in the plan.

The VIII. Corps lost 2,206 men and 125 officers out of its 15th Division, 841 men and 59 officers out of its 16th Division, 30 men and 50 horses of its corps artillery, 13 men and 3 officers of the divisional artillery. None of these guns took part in the insane advance across the valley. The total infantry loss was 3,047 men and 184 officers.

The VII. Corps lost, in infantry, 715 men and 29 officers, in artillery 103 men and 183 horses.

The 1st Cavalry Division lost 7 officers, 88 men and 177 horses.

A justification of such losses as took place in the VIII. Corps infantry might be pleaded on two grounds, and these are pleaded in the German Official Account:—(1), the demoralization of the enemy; (2), the "containing" of the French left wing, thereby preventing any part of it being detached to the north to help at St. Privat and Roncourt. Neither plea is, in this case, valid. The French left wing at nightfall held, with good reason, the opinion that they had gained a victory; and it would have been just as well "contained" by more cautious action. The French Guards, in rear of the left wing, were available for reinforcing at the further flank, and were not used till it was too late. Can anyone suppose that, had von Steinmetz contented himself with artillery fire and threatening with his infantry, the French would have felt themselves free to remove part of Frossard's Corps (2nd) to the north flank? Can anyone suppose that von Steinmetz would have escaped censure, if the attack on the French *right* flank had failed and the enemy had advanced to attack at dawn next morning?

(2). GERMAN CENTRE.

The IX. Corps had orders to take in hand that part of the French front extending from La Folie Farm to Amanvilliers, which turned out to be the centre of the hostile position; but von Manstein, commanding the corps, had instructions to attack only "if the enemy's right flank was at La Folie." But when the advanced guard was checked by fire from Chantrenne Farm, and von Manstein went forward to reconnoitre, he saw

GUARDS and IX. CORPS on 18th AUG. 1870, about 3 p.m.

PL. XIV.

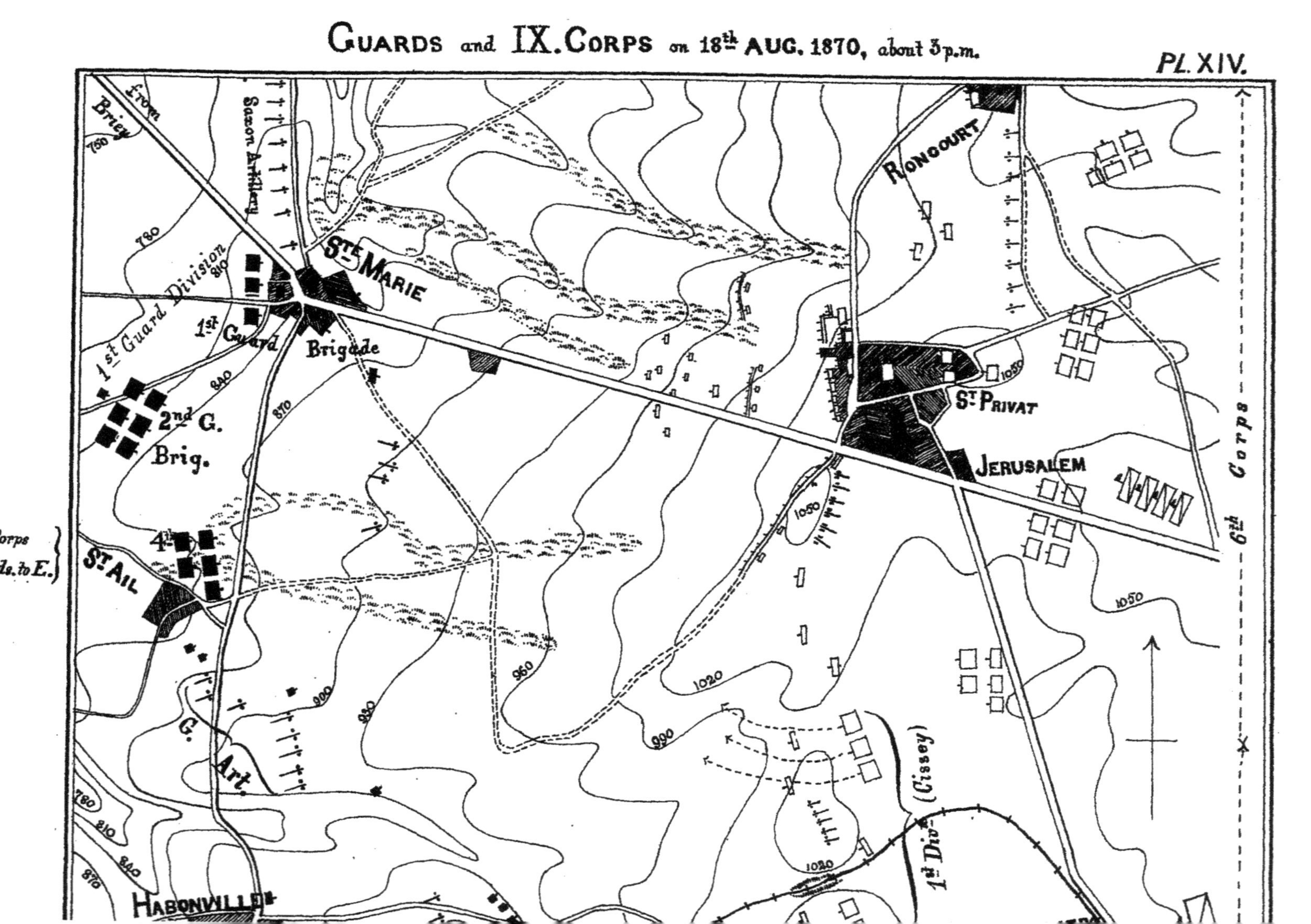

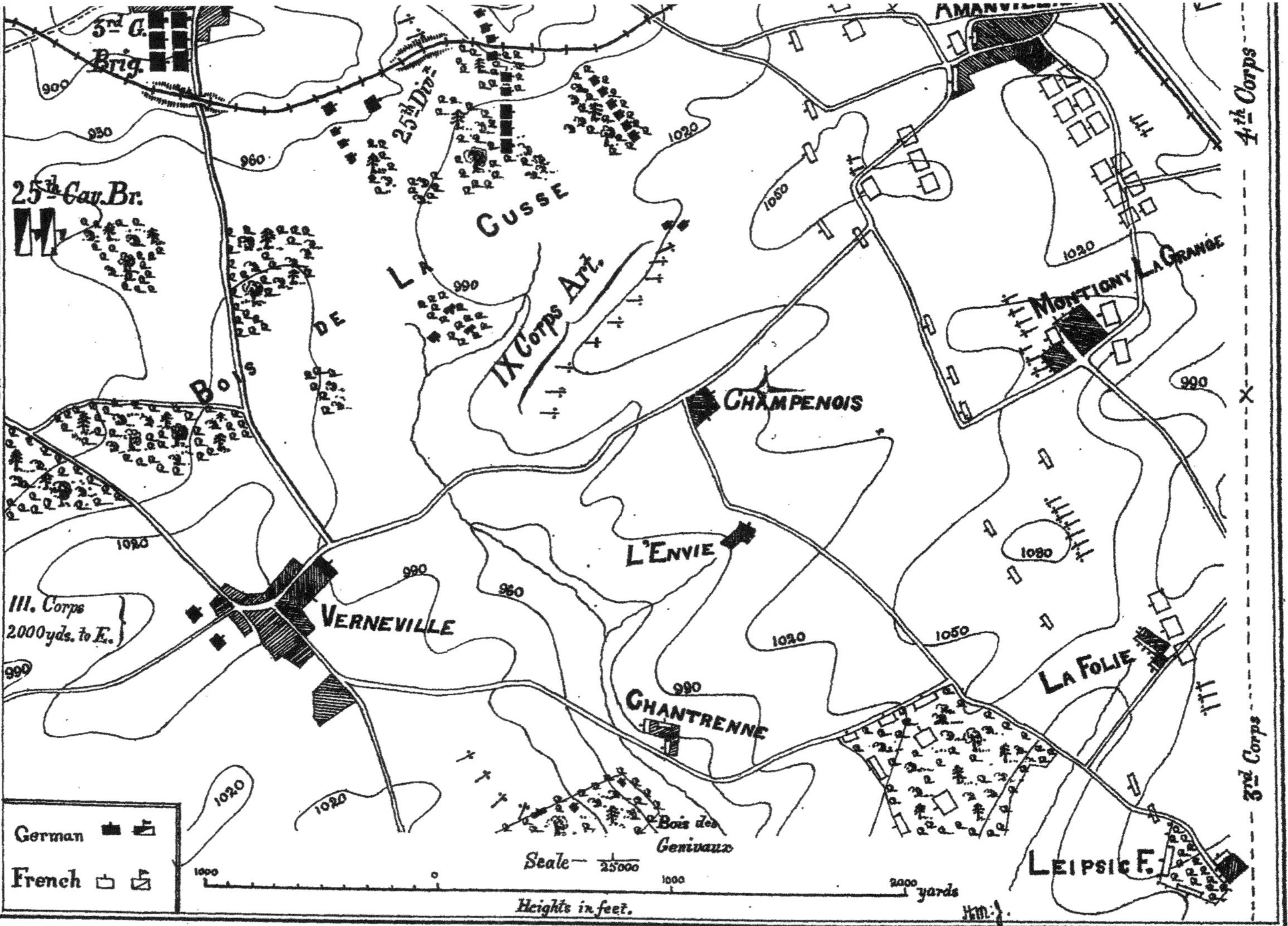

Each gun represents a battery.

the French camp about Amanvilliers apparently in a state of heedless unconcern. Thinking this too good a chance to miss, he proceeded to the attack on his own responsibility.

At 11.45 only the single battery of the advanced guard had arrived, and it opened fire from the height immediately to the east of Verneville; but the range was found too long, and as the corps artillery arrived, it formed up on the long spur that comes out south-westwards from Amanvilliers, and opened fire. *v. Plate* XIV.

The 4th French Corps was opposite; it was to some extent surprised, but quickly took up its line of battle and brought its powerful artillery to bear, 12 batteries and 3 mitrailleuse batteries.

The Prussian batteries at once began to find themselves in difficulty; they were completely exposed on the open ground, while many of the French had the shelter of walls or emplacements. The guns were taken in flank from the north of Amanvilliers, and very soon officers, men and horses began to fall.

Isolated advances of detachments of French infantry caused still further trouble, and the Prussian main body had not yet arrived; so, about noon, von Manstein made the attempt to cover the left flank of his line of guns with cavalry. But these had to be quickly withdrawn. Some infantry, now arriving in the Bois de la Cusse, was sent forward to the guns, and a battalion from Verneville moved out and captured the L'Envie Farm, which had been held by a handful of French. The latter, in retreating, was met by an advance of strong supports, and a strenuous attempt was made to retake the farm, but the single battalion held on successfully to what it had won, though with great loss.

Meanwhile the nine batteries on the forward spur were in a most precarious position. A French mitrailleuse battery had come out from Amanvilliers to within effective range of the left flank guns. The Prussian heavy battery on that flank had already suffered terribly. In a few minutes more it had lost several officers, 5 gun commanders and 40 men, while nearly all its horses were rendered useless. Strong detachments of hostile infantry made a rapid dash towards the battery, which only succeeded in getting away two of its guns. A fresh battalion having reached the nearest copse of the Bois de la Cusse moved out towards the abandoned guns, but suffered so badly that it had to be withdrawn; but the enemy's infantry advance had been checked.

Although sorely pressed, the Prussian artillery maintained an unbroken and vigorous fire; but as the state of the contest and the form of the ground rendered it difficult to replenish the ammunition in good time, especially on the left wing, the batteries of the corps artillery were barely able to continue the struggle at 2 o'clock.

The position was now as follows:—

The combined artillery on the long spur was still maintaining its positions, though with difficulty after the annihilation of the left flank battery. L'Envie Farm was occupied, but Champenois was still in French hands. Of the infantry only six battalions had been as yet engaged. The parts of the Bois de la Cusse nearest the enemy were held by two and a-half battalions; half a battalion was in L'Envie; two and a-half were engaged in a stationary contest near Chantrenne.

The rest of the artillery of the IX. Corps (five batteries) was in front of Habonville, firing towards St. Privat; thus there was divergent fire, and Amanvilliers, between the two directions, was left almost untouched.

What are we to say of von Manstein's operations? He knew that the Guards were coming up on his left, and the Saxons still further to the left, and he had instructions to withhold his real attack, if the enemy stretched far away beyond him. We are told that he could not let slip the opportunity of taking the enemy by surprise; but is it a surprise of any value when you have only a single battery at hand, and the enemy is encamped close to a prepared position and in great strength? This kind of surprise simply results in giving the enemy ample warning.

Then, when the corps artillery arrived, its action was bold to the extreme of rashness. The Germans never shirked losses; but was there in this case any sufficient advantage gained? Why should the 14 available batteries not have turned their whole attention to the French 4th Corps from Amanvilliers to La Folie, keeping at first to a longer range, and keeping out of reach of any French guns north of Amanvilliers? Von Manstein seems to have forgotten for the time the principle of concentrating superior fire on a chosen part.

The great steadiness of the German gunners was most laudable, but the commanders in this case exposed them to useless risk. The battle was eventually won; and no doubt the indomitable pluck of the gunners had a great effect on the *moral* of the enemy in the future; but the troops of the French left and centre were far from being demoralized by the herculean efforts of von Steinmetz and von Manstein. They were, on the contrary, when evening fell, jubilant with success, and many hoped to receive orders to advance at dawn. In their 2nd and 3rd Corps demoralization set in with the order to retreat among the forts. What if the French had pushed their counter-attacks with more combination? What if the Germans had failed to carry the day? The terrible losses of the artillery here, of the Guards' infantry at St. Privat, and the still more useless slaughter of von Steinmetz's men at Gravelotte—these losses would have been severely felt, if the French had held their ground for the day, and had attacked in force at dawn next morning. Should we then have heard so much praise of the "splendid co-operation" of the German commanders at this fight?—or of the brilliant work of the artillery?—perhaps of the *personnel* of the batteries, but certainly not of the commanders who drove them to useless sacrifice.

The French missed many chances of making an end of German batteries; their counter-attacks were isolated, and never really pushed home. A half-hearted counter-attack is seldom of any value.

Towards 3 p.m. von Manstein withdrew the guns from the spur, by batteries from the left. The French, seeing the movement, sent infantry forward, and one of the batteries had to halt and fire case. One gun had to be temporarily abandoned. The five batteries on the other (north) side of the Bois de la Cusse had also to leave five wagons on the field, so great was the loss of horses.

One proof of German prudence was shown when von Manstein set his pioneer battalion to fortify Verneville.

(3). German Left Wing.

Towards 5 o'clock the Guard Corps had assumed the following positions:—

The 1st Division occupied Ste. Marie with seven battalions, its other troops being to the west and south-west of the village. The 2nd Guard Division had given up a brigade (the 3rd) to the IX. Corps, the other was deploying at St. Ail, while one

1ST GUARD BRIGADE on 18th AUG. 1870, about 5·30 p.m.

PL. XV.

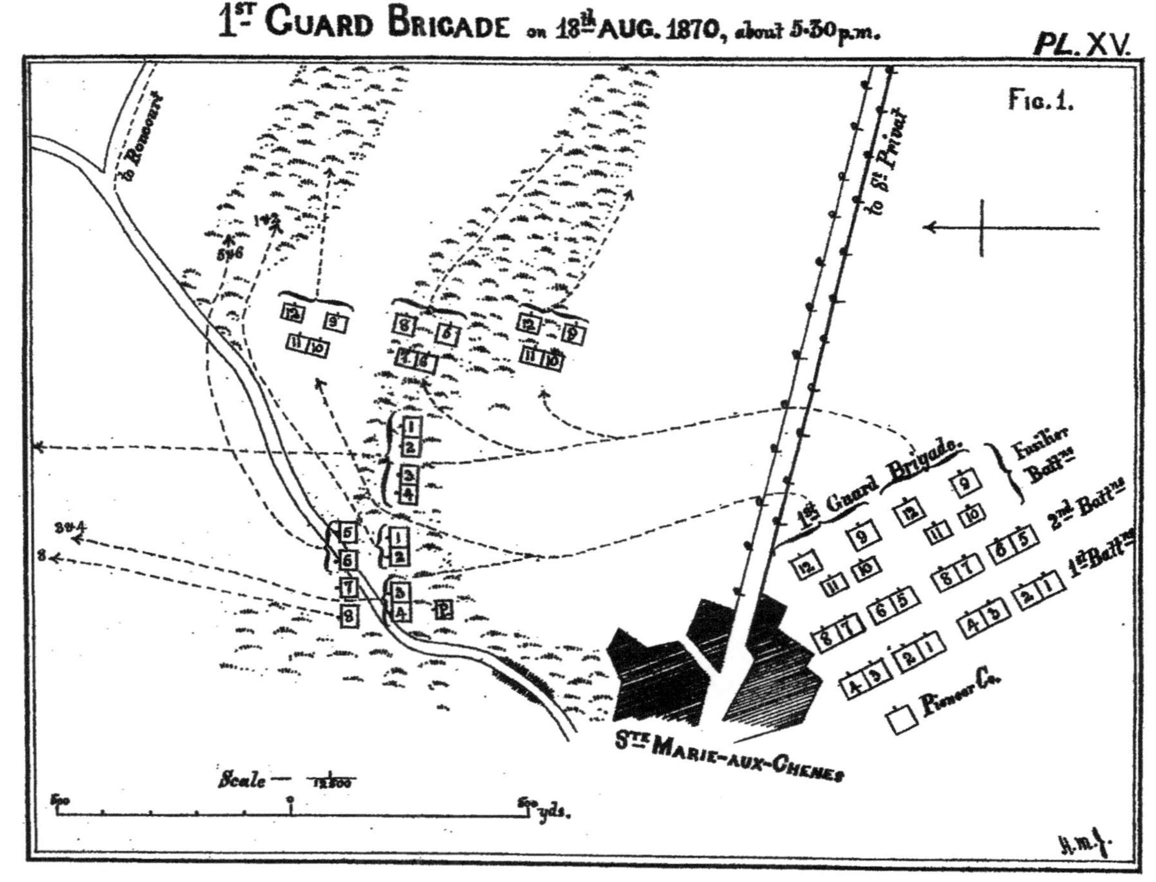

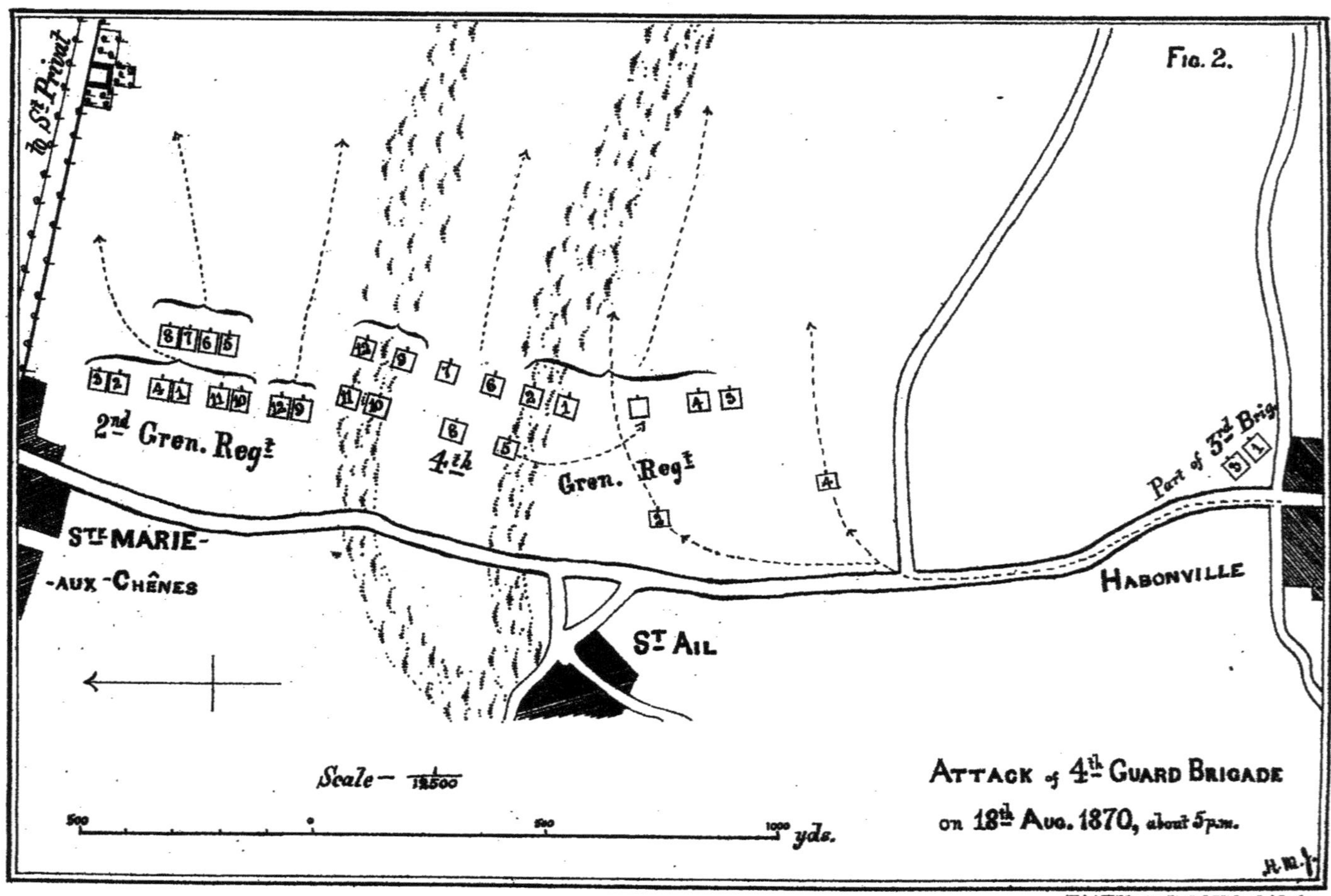
Fig. 2.
To St. Privat
2nd Gren. Regt.
4th Gren. Regt.
Part of 3rd Brig.
Ste. MARIE-
-AUX-CHÊNES
St. AIL
HABONVILLE
Scale — 1/12500
500
0
500
1000 yds.
ATTACK of 4th GUARD BRIGADE
on 18th AUG. 1870, about 5 p.m.
H.M.

battalion protected the artillery between Habonville and St. Ail. To the south of the latter were posted eight, to the north of it four, batteries of the Guard. Between the Bois de la Cusse and the left of the whole line there were altogether 180 German guns in action, the Saxons having 12 batteries to the left, and the IX. Corps six batteries to the right. The French guns opposing these, outnumbered, had for the time almost ceased fire. Thus, before 5 p.m. there was a lull along the entire front of the IInd Army, the German artillery keeping up a moderate fire. The pause was for the purpose of awaiting the completion of the turning movement of the Saxons.

The commander of the Guards now saw signs that the full attack of the XII. (Saxon) Corps would not be long delayed, so he determined to commence his frontal attack. The 4th Brigade accordingly advanced. The O.C. Guard Corps betook himself to the 1st Division, and ordered its commander also to advance. The latter pointed out that the Saxon flanking column was not yet in sight, and that success would be almost impossible without powerful artillery preparation. But, as the 4th Brigade was seen to have commenced its movement, and appeared to be getting along well, the order was carried out forthwith.

Attack of 4th Guards Brigade on St. Privat.

I shall take the attack of the 4th Brigade first. It had deployed in two lines, with skirmishers in front along the St. Ail-Ste. Marie road to the left of the long line of Guards artillery. The 2nd Grenadier Regiment had the left of the line, between the high road and a grassy hollow, which ran from St. Ail to St. Privat height.

The 4th Grenadier Regiment was to advance along the hollows and between them, but they were found to be as much under the enemy's fire as any part of the ground.

Dense swarms of French skirmishers covered the slope and the ridge. On the latter they had ensconced themselves in shelter-trenches and behind a hedge that lines the south-west road from St Privat, whilst the water-channels intersecting the fields at distances of 20 paces offered the assailant a very meagre cover.

v. Fig. 2, *Plate* XV.

Even during its deployment at St. Ail the Prussian Brigade found itself overwhelmed by a shower of bullets; the ground was so hard that the chassepôt bullets ricocheted freely. The French infantry must have been firing at about a mile range.

The 2nd Grenadier Regiment lost its commander, and two battalion commanders, at the very start. The 2nd Battalion broke up its outer company-columns into skirmishers on starting, and the example was soon followed by the other battalions. Advancing by rushes, they all gradually bore off to the left towards the high road, but before long the advance was checked. Most of the regiment were now in the road ditches, at about 400 yards from the French skirmish line, and were keeping up an exchange of musketry fire, when the enemy increased their troubles by turning some guns to sweep the road. Only one half-battalion, viz., the 9th and 12th Companies, kept straight to its front, moving on the hedge before mentioned.* By this time the regiment had lost nearly all its officers, and the companies were reduced to mere knots of men. It was plain that there was not here sufficient strength to do anything further.

The other regiment, the 4th, had advanced at the same time, the Fusilier or 3rd Battalion on the left in the north grassy hollow, half of the 1st Battalion in the south hollow; half of the 2nd Battalion presently came into line to fill the gap between them, and the front was extended to the right by the advance of the other half of the 1st

* *i.e.*, towards height 1050—*v. Plate* XIV.

Battalion. (For the numbering of the battalions and companies of a regiment, see *Fig.* 1, *Plate* XV.). Two companies of the 3rd Brigade from Habonville also joined in.

The attacking line was now as shown in *Fig* 2, *Plate* XV. At a signal, the whole charged forward, but were met by such a fire (volleys) that the frontal attack along the north hollow broke down almost immediately; but the rest of it was so far successful that the enemy on the south part of the spur abandoned it and withdrew to Jerusalem.

v. Plate XIV.

The portion of the spur thus won, about 1,200 yards from St. Privat and 1,800 from Amanvilliers, was now subjected to a converging fire from the direction of these two places, and strong columns of the 1st Division, 4th French Corps, were seen moving towards it from the south-east. The four companies on the spur, much mixed up, formed into small groups and met the advance by rapid file-fire, while the Prussian artillery saw the danger and turned its attention to the French counter-attack. But the position of the advanced infantry was still so critical that two Guards' batteries galloped forward to short range, but had the greatest difficulty, owing to rapid losses, in opening fire. When they got set, however, aiming at the counter-attack, the latter wavered and then halted, and the fight became stationary.

Meantime, the six companies moving between the hollows were easing the situation for those on the road by pouring a flanking, almost enfilade, fire on the defenders of the hedge. Towards 6 p.m. the French abandoned the ridge entirely, leaving 200 unwounded prisoners in the hands of the Prussians; but no further advance was possible.

The attack on the north side of the high road began about half an hour after the commencement of that just described.

Attack of 1st Guards Brigade on St. Privat.

The conditions of the attack were far from favourable; everywhere there was a bare slope rising gradually towards St. Privat and Roncourt, with only a few potato patches and a few isolated trees and shrubs growing on it. The foot of the slope was bordered by a slight valley, trending off from Ste. Marie to the north-west. Out of it stretch towards the east two narrow strips of meadow 500 paces apart; the ridge between these is like a broad undulation.* Towards the top of the slope, close to the village, were several parallel walls high enough to give kneeling cover, and a few trenches had been dug. These lines commanded one another, and were filled with riflemen; while immediately in rear and higher up stood the solidly-built village, whose houses were manned in every tier by rifles.

The French artillery in front of St. Privat and to its north had already been silenced and had retired; but from the south of St. Privat heavy French batteries swept the ground of advance most effectively, and there were masses of infantry between Roncourt and St. Privat. The Saxon batteries were now engaged with Roncourt, and the Guard artillery with the French on the south of St. Privat, so that the coming frontal attempt on the latter seems a most foolhardy proceeding.

Fig. 1, *Plate* XV.

At 5.45 p.m. the six battalions of the 1st Guard Brigade were as shown, in three lines. They wheeled after crossing the road, the Fusilier battalions immediately throwing out their skirmishing *züge*, which however made no progress. The north movement was kept up for 500 yards. Then the wheel to the right began, but the troops were already suffering from chassepôt fire at 1,200 yards, and that of the heavy

* Ste. Marie to St. Privat, 2,500 yards (German pace about 30 inches).

guns already mentioned. So the battalion at the pivot moved forward at once and then the next, and the advance was made towards St. Privat in échelon from the right.

The Fusilier battalion of the 3rd Regiment thus got started first, and reached to within 750 yards of St. Privat, but with great loss. Its commander, keeping on horseback, had been killed at the very start, and the C.O. of the regiment was wounded. Half of the battalion was following in close order, and managed to press forward thus to the skirmish line; but it had then, under the stress of heavy losses, to break up into extended order. The captain who was now in command rose and bore forward the thinned ranks in another attempt, but he fell at once mortally wounded and the remnant of the battalion was shattered.

On the left of this battalion, the 2nd had also tried to advance in the same manner, but had retired behind a line of skirmishers and assembled in one of the grassy hollows. But the cover here was found to be quite ineffective; the battalion commander was severely wounded. Soon the remains of the battalion were under the command of a lieutenant, who got the men up and charged the first line of the enemy and drove it in. He then established his men where he was as a skirmishing line, lying down.

To the left, again, of this battalion, the Fusilier battalion of the 1st Regiment had completed its wheel and advanced. When its skirmishers reached within 500 yards of the French advanced line, they started firing, followed by the rest of the battalion in close order. But the whole began to lose men at an appalling rate. The O.C. battalion was early wounded severely, and the battalion gradually lost all its officers. The left flank company was reduced to a small knot of men; the other three had only enough left to form a thin line of skirmishers.

The 2nd Battalion of this regiment had moved north behind the three battalions already engaged, marching in line. It made a wide sweep to the north before turning east, and was followed by the 3rd line of the brigade. But the O.C. 1st Regiment, seeing his Fusilier battalion in sore straits, wheeled half his 1st Battalion up on to the left flank of his Fusiliers, in concert with whom it made several fresh attempts to get forward, but with no success. About 500 yards in front of the French position, the remains of these six companies, in a state of exhaustion, threw themselves down in a trench the enemy had evacuated earlier, and there remained.

As these movements went on, the 2nd Battalion of this regiment completed its wheel and found it was being fired on from Roncourt. The O.C. Brigade sent up two companies (half a battalion) out of the north grassy hollow to take up a flanking position towards Roncourt, one company to support the left flank of the troops facing St. Privat, while the other company joined the 3rd line of the brigade moving on to the left. The O.C. 2nd Battalion was killed during this movement.

The whole of these events had occupied about half an hour; at 6.15 p.m. four and a-half of the battalions of the brigade were within 500 to 700 yards of the object of attack; the remainder had gone north and joined the Saxons moving on Roncourt.

In the instinctive effort to avoid the frontal fire of St. Privat, these troops had edged off to the left, and there was a gap of 600 yards between their right and the road. To fill this gap, the 2nd Guards Regiment of the brigade was brought up from behind Ste. Marie; it crossed the road 500 yards forward from the village in two lines of half-battalions, wheeled towards St. Privat, extended the front line into company-columns at open intervals, and threw out the regulation skirmishers. This was all strictly regulation. In a few minutes the O.C. Brigade, the O.C. Regiment and the O.C. 1st Battalion were severely wounded. This battalion gradually lost all its officers, but it worked up into position abreast of the line of battle. On the left of the 1st Battalion, a major took forward the 2nd Battalion rather more successfully, but still it lost heavily; and on the right the Fusilier battalion with great difficulty came up on both sides of the road.

The attack was now at a standstill, nothing decisive having been effected, and thousands of killed and wounded strewed the blood-stained slopes; but the troops, true to their training, showed no back to the enemy, and clung to their dearly-won ground. Now was the time for a French counter-attack, but no such move was made. Only once a French cavalry regiment from behind St. Privat came out towards the left of the 1st Guards Regiment, as if meditating a charge, but it was driven back with heavy loss by the skirmishers. A German Lancer regiment, on the look-out for this, came out from behind Ste. Marie, but was not able to close, and retired with loss.

The 4th Guards Regiment was now brought along the low ground northwards, and wheeling came out in front on the north-west of St. Privat, losing its commander. The divisional commanders and the O.C. Guards Corps now began to see that nothing could be achieved without a considerable artillery preparation, and arrangements were made for this. Twelve batteries of the Guards had been all this time deployed between Ste. Marie and Habonville, firing between St. Privat and Amanvilliers.

There was a general pushing forward of guns, four batteries going up to within 1,000 yards. This bold action, and the development of the Saxon attack on Roncourt, soon put a new complexion on affairs. At 8 p.m. the Germans were in possession of the village, and captured 2,000 unwounded prisoners. The place was in flames.

The 1st Division of the Guards, part of whose operations we have seen, lost 66 officers and 1,214 men killed, 95 officers and 2,876 men wounded—total put *hors de combat*, 161 officers and 4,177 men. Most of the damage was due to attacking, without proper artillery preparation, a position of great natural strength, which at this stage could only be attacked in front. The question is whether St. Privat would not have quickly fallen by the mere capture of Roncourt, which *could* be taken in flank. But it must be remembered that night was coming on, and that it was held essential to achieve something definite and decisive before dark.

CHAPTER XIII.

BATTLE OF BEAUMONT, 30TH AUGUST, 1870.

ON the evening of August 29th, German cavalry reconnaissance showed that Beaumont was occupied by the French, and that a part of their army had crossed the Meuse, north of that place. As a fact, the 5th French Corps was at Beaumont, acting as flank guard to the other three corps, which were aiming for Mouzon from the west.

v. Plate X[illegible].

On that same evening, on the German side there stood at Nouart (seven miles south of Beaumont) the bulk of the XII. or Saxon Corps—23rd and 24th Divisions—with detachments and cavalry along the road north-east as far as Stenay, which is seven miles south-east from Beaumont. The IV. Corps bivouacked three miles south of the XII., and the Guard Corps were at Buzancy, a few miles to the east.

The country between these corps and Beaumont is a tangled mass of forest and ravine, impassable, for the most part, even for infantry, with very bad roads; but from Buzancy a passable road runs through Sommauthe to Beaumont, while the good high road goes from Nouart by Stenay to Beaumont.

The object of the Germans was to destroy the French flank guard before it could cross the Meuse.

30th August.

At 6 a.m. on the 30th there was sufficient information to enable a plan to be made. The XII. and the IV. were to advance on Beaumont in four columns—1st column, consisting of the cavalry division, 23rd Infantry Division and the XII. Corps Artillery, to travel by the high road, so as to reach Beaumont from the south-east; 2nd column, 24th Division, to move by the road through Forêt de Dieulet on Belle Tour Farm, where it would easily line up with the 1st column; 3rd column, 7th Division of IV. Corps, to take a forest road due north from Nouart, and then work off to the right towards Belle Tour; 4th column, 8th Division of IV. Corps and the corps artillery, another forest road to the left through Bois du Petit Dieulet.

As a supplement to these movements, the 1st Bavarian Corps was to advance by the Sommauthe road, and the Guard Corps to move in front of Nouart as a general reserve.

Each of the advancing columns, on emerging from the forest, was to await the arrival of the others, and open the attack with artillery alone.

The French 5th Corps committed the unpardonable sin of neglecting to watch the roads between their camp and the enemy. They knew he was only a few miles distant, but they *assumed* he was marching on Stenay in order to cross the river there. Accordingly the commander, wishing to give his troops a rest, let them stay in camp, cooking, with the intention of moving on Mouzon after noon. A few sentries, stationed a short distance from the camps, formed the sole "service of security."

"The town of Beaumont, lying in the middle of a ravine, is surrounded on three sides by thick woods, which encompass it in a semi-circle to the southward, their

ERRATUM.

Page 111, in the margin, *for* "*Plate* XVI." *read* "*Plate* X."

borders lying about two miles from the town. Between the woods and the town there is open hill country, which allows of free movement for all arms. On the east there are steep declivities to the Meuse."

A French officer, writing recently of the opening of this battle, speaks as follows :— "During the night of 29th and 30th August, 1870, the 5th French Corps arrived at Beaumont, harassed by the fatigues of a series of painful marches and counter-marches, and by the engagement it had had the day before at Nouart with the hostile advanced guards which pressed it closely in its first (attempted) march on Stenay. Without any previous study of the ground, in this *funnel* of Beaumont, de Failly's staff fixed the camping positions of the corps. The 3rd Division was established to the east of the Mouzon road, a brigade in 1st line, the batteries and mitrailleuses behind, and a brigade behind that. In the dark night, the 1st Division took a similar formation, north of the Beaumont-Besace road and west of the Mouzon road ; but one brigade was south of the former road.

"The reserve (corps) artillery—two heavy batteries, two light, and two of mitrailleuses, with the 'park'—halted south of Beaumont, covered on the rear (*i.e.*, towards the enemy) only by the camp-guard, massed at scarcely 200 yards' distance, facing the great woods.

"The proper rear guard . . . Maussion's single brigade, with a battery and some mitrailleuses, arriving towards 4 a.m., found the ground about Beaumont occupied, passed through the sleeping town, and went on to establish itself on the heights situated to the east of La Harnoterie.

"On the 30th, about midday, the German advanced guard batteries opened fire unexpectedly on the reposing camps, and threw them into an indescribable panic. The reserve batteries succeeded only with the greatest difficulty in harnessing and getting clear of the dangerous position where they were exposed to the view and the blows of the enemy, and in carrying themselves on to the heights north-west of the town.

"Already reduced to five batteries on account of these first losses, they opened fire on the edge of the Dieulet Forest, from which at that moment were debouching in mass the German infantry and artillery.

"The 3rd Division batteries, and still more those of the 1st Division, more exposed as they were to the enemy's sight, only escaped from their bivouac with serious losses and went off to take position astride of the Mouzon road, north of the town.

"This retirement, so harshly imposed on these guns, most clearly indicates the fighting position on which the batteries *should have bivouacked* the night before.

"I shall not," he continues, "enlarge any further on the senseless situation in which the reserve artillery and the 'park' had been placed, fooling in the rear of the column, although the 5th Corps was moving in retreat. As I have said before, nothing had been arranged in time for the. emplacing of the troops in bivouac, nor for the order of departure next day, nor for the ensuring of such a zone of manœuvre as would be necessary in order to sustain the fight which ought to have been expected from moment to moment, since it was known that the Prussians were following.

"The heroic conduct of the 4th Battalion of Chasseurs and the self-sacrifice of the regiments of Saurin's Brigade permitted our reserve batteries to escape, though in inexpressible disorder, from complete and useless destruction."

The 8th Prussian Division was the first to reach the edge of the forests, about noon. The leading battalion halted on the slope behind Petite Forêt Farm, and sent a company quietly to occupy it. A French camp was seen about half a mile distant, and another north-west of the town, and the occupants seemed quite ignorant of what was impending. The commander of the division, as usually happened with German generals in this war, could not bear to let such an opportunity slip; so, without waiting for the other columns, he sent the whole of the leading battalion on to the height where the company was, ordered the 16th Brigade to deploy behind the height, and the Divisional Artillery to move up on to it. The battalion and the batteries of the advanced guard were in position before the French were aware, but something alarmed them before the 16th Brigade was free of the woods. At 12.30 the batteries opened fire on the camp, which had the appearance of a hive of startled bees, and was sending out rapidly a swarm of skirmishers.* French guns appeared at once to the south and west of Beaumont, and on the heights to the north an ever-increasing line of artillery began to form up.

The question arises whether this sort of surprise is of any value. Inter-communication between the columns had long been lost, owing to the extreme density of the forest undergrowth, and the other columns might be long delayed. To us it seems that the commander of the 8th Division was *preventing* a really useful surprise—*i.e.*, a sudden appearance in such strength that the hostile camp might have been "rushed" straight away.

The small force that opened the fight was at once in trouble; the guns were very soon being served by only two or three men each.† Towards 1 p.m. the French counter-attacked, skirmishers followed by closed supports; the Germans fired volleys with such effect that the skirmishers lay down and the supports retired out of range. Three German battalions were now available for firing line, and these pushed forward to the front edge of the height, and were soon joined by two more battalions; these five battalions effectually prevented any further attack.

The corps artillery was now deploying from the wood; its escort of three battalions (71st Regiment) lined the wood near the point of issue. Before these guns were able to join in, the advanced infantry again moved forward to the attack, helped, this time, on their right by the leading regiment of the 7th Division. The commander of the latter, hearing the firing on his left front, had determined to join in without waiting. The four divisional batteries moved up to the left side of the height in front of Belle Tour, where they suffered very severely from the chassepôt fire directed from the opposite ridge. So all the troops clear of the woods were pushed forward into the fighting line, and these succeeded in throwing back the French skirmishers.

Thus, about 1 p.m., five battalions of the 8th Division, three of the 7th Division, and eight batteries constituted the firing line. The French were bringing down some supports from Beaumont, and with the help of these they made so determined an attack

* Saurin's Brigade, mentioned in the last paragraph of the long quotation given above.

† The same French officer previously quoted says:—"The Prussian divisional batteries engaged with such precipitation that they found themselves, from the very beginning of the action, in the very skirmish line and suffered much. . . . In spite of exceptional conditions of fatigue, of moral depression, of complete surprise, the French troops inflicted serious losses on the heads of the German columns. In circumstances less unfavourable, they would have crushed and hurled back into the woods the gunners and infantry arriving prematurely and debouching in succession without proper covering."

On the other hand, the Germans only attacked in haste because they saw the French unprepared.

on the right half of the 8th Division that they reached to within 50 yards of their enemy; they were finally checked and driven back by a bayonet-charge, the whole German skirmish line reaching, this time, the skirts of the hostile camp. Seven guns were captured here. The camps were soon cleared of the enemy, who retired mostly through Beaumont. Three battalions followed close on the heels of the French, and entered the deserted camp north of the town.

It was 2 p.m. when the town was taken; the losses of the attackers had been very great. The 66th Regiment, on the right flank, which had had to deal with flank fire from the Stenay road and the heights beyond, lost 20 officers and 500 men. The first battery at Petite Forêt lost 3 officers, 26 men and 34 horses. The French guns had been speedily silenced.

The IV. Corps Artillery, now clear of the forest, had had its fire masked by the victorious advance of the infantry on the heels of the enemy. It now followed the infantry, and chose new positions. The whole of the French artillery at 2 p.m. was forming a line from La Harnoterie to the heights above Létanne. During the events just described, the Saxons and the Bavarians had joined in the struggle. The 24th Division had found its proposed road impassable, and had taken one to the right (*v.* plan). Again, it had been badly delayed in crossing the swampy brook, and a bridge had to be made. Its cavalry and artillery went to the right on to the Stenay road, and inserted itself in the column there.

Saxons.

The leading regiment of the 23rd Division reached the bridge at 1 p.m. It was immediately under fire from the copse on the right, which it attacked and occupied. It then proceeded to drive off the French detachments which, from the right of the road, had been harassing the IV. Corps with a flanking fire. Then the two light Saxon batteries shown in the plan took post to the right of the road and opened fire. Soon more batteries arrived, and the bulk of the Saxon divisional and corps artillery moved forward to the heights east of Beaumont, to join in the attack on the French 2nd position, escorted by some infantry battalions which posted themselves below the guns on the forward slope. Two regiments advanced on Létanne.

1st Bavarian Corps.

The Bavarians, coming up from Sommauthe, began to arrive after 1 p.m. Two batteries, pushed well on in front with a cavalry escort, were in time to fire on the French retreating into Beaumont (*v.* plan). The commander of the leading division, the 2nd, planned to make his main advance by Thibaudine, which promised to bring him well on the hostile right flank. The leading regiment accordingly moved off in the wood to the left of the two mentioned batteries. Just then the cavalry regiment was ordered to charge an apparently isolated mitrailleuse battery south of La Harnoterie; but it was received by a brisk musketry fire from the copses there, and on its return, *re infectâ*, it was fired on from La Thibaudine.

By 2.15 the Rifle Battalion had established itself on the embankment of the high road east of Thibaudine, and part of the regiment in the wood had reached its north edge and was firing on the farm. At this moment, unexpected fresh troops of the enemy appeared on the left flank from the direction of La Besace; these were part of Dumesnil's Division of the 7th Corps, making for Mouzon. They immediately attacked the German regiment in the wood and to the north of it.

At this time no less than 25 German batteries were in position—to the right of Beaumont, 12 Saxon and 4 Prussian—to the left, 6 Prussian and 3 Bavarian. The

French guns did not long sustain the fire, the mitrailleuse batteries in particular retreating almost at once. The French infantry meanwhile were taking up a new position from Yoncq to the farm in the Bois de Givodeau, and by 3 p.m. the French artillery had altogether retired. This was an unfortunate move, for the new position had no scope for gun positions.

Meantime, the IV. Corps infantry had been resting and re-forming. It then advanced, assisted on the right by such troops of the Saxon divisions as could find room to deploy.

The Bavarians had been drawn into a separate engagement to the left, which I do not propose to describe. An indication of its nature is shown on the plan. The Bavarians had the better of a tough fight, and halted at 4.30 p.m. in their pursuit on the Besace-Yoncq road. They had prevented Dumesnil's Division from reinforcing the French 5th Corps.

The Crown Prince of Saxony, commanding the Meuse Army, arrived at Beaumont towards 4 p.m., and his aim now was to cut off the 5th Corps from Mouzon. In this he failed.

At 3 p.m. the 7th Division advanced with a brigade (13th) and 3 batteries on the east of Beaumont, the other brigade (14th) taking the west side. The 8th Division prolonged the line to the left, and part of it captured La Harnoterie about 3.30 p.m. When the French abandoned their position on the heights, they were temporarily lost to view; so the O.C. IV. Corps ordered forward the corps cavalry, the 7th Division to move on the Bois de Givodeau, the 8th on the left side of the high road. The cavalry were soon under fire from the heights east of Yoncq and from the wood, so they were withdrawn. The regiment that had stormed La Harnoterie was checked in attempting to reach (905). The 14th Brigade and the 8th Division were now to make for (918).

De Failly, commanding 5th French Corps, was leaving a rear guard on (918) and in the Bois, and trying to rally his troops on Mont de Brune and at Villemontry. The O.C. 12th (French) Corps, already across the river at Mouzon, sent a brigade and three batteries to Villemontry to help. These troops deployed about 5 p.m. A division of the 12th occupied the other bank of the Meuse from Alma northwards, as shown in the plan, and prevented the attempted advance of some companies from Le Fays Copse along the river bank.

With some loss the farm and the edge of the Bois de Givodeau were rushed by the 13th Brigade, who then found the wood so dense that all cohesion between the advancing parties was lost. As they arrived at the north edge and attempted to advance in the open, they were several times driven back by heavy chassepôt and mitrailleuse fire from Villemontry and its copse, and presently some French battalions broke forward from the village. This attack was helped by guns on the other bank of the Meuse, but failed to drive the Prussians within the wood.

By this time (5 p.m.) the 14th Brigade was in possession of the wood to the left of the high road. Assisted by the 8th Division, it had first captured (918), then the wood beyond, taking in the course of operations eight guns and four mitrailleuses abandoned by the enemy. Two companies reached the low ground under Mont de Brune. A large part of the 8th Division moved into the valley at Grésil, with a view to enveloping Mont de Brune. Soon after 5 p.m. three batteries were on (918), shelling the rapidly retiring enemy; and a little later, three more batteries filled all the available space and began long-range fire at Mont de Brune. The Villemontry Copse had been occupied, but it was now evacuated, as strong attack was expected from Mont de Brune. The enemy re-occupied the copse.

At 6 p.m. there was a pause in the IV. Corps engagements. The 13th Brigade held the north edge of the Bois de Givodeau, but could not make further progress, it being impossible to bring up the guns to prepare the way.

Meanwhile the Saxon 24th Division had been following on the extreme right. Attempting to advance from Le Fays Copse along the river, it was received by very heavy fire from Alma and the wood above it. Saxon batteries, accordingly, to the number of five, took post on (768), and by 5.45 two heavy batteries of the 23rd Division were firing from the top of the slope near the Givodeau Wood Farm. These seven batteries succeeded in attracting much of the French fire away from the wood. Some regiments of Saxons thus succeeded in reaching the north-east corner of the Bois, but there all advance ceased about 6 p.m.

An attempt was made, between 5 and 6 p.m., to shift the enemy from the other bank by using the Saxon cavalry division which was at Pouilly; but the attempt failed.

The French on Mont de Brune—several battalions and squadrons and three batteries—were observed to be facing east, when the 14th Brigade attacked from the south, and part of the 8th Division from the south-west. There was a sharp fight, in which the French horse attempted a charge. The hill was taken, along with some guns.

Again, from Ponçay Mill, the French tried a cavalry charge. The Prussians met it with great coolness, the cavalry regiment lost 11 officers and over 100 men, and fled in wild disorder to the Meuse.

But further advance in the open against the masses occupying Mouzon was impossible. Batteries gradually arrived on the height and in the low ground to the right (*v.* plan), and the rest of the fight consisted for the most part in shelling the disordered masses of infantry, artillery, cavalry and wagons struggling across the bridges and fords about Mouzon.

By dark, the enemy was entirely across the river, and the IV. Corps held the final positions shown in the plan—*d, d, d.*

This was the first battle of the IV. Corps. It lost 126 officers, 2,878 men and 248 horses, while the Saxon loss was only 85 men.

SEDAN. 1st SE

Champ de la Grange
Fleigneux
9th Div.
V. Corps
10th Divn
V. C
St Menges
Vrigne-aux-Bois
from Mezières
Maison Rouge
43rd Bn
XI. Corps
XI. Corps Artillery
Floing
Cazal
7th C
Sedan
4th Cav. Divn
canal
Wurt. Division
Torcy
Donchery
field bridges
R. Meuse
French —
German —
2nd Bavarian Corps
Frénois
Cheveuges
from Chéhery
Each gun represents a battery.
Scale
yards
Heights in feet.
St Menges
road
XI. Art.
Fr. guns
Cazal
No I.
Heights in these Sections are exaggerated 7 times.
V. Art.
road
Fr. guns
Bois de la Garenne
No II.

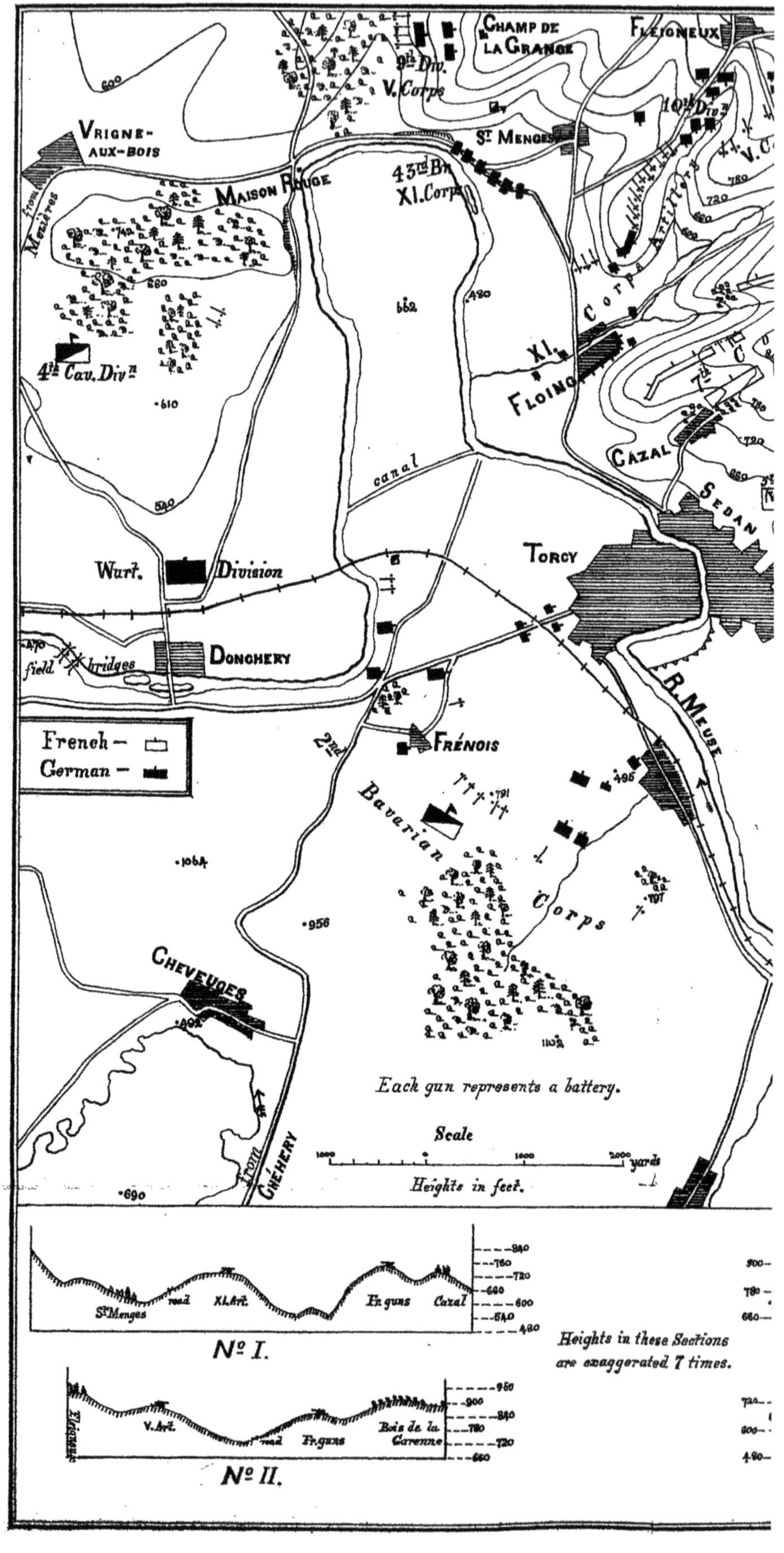

La Chapelle

Illy

G. Cav. Divn

Givonne

1st G. Brigade

Villers Cernay

Haybes

Guard Arty

3rd Guard Brigade

Bois de la Garenne

Corps (Wimpfen)

Fond de Givonne

Daigny

12th Corps (Lebrun)

Francheval

La Moncelle

Rubécourt

Balan

5th Bav. Brig.

XII Corps Arty

7th Divn

Lamecourt

6th Bav. Brig.

Va Beurmann

4th Bav. Brig.

1st Bav. Br.

Bazeilles

2nd Bav. Brig.

to Carignan

Dorival

Douzy

Cuir. Brig.

12th Cav. Divn

R. Chiers

field bridges

IV. Corps Arty

Remilly

Mairy

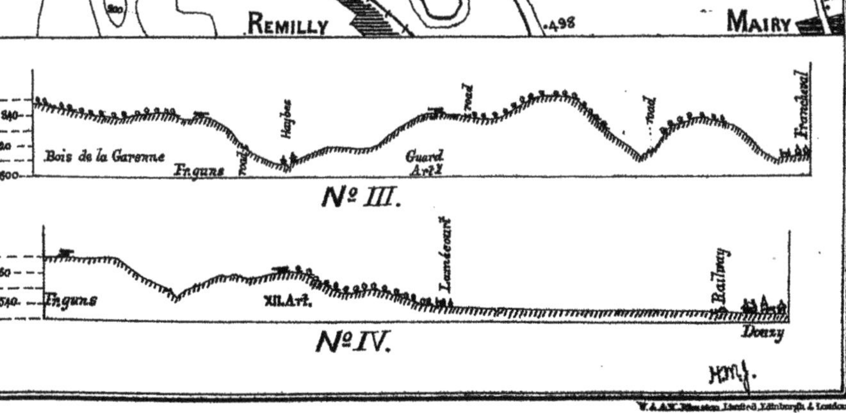

CHAPTER XIV.

SEDAN.

THE battle was fought on September 1st, 1870. Marshal MacMahon, driven to his ruin by Governmental interference from Paris, was making the attempt to pass round the German armies by their right flank, *i.e.*, between them and the Belgian frontier, with the idea of going to the succour of Bazaine, who was now shut in Metz.

MacMahon had four corps, the 7th, 5th, 1st and 12th, of a nominal strength, including 5,000 independent cavalry, of 140,000 men, 67 batteries and 14 mitrailleuse batteries; but the 7th and 5th had lost heavily in the fighting about Beaumont on August 30th, and the 5th in particular was exhausted and demoralized.

The King of Prussia, leading the IIIrd Army and the "Army of the Meuse," had considerably larger forces. One corps, the VI., far away to the south-west of Sedan, took no part in the fight, nor did the bulk of the cavalry divisions. The rest, available for the battle, numbered 164,000, and could put 116 batteries into line.

The orders were that the Army of the Meuse—Guards, XII. or Saxon Corps, IV. Corps, as they stood from right to left—should prevent the French from escaping eastwards from Sedan; and that the IIIrd Army—1st Bavarians, 2nd Bavarians, XI. Corps, V. Corps, Würtemberg Division, also from right to left, should attack from south and east.

The French troops encamped on the high ground above Sedan. What MacMahon had in his mind, beyond confusion and uncertainty, is not clear; but it is clear enough that he committed the unpardonable sin of allowing a field army to be surrounded. And the sin was doubly black from the fact that this was the only free army left to the French. Even if he succeeded in breaking through, sword in hand, he could only hope to get clear with part of his force, and without any train; and where was such a body to make for? Metz was six marches distant, and the addition to Bazaine of 40,000 or 50,000 hunted men, marching light with no reserve of ammunition, would have been a poor addition to that fortress, even if the German investing army should fail to stop them. In the other direction, the nearest fortress capable of sheltering an army was Lille, and that was 108 miles distant. To the north, the way was yet clear, but that meant the humiliation of disarmament in Belgium. (It is worthy of note that the German Corps had instructions to follow into Belgian territory, if there was any sign of immediate disarmament not taking place).

Plate XVIII.

If cutting a way out, and inflicting as much damage as possible, were to be resorted to, it looks as if the south would have been the most promising direction; but the idea seems to have occurred to none of the leaders—perhaps they thought the bulk of the Germans were still between the Beaumont-Buzancy road and Sedan. But it is worthy of consideration whether, when the whole of the Meuse Army and most of the

Bavarians were across the river on the east of Sedan, and the whole of the XI. and V. were across on the west of Sedan and committed to a northward march—whether at that time, say 8 a.m. on 1st September, the massing of two Army Corps at Torcy for the purpose of breaking out might not have proved successful. The other two corps would have retired fighting, more or less rapidly according to the success of Torcy. But such an operation could only have been carried out by an army in a high state of discipline, ably commanded and ably staffed.

It is not argued that any sort of victory would have been thus achieved; but it seems as if success in getting rapidly through the 2nd Bavarians would have landed the French in the midst of the German *étappen* lines, where they might have been able to work incalculable mischief, and perhaps after all escape towards Paris.

On the night of August 31st there was a fight between the 1st Bavarians' advance guard and the defenders of Bazeilles, ending in the retirement of the former, who however held on to, and barricaded, the railway bridge.

The same night, the XI. and V. and Würtembergers started crossing about Donchery.

MacMahon seems to have issued no plan for the morrow; it can only be supposed that he did not know that as many as seven and a-half hostile Army Corps were so close at hand. His own 7th Corps held the ground about St. Menges, Floing, Illy; 1st Corps towards Daigny and Francheval; 12th towards Bazeilles; 5th on the heights immediately north of Sedan. Nevertheless, the German headquarters did quite right to assume that MacMahon would try to escape, if he discovered his danger during August 31st—that he might even attempt it by a night march.

The battle may be conveniently, for purposes of description, divided up into separate attacks, but it must be understood that these can by no means be looked upon as isolated attacks, for they were carried out in such a manner as to assist each other most notably. At the same time it will be oonvenient to consider them as follows:—(1) Operations of the Bavarians and IV. Corps against Bazeilles and La Moncelle; (2) of the Guards and Saxons against the front, Givonne to La Moncelle; (3) of the V. and XI. Corps against the front, Floing to Illy.*

September 1st.

Von der Tann, commanding the 1st Bavarian Corps, had received the order to join in the battle "when the Army of the Meuse shall have sufficiently advanced"; but a verbal addition had been made to the effect that he might use his discretion in attacking earlier, "if the enemy could thereby be detained in his present position." An officer of the General Staff had during the previous evening reported that his observations led him to suppose that the French were planning to march off at daybreak to Mezières—hence the latitude allowed to von der Tann. The latter decided to attack at once.

Description of Bazeilles.

The extensive village of Bazeilles, surrounded with gardens and park enclosures, consists mainly of strongly-built houses. The roads from Balan and Douzy, meeting at an obtuse angle in the village, divide off the north-east part from the rest. In the latter part is an extensive market and a large stone church, and at the extreme south-east border lies Château Dorival with its park. From Villa Beurman, situated in the angle between the Balan and Daigny roads, the main road can be enfiladed in its whole length. To the north-east lies the park of Monvilliers, plantations and meadows and

* The Map, *Plate* XVIII., should now be carefully studied, along with the four sections of ground given below it. The 1st is on the line St. Menges-Cazal, the 2nd Fleigneux to the centre of the Bois de la Garenne, the 3rd from the Bois to Francheval, the 4th from the north end of Fond de Givonne to Douzy.

orchards; its north-west boundary was a strong hedge and ditch; on the east it had a high wall, with only one entrance. The Givonne, flowing through the park in two branches, had only two bridges, and was there almost unfordable. North of the park it was negotiable; and the ground hereabouts up to La Moncelle was practicable for all arms.

Ordinarily the attack of such a place would have been prepared for by artillery; but the intention was rather to entangle the French 12th Corps in a struggle than to hasten its departure, if their move to Mezières was really going to be made. Moreover, the attack began in the dark, and the thick morning mist did not permit of gun fire from the left bank till 6 a.m.

Comments on the Bazeilles Attack.

Von der Tann's idea was to rush the village under cover of the obscurity, and to hold its north and west edge in such a manner that he could debouch in strength towards Balan and Fond de Givonne, if the French showed signs of evacuating the whole position. But this commander cannot, in my opinion, be relieved of the responsibility of having inflicted huge losses on his troops with insufficient reason. If the expectation was that the French were about to evacuate to the west, the establishment of a division on the railway embankment, and of another east of the village and towards La Moncelle, the artillery taking post on the heights to north-east of the village, would have done all that was required. The argument is that the attempt to surprise the village was right enough, but that, as soon as it was found that its capture in this manner could not be done, the infantry should have been withdrawn from the village and posted *en potence,* while the artillery rendered the place untenable.

On the other hand, there is this to be said for it. If the Saxons should find the French position opposite La Moncelle impregnable, the seizure of Bazeilles would help them by bringing the Bavarians on the right flank of the French 12th Corps. Those who seek to justify outright the desperate struggle for Bazeilles argue that von der Tann "could not have been expected to remain on the other bank of the Meuse and look on idly while the Saxons prepared a way for him." Between this absurdity and a desperate attack on the village there is a happy mean, which I have suggested.

The Attack on Bazeilles.

At 2.30 a.m., in darkness and mist, von der Tann and his staff arrived at Allicourt. The decking of the pontoon bridges was noiselessly completed.

The whole of the 1st Infantry Brigade (six battalions) was to cross by the pontoons; the 2nd Brigade and the divisional troops (two battalions, a 4-pr. battery, and a cavalry regiment) were to follow by the same route. The whole of the 2nd Infantry Division was eventually to cross at the same place, but to leave for the present the 4th Brigade and two batteries in reserve near the bridges. Similarly, three battalions of the 2nd Brigade were left to guard the railway bridge.

The artillery had orders to hold their fire.

Just before 4 a.m., and simultaneously, the columns began to cross. One company preceded the advance in skirmishing order, and two battalions of the 2nd Brigade got well into the streets before the enemy was awake to what was going on; but there were barricades, and a savage engagement at close quarters at once began. The French Marines were in the place, and they had loopholed the houses to good effect. The now-scattered companies in the village began to suffer severely.

A battalion of the 1st Brigade was posted to hold the station, so as to command the exit from the town on the east; another was ordered to pass along the eastern side of the village, and endeavour to take the defenders on their left flank; but, probably owing to a misunderstanding, it went along the whole length of the park wall of Monvilliers, and, not finding any gap, eventually rushed on to La Moncelle under a heavy fire. There it established itself in the first houses.

The rest of the 1st Brigade had meanwhile plunged into the village, and become entangled in the desperate struggle raging there. Some companies got as far as the junction of the Balan and Douzy roads; and a certain major, avoiding a barricade by a side lane, had penetrated to a house on the north outskirts, and established himself in it with a few men. This house commanded the road by which French reinforcements were pouring in. But all the other German detachments were pushed back by the enemy, and the major and his men were captured and sent into Sedan.

By 6 a.m. the whole of the 2nd Brigade was in the fight, but no progress was made beyond the level of the church, and the detachments were hard pressed.

The whole of the 1st Infantry Division was swallowed up in the fight, except the three battalions at the railway bridge. In an effort to prevent further French reinforcements reaching the village, the division commander ordered a 6-pr. battery on to the heights east of La Moncelle, where it maintained itself for hours against great odds.

It was now 7 a.m., and the fight in the village was not going well. The mist had cleared off, and the corps artillery from the other bank were able to deal with the columns of French reinforcements as they appeared. Marshal MacMahon was wounded at this time by a shell, and was carried off the field.

The 2nd Infantry Division now arrived at the village. Its leading brigade, the 3rd, went along the park wall. Two battalions were to push to the left through the park, and attack the houses on the north side of the village; but by some mistake only one and a-half companies entered the park. The rest went on, just as a battalion had done earlier, pushed through to Petite Moncelle, and ejected thence a French detachment. The one and a-half companies were presently reinforced by four more companies, and the rest of the 3rd Brigade distributed itself along the stream from the north end of the park to La Moncelle.

If the essence of good tactics is to bring a preponderance of force at a chosen point, these scratch proceedings cannot be commended.

The 2nd Infantry Division now sent a 6-pr. battery to the heights.

The Villa Beurman was going to be the decisive point. At 7.45, two 4-pr. guns were brought into the town, and at 70 yards range battered a house near the junction of the main roads. This drove the French out, and another advance was tried to the Villa Beurman, assisted by the two guns; but only 12 rounds were fired before nearly all the gunners were down, and infantry had to drag the guns into a side street.

The 4th Brigade was now across, three battalions moving from the railway bridge, and the rest from the pontoons.

Von der Tann, assuming that there was now no chance of his being driven back across the Meuse, nevertheless needed an intact reserve at hand; so he requested the commander of the IV. Corps to send his men across. Accordingly the 8th Division and its artillery arrived on the scene, and marched towards the gap between Bazeilles and La Moncelle. The French were pouring down reinforcements in this direction in such a manner as to indicate that they might be intending to break through at this point.

Bazeilles was on fire soon after 8 o'clock, and the smoke and flames were slowly but surely forcing the French to abandon the houses; but the position of the Bavarians was becoming critical. Until the 8th Division arrived they had no supports, and many of the men in the struggle were coming to an end of their cartridges. Some of the leading Saxon troops were up by this time, and were mixed up with the Bavarians along the park and to La Moncelle. Reports began to come back from the firing line that they could not hold out unless reinforced. One of the 6-pr. batteries had been forced to retire, and things were looking bad, when the 4th Brigade brought up its last men and guns, who plunged into the fight, partly in the village and partly in the park. Two batteries climbed the heights and opened fire; and a whole Saxon brigade dashed into the fight north of the park.

The last of the 1st Bavarian Corps now arrived, and, wisely avoiding the town, joined in the fight in the park. By working thence into the town and towards Villa Beurman, the French were gradually driven out; and soon after 10 a.m., the villa had been taken and all the rest of the village.

Of the 1st Bavarians 15½ battalions and 2 guns had taken an immediate part in the fight for the town.

The troops between La Moncelle and the village now took the offensive, assisted by the Saxons, and after some close fighting drove the enemy from the heights above the park.

The Bavarians now re-formed and served out ammunition; and the corps artillery was ordered from Remilly to the heights east of La Moncelle. The Saxons who had helped now sheered off to the north to join hands with the advance of the Guards.

Von der Tann was arranging the holding of the village, and beginning to move forward from it, when the Crown Prince of Saxony requested him not to go further towards Fond de Givonne until the Saxon and Guards' attack should have fully developed.

The general therefore continued for the present with artillery alone; the 1st and 4th Brigades held the north and west outskirts, 3rd near La Moncelle, 2nd in reserve at Bazeilles.

Saxon Attack.

The first guns on the heights east of Moncelle had been a Saxon advanced guard battery, which opened fire at 6 a.m. This battery was risked in order to draw off the superior French fire from the Bavarians; it was joined an hour later by the two Bavarian batteries already mentioned. The advanced guard drove the French out of La Moncelle without much difficulty, but the combatants remained at points within 50 paces of each other, ensconced in ditches and hedges. A powerful French infantry was on the opposite crest, and from it detachments were seen making their way to occupy some solid detached houses on the slope. Two or three Saxon companies made a dash for the houses, and hastily put them in a state of defence. Here, under a hail of shot and shell, they held out, and the struggle for a long time centred round these houses. But no reinforcement could be sent; for just at this time the French 1st Corps were pushing forward strong bodies into Daigny, some of which pressed forward nearly to the Bois Chevalier,* where there was only as yet a cavalry regiment. The whole of the rest of the Saxon advanced guard had to attend to this movement, and then the leading division, 24th, of the corps had to take the same direction.

The divisional batteries lined up about 7 a.m. to the right of the three already engaged, and at once found themselves taken in flank from the direction of Daigny by

* The large wood between Daigny and Rubécourt.

gun and mitrailleuse fire. They accordingly wheeled to the right and opened fire. Soon after 8 a.m. the Saxon corps artillery arrived, and by 8.30 ten Saxon and two Bavarian batteries were firing from the heights. They suffered severely from the French musketry, but it was held necessary to endure this for the present, as the French infantry looked menacing and far outnumbered the Germans for the moment. About 9 a.m. French infantry was seen advancing on Bazeilles and La Moncelle in strength.

French Change of Commanders.

The French Army had been suffering from a change of commanders. About 7 a.m. Ducrot of the 1st Corps had taken over the command. It is probable that he knew nothing of the XI. and V. and Würtembergers crossing about Donchery, and he issued orders for a concentration of the army towards the heights between Floing and Illy. The 1st Corps and the left division of the 12th actually began the movement, and the Division of the 12th engaged at Bazeilles began to prepare for gradual withdrawal, helped by an advance of another division towards La Moncelle.

But before anything substantial was effected, General Wimpffen, the new commander of the 5th Corps, claimed the command-in-chief, and Ducrot handed over. Wimpffen objected altogether to Ducrot's plan, and issued the necessary counter-orders. All of this happened before 8 a.m.

But the advance Ducrot had ordered towards La Moncelle continued, and was conducted with vigour. The Saxon corps artillery, which had been dealing with the French guns opposite, was compelled to turn its attention to the infantry, and eventually had to be drawn back, when the French skirmishers got to within 300 yards. But the French advance was slackening, and the guns, along with fresh arrivals, taking up new positions, opened fire again, and were shortly able to resume their old ground. But the infantry in La Moncelle were hard put to it; the 24th Saxon Division was employed towards Daigny, and the 23rd had only reached Le Rulle* at 9 a.m. The help of some battalions of the 4th Bavarian Brigade, as previously described, was very welcome.

The leading brigade of the Saxon 23rd Division was to march to Bazeilles; but on its way the report came that the troops towards La Moncelle were coming to the end of their ammunition; so the brigade swerved to the right towards the north end of the park.

Skirmishing and gaining ground here along with the Bavarians, these reinforcements helped materially in clearing the way to the north edge of the town.

Also more Saxon batteries were arriving, and soon after 9 a.m. 13 Saxon and 3 Bavarian batteries were in action on the heights, the line stretching from the copse above Monvilliers to opposite Daigny.

The head of the 8th Division, IV. Corps, was now reaching the field. The advanced guard was at the railway station about 10 a.m. The 7th Division reached Lamécourt at the same time, and the Saxon cavalry division was concentrated at Douzy.

Saxon Right Wing.

We saw that the leading Saxon division, the 24th, had to move towards Daigny, as it was found that the French were across the Givonne at this point and seemed to be pushing forward. In fact, a division of the 1st Corps had crossed early. There is a road along the front of the Bois Chevalier, and the Saxons deployed along it, battalion by battalion, as they arrived from Lamécourt. Several battalions of French, accompanied by mitrailleuses, attacked in skirmishing order, and the Saxon right flank was in danger—

* Off the Map.

the more so, as four battalions there had left their knapsacks behind by order, and had omitted to remove the cartridges from them. The guns on their left were also in danger, and it was at this critical moment that the temporary retirement of the latter took place.

The situation on the road was becoming dangerous; some companies on the right had to be drawn back, with their last cartridges in their rifles, the French were edging round the right flank, and the Saxons were preparing to dispute their position with the bayonet, when reinforcement arrived, in the nick of time and at the point of danger. While a Saxon regiment arrived on the left, a curious piece of good fortune favoured the Germans. On marching off from Rubécourt, the 24th Division had detached a battalion towards Villers Cernay to gain touch with the Guards. Finding the great wood on its left more open as it advanced, the battalion entered it and proceeded north-west in company-columns. It emerged from the wood beyond the right flank of its hard-pressed comrades, just as the French were preparing to close. Without waiting to change formation, the battalion doubled forward, and the enemy retired at this point, not knowing what further surprises there might be in the wood. The battalion continued its charge and captured two guns and a mitrailleuse. The whole of the 24th Division now pushed forward towards Daigny, from south-east and east. The French made a slight stand on the edge of the ridge to save the mitrailleuses they had brought up with them, and then retired rapidly into Daigny and the enclosures about it. Six battalions, and a detachment of Bavarians who had worked up northwards from La Moncelle—to such an extent was mixing up going on—pushed on for Daigny, drove the French across the valley and got possession of the bridge and the whole of the village. The fight here now became a stationary exchange of musketry.

The Saxons whom we left in the two detached houses in front of La Moncelle were by now in great jeopardy, having almost finished their cartridges, while French skirmishers were within 60 yards of them. But help was now available; small Bavarian detachments moved up, and then a substantial force of the 8th Division crossed the stream. Bavarians and Saxons then attacked towards the hollow road about the height "635," and—but not without loss—won the road, and thus the watershed of the great spur, capturing two or three abandoned guns. But further advance here was checked by the enemy's fire from the right front, and the skirmish line alone held the edge of the heights.

It was now after 11 a.m. By 12 noon a general pushing forward of Bavarians and Saxons towards Balan had taken place; the French artillery, which had done good work in covering the retreat from Bazeilles, withdrew behind "656," and two Saxon batteries crossed the ravine at La Moncelle and came on to the heights. Detachments were forward pushing the retreating enemy through Balan and towards Fond de Givonne; 10 Bavarian batteries were beyond Bazeilles to the north-west; the interior of the village was a mass of flames; a brigade of the 2nd Bavarian Corps was below the road to the west of Bazeilles, the 4th Bavarian Brigade held the Villa Beurman and the park, with Saxon battalions and guns up the slope in front; the 1st Bavarian Brigade had been withdrawn east of the park, the 2nd, with its guns, posted in reserve between the station and the Douzy road; on the other side of the railway the cuirassier brigade and its horse-batteries; the 3rd Bavarian Brigade garrisoned La Moncelle, with the bulk of the 8th Division, IV. Corps, up the slope in front; the 7th Division still in reserve at Lamécourt; the Saxon 24th Division in Daigny, with 15 batteries along the heights behind; their 23rd Division was mixed up with the Bavarians from La Moncelle to "635." The artillery of the IV. was in reserve on the left bank of the Meuse; the bulk of the 2nd Bavarian Corps was occupying Wadelincourt and Frénois and working to Torcy, and firing with artillery across towards Balan.

When Prince George of Saxony, now commanding the Saxon Army Corps, saw that the heights opposite La Moncelle were taken, he began to draw off the Saxon 23rd Division northwards; for the Guards on his right were to go round the north of the French positions to Illy, and link on with the IIIrd Army, which was believed to be reaching Fleigneux. To assist this movement and leave no gap, it was necessary for the Saxon Corps to leave the business at Bazeilles and La Moncelle to the Bavarians and the IV. Corps.

Guards' Advance.

At an early hour the O.C. Guards' Corps at Francheval heard the firing at Bazeilles. To avoid the Bois Chevalier, through which there was no direct road, the corps had to move by Villers Cernay. The 1st Division, and the corps artillery, was to move to Givonne, with the cavalry on the right flank, the 2nd Division to halt for the present at Cernay. The advanced guard drove a small detachment from the village, and pushed back another further on. A battalion got down to the road between Haybes and Daigny; but when more troops cleared the woods and came out on the open, they were met by a storm of shell from the opposite heights. A dash was made for Givonne and part of the village was cleared of the enemy. The French say that Ducrot's withdrawal that we have noticed rendered this possible. The Guards' artillery found the roads very bad—and only came up slowly; by 12 noon, however, two batteries were firing from "1023" and seven from the ridge between Bois de Villers Cernay* and Chevalier. Sniping was experienced earlier from La Chapelle; so a flank guard took post towards that village. The Saxons not having yet decided their contest at Daigny, the 4th Guard Brigade went towards that place to help. When, at 10.30 a.m., the Saxon 24th Division, assisted materially by the fire of the Guards' artillery, had occupied Daigny completely, the said 4th Brigade dashed into the valley and captured seven guns and three mitrailleuses and made some prisoners. Meantime, the original vanguard held on to Givonne, though the French fire rained upon it from three sides.

It seems that the 1st Corps had withdrawn all but rear guards from Daigny and Givonne. Now, under Wimpffen's orders, they were coming back in strength, but the Germans had so established themselves in the valley, that the French efforts were in vain, and they had to content themselves with lining the heights to the west with guns, the infantry lying down in dense masses behind the guns.

The O.C. Guards now turned the attention of his artillery towards the north end of the Bois de la Garenne, for his orders indicated that he was to join hands with the two corps of the IIIrd Army who were working from the west towards Illy. From the height "1023" he could see the XI. Corps fighting at St. Menges. He wished to go through Givonne, for which purpose it was necessary to prepare the way with artillery.

Only the north end of Givonne was actually occupied, and about noon the French made a foolhardy attempt at this point. Ten guns and mitrailleuses galloped, in spite of the hostile fire, into the village, across the bridge and into the south part of the

* The wood between Givonne and Villers Cernay.

village. It may be supposed they wished to get in some short-range fire against the nearest Guards' guns; but the whole were captured by a fresh advance of German infantry before they had even unlimbered.

By noon, Haybes and Givonne were occupied, the bulk of the 1st Brigade was in Villers Cernay Wood, the 3rd Brigade behind the seven batteries, the 2nd Brigade in front; the cavalry division on the road to La Chapelle from Givonne.

IIIrd Army.

The first orders to the XI. and V. Corps were that they were to reach the Sedan-Mezières road from Donchery, the XI. on the right to Vrigne-aux-Bois, the V. on the left to Vivier-au-Court. It is perhaps worthy of note that the V. had to start unfed. It was soon known that the French had not come west from St. Menges, so the two corps were ordered to wheel to their right. Each of them placed its corps artillery close behind its advanced guard. To the XI. Corps, St. Menges and Floing were assigned as point of attack; the V. was to make for Fleigneux. Now, there is only one road round the bend of the Meuse, and why the French were not commanding it is a marvel. It was advisable that the artillery of the V. should come into line as soon as possible on the left of that of the XI. against the French 7th Corps on the heights from Floing to Calvaire d'Illy;* so as soon as the advanced guard of the XI. drove the French out of St. Menges, and the guns of that corps passed the defile, the infantry divisions of the corps were taken off the road at Maison Rouge to let the V. Corps guns past. These then wheeled to the left and mounted the slope to Champ de la Grange Farm, whence they trotted on, keeping as much out of view of the French position as possible, to position in front of Fleigneux. They had to borrow an infantry escort from the XI. Corps, to guard their left flank, and some cavalry regiments went there also.

v. Sections I. and II., *Plate* XVIII.

The guns of the XI. had meanwhile started a lively battle with the hostile artillery above Floing. The deep valley of the Floing rivulet runs from Illy into the Meuse, through Floing. A branch comes down from Fleigneux. Between the two streams, but mostly well back towards Fleigneux, was the position of the ten batteries of the V. Corps. The average range to the crest of the opposite heights, the French main line, was less than a mile. This mass of guns was prolonged to the heights above St. Menges by the XI. Corps guns, firing at much the same range.

When this great mass of artillery began to get some work in, preparations were made to attack Floing and the heights beyond. Liébert's Division of the 7th Corps had prepared the village for defence and done some entrenching up above in rear. Dumont's Division held along to the Calvaire d'Illy, and, in conjunction with Dumesnil's immediately behind the centre of this front line, had prepared a farmhouse and entrenched in its neighbourhood. The advance over the open from St. Menges to Floing cost men, and the village had to be hotly fought for. Above it, the slopes are steep, and the earth is held up in irregular terraces by retaining walls. The German guns beat back the French skirmishers from the edge of the slope, and the companies began to emerge on the top one by one. As this fight progressed, the commander of the V. Corps sent in the 19th Brigade against the centre of the French front as a diversion; but it is noticeable to what an extent the attack from the north was an artillery attack, the result being a comparatively small loss on the German side. The 19th Brigade was badly treated by the French fire as it came down the other slope of the valley and crossed the brook and the road; but it deployed its whole four battalions

* A point between Illy and Bois de la Garenne.

and gained the crest without much more loss. This rather took the pressure off the troops attacking Floing. Before this time the French had made several attempts to send infantry against the monstrous line of guns that was galling them, and that had driven many of their own guns and mitrailleuses into the Bois de la Garenne for shelter. As many as six battalions in a single line had tried to get forward, but had been stopped every time on the open slopes by the German guns alone. About one o'clock they also made a desperate effort from the wood towards the Guards at Givonne, but the V. Corps guns reached this movement in flank, and had a great hand in the checking of it.

There were now plain signs of the French giving way. Crowds of German companies were pushing on to the top of the ridge, when some interesting attempts were made by the French cavalry, and the unaccustomed sight of volley-firing was manifest.

Von Heydekampf.

A German chronicler of the deeds of the V. Corps thus describes these episodes :—"Two squadrons of lancers, coming out of a ravine to the south, charged on to the height, one hurling itself against the 5th and 7th Companies, the other against the 1st Battalion. Their charge was broken against the line of skirmishers, which delivered a concentrated fire; it turned aside, passed along the line, and hurled itself partly to the foot of the slope and partly through Floing. The 2nd and 3rd Companies had just established themselves near the church; they threw themselves into the courtyards and the side streets and delivered a murderous fire on the passing horsemen."

And again :—"At the moment when the companies on the top were re-forming and preparing to carry the last trench, some squadrons charged resolutely the 2nd, 3rd and 5th Companies; the last company, having formed up in time, received the charge with volleys, the other two, still scattered for the most part, by a rapid independent fire at short range. The first rank of the horsemen was destroyed; the others fled at speed, but a few of them broke through."

Immediately afterwards another cavalry charge was made; "a part of the skirmishers were ridden down, but the charge broke itself on the supports, and swerved to the right against the flank and rear of another company; but this one had by now formed square and easily repulsed the attack." A few moments later, another company stopped a charge at 400 yards.

These troops were preparing for further advance, when several squadrons attempted another charge; but this time the pace was so slow, and the German infantry so steady, and so sure of their ability to repulse it, that they coolly held their fire till the cavalry were quite close. The latter swerved to the right and came on the rear of another battalion, which coolly turned and repulsed them. This was the last attempt of the kind.

It was just after this that the 19th Brigade completed its advance, and took the farm and the trenches about it. It was now past one o'clock, and from every side the Germans were now engaged in driving back the enemy on Sedan, until, late in the afternoon, the white flag fluttered on the citadel. A glance at the map that shows the positions at noon demonstrates to what an extent the attack from the north was prepared by artillery. The losses of the different corps will show this also.

They were :—

1st Bavarians.—121 officers, 1,988 men, of which only 14 were gunners.

2nd Bavarians.—93 officers, 1,888 men, of which all but 50 were in the 3rd Division, which alone crossed the Meuse.

Thus in three divisions there was a loss of nearly 4,000, or nearly 10 per cent. of the infantry.

Saxon (XII.) Corps.—62 officers, 1,365 men, of which 48 were gunners. A great part of the Saxon infantry loss took place near Bazeilles; they lost about 5 per cent. of their infantry.

Guards.—25 officers, 424 men, of whom over 60 were cavalry and gunners. This corps was greatly assisted by the guns of the V., and was also the last corps to arrive in line.

V. Corps.—47 officers, 973 men. About 4 per cent. of their infantry.

XI. Corps.—100 officers, 1,456 men. Thus, much more than the V., owing to the close fighting at Floing.

The total loss in the whole battle was 466 officers, 8,459 men.

The French lost much more heavily in killed and wounded.

CHAPTER XV.

INVESTMENTS IN 1870–71.

Investments. I SHALL now refer to the great investments of the 1870–71 war. So great were they—I refer to Metz and Paris—that one looks in vain for any precedent with which to compare them. The famous investment of Alesia by Cæsar, when, with 70,000 men, he blockaded Vercingetorix with 80,000, is the nearest approach we can find. At Ulm, in 1805, 25,000 Austrians were blockaded, but the capitulation came in a few days Sebastopol was a remarkable case, but the investment was not complete, and the operations therefore took the form of very active siege. Saragossa, besieged in the Peninsular War by the French, was a populous place, but the number of regular troops in it did not probably exceed 25,000 at any time; and, here again, the French did not merely invest, but engaged in continuous and most fierce assaults.

In Metz the problem was to keep in a highly-trained army of 150,000 men. In Paris there were 50,000 or 60,000 trained troops of inferior quality, but in addition a vast population of great pluck and high spirit.

The task was to maintain the investment, in spite of its great extent, so as thoroughly to prevent the entry of all supplies and cut off all communication; to post the troops in such a manner as to prevent all attempts at breaking out.

In order to deprive the enemy as far as possible of grazing ground, etc., the investment line had to be drawn in as close as possible, and this also tended to shorten the line. The French, on their side, pushed out their outposts as far as possible, so that there was pretty close touch all the time, and the superior range of the chassepôt often caused great annoyance to the Germans. As far as possible, the outposts got shelter from the weather in villages; where this was not possible, huts were built; where the ground was in view of the heavy fortress guns, underground bombproofs were constructed. The picquet line had usually a strong defensive line prepared immediately in rear, but sometimes the picquet line itself was prepared as the 1st line of resistance. As time went on, a 2nd and a 3rd line of defence were prepared. Defensive works were rather a novelty for the German troops, and at first these were not undertaken with that energy and industry that the occasion demanded. The main lines of investment were marked out from the first by the supreme command with conspicuous ability, but the details of work in each section were left entirely to the troops of the section. Corps that had won victory after victory by a vigorous offensive in the open, did not at first take kindly to the work, and there were sometimes weak points where two sections joined. The technical execution of works directed by the engineers was good, but sometimes true tactics were disregarded.

But soon the Germans took to the work in earnest, and the daily labour of thousands converted villas and châteaux into fortresses and woods into impassable obstacles.

In each place, the first entrenching done was separate rifle-pits, deep enough for a

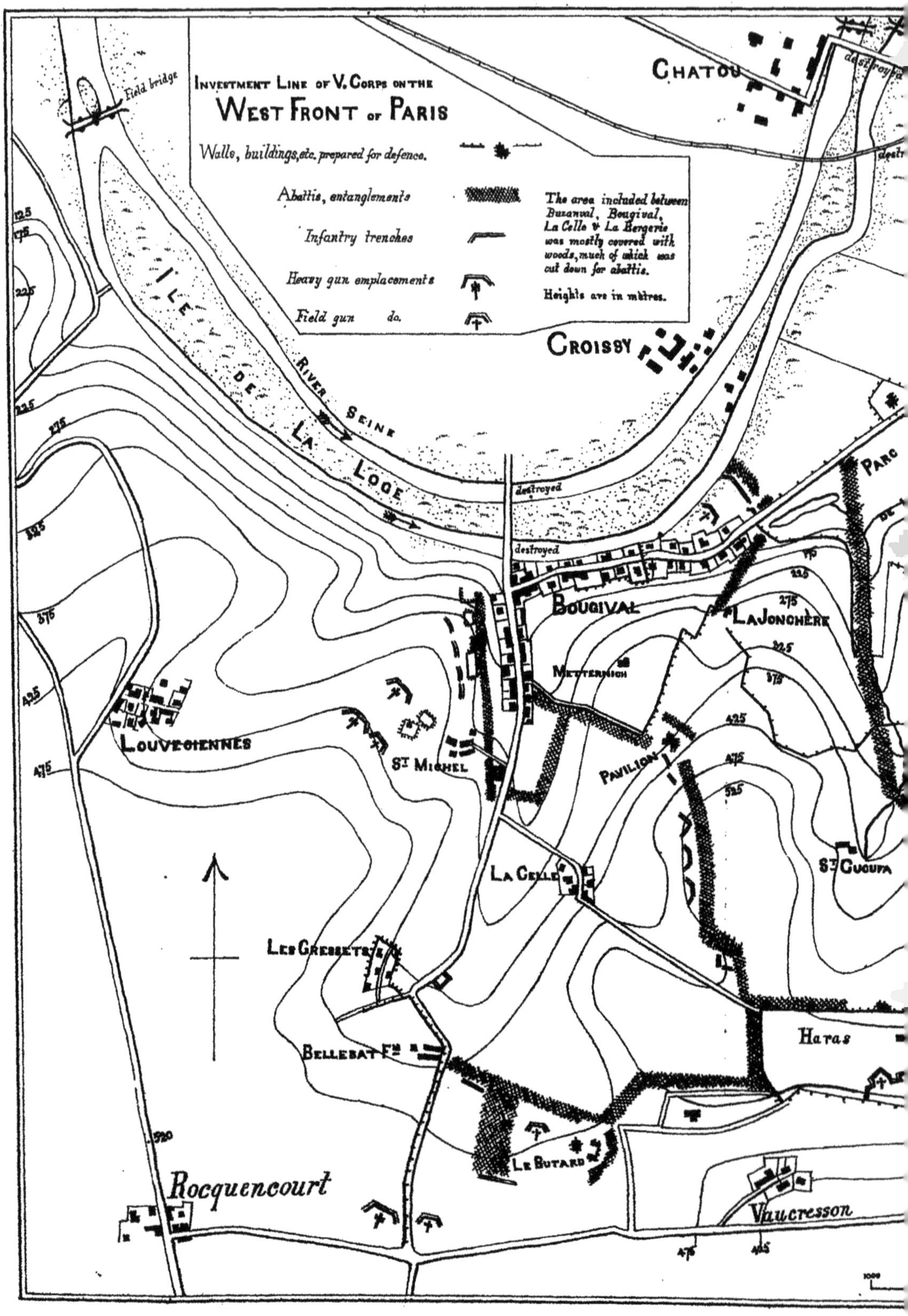
INVESTMENT LINE OF V. CORPS ON THE
WEST FRONT OF PARIS
Walls, buildings, etc. prepared for defence.
Abattis, entanglements
Infantry trenches
Heavy gun emplacements
Field gun do.
The area included between Buzanval, Bougival, La Celle & La Bergerie was mostly covered with woods, much of which was cut down for abattis.
Heights are in mètres.
Field bridge
CHATOU
CROISSY
ILE DE LA LOGE
RIVER SEINE
destroyed
BOUGIVAL
LA JONCHÈRE
METTERNICH
LOUVECIENNES
ST MICHEL
PAVILION
LA CELLE
ST CUCUFA
LES GRESSETS
BELLEBAT FM
Haras
LE BUTARD
Rocquencourt
Vaucresson
PARC

PL. XVII.

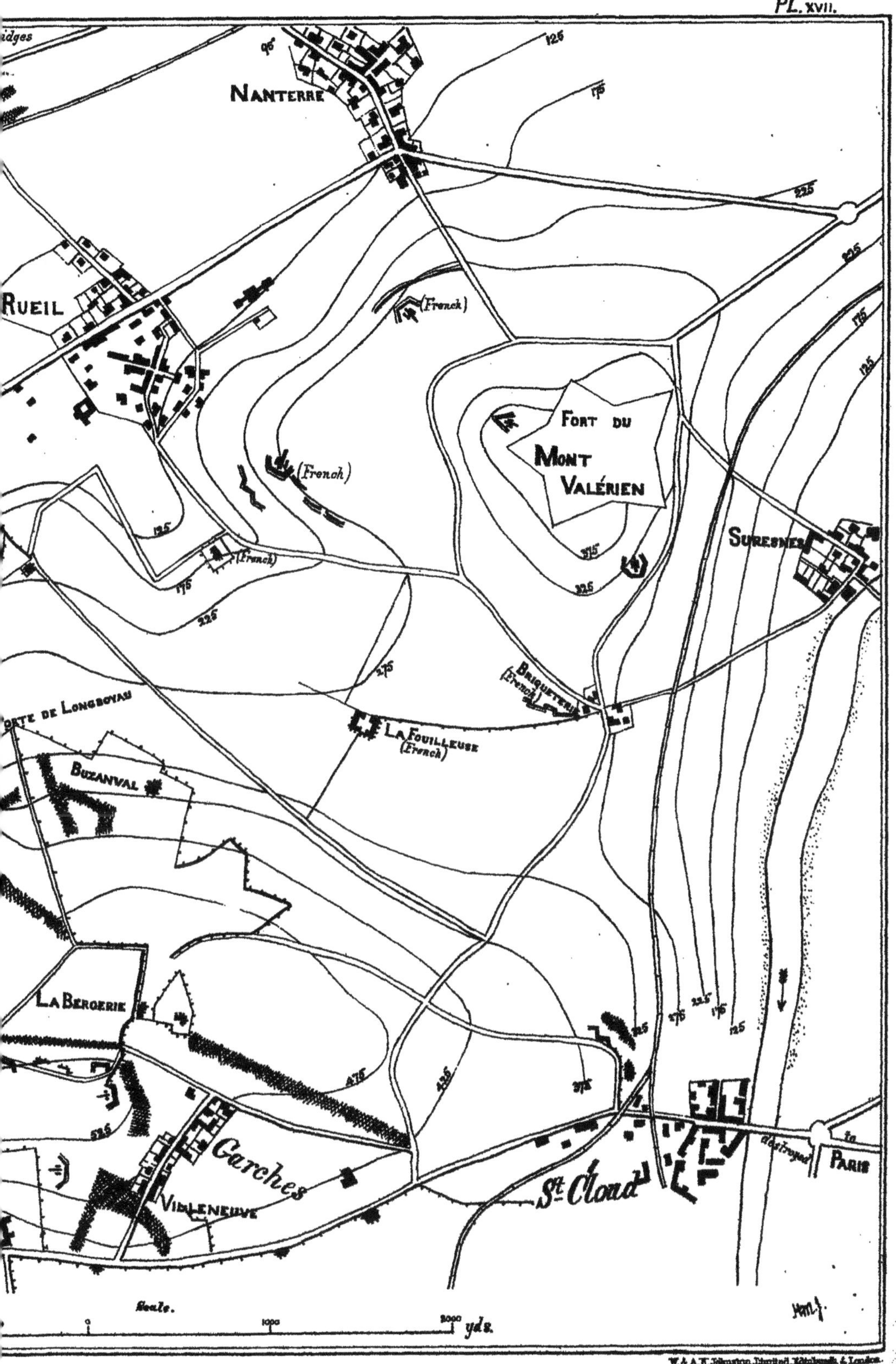

W. & A. K. Johnston Limited, Edinburgh & London.

man to stand in. These would be later joined up. The shelter-trench was usually found better than the redoubt. Field guns were still put into redoubts, but the idea grew that they were better placed in emplacements, and that shifting them from one to another gave them a better chance against the heavier guns of the fortress. Especially at Paris, the weight and power of the fortress artillery were found troublesome. At the start, Paris had 576 field pieces and 2,627 heavier guns available; private enterprise in the city kept adding to these guns of heavier weight and longer range, until there were at last 648 field guns and 3,192 heavy guns, some of these achieving ranges of nearly 10,000 yards.

Abattis played a great part in the investing works; huge belts of this obstacle were constructed, as much as 100 yards wide at places, and these were placed, as far as possible, out of sight of the French artillery, while being under close fire of the Germans, but not *in* the German defence line. Acres of *trous de loup* were constructed, and sometimes even inundation was resorted to, to narrow the ground available for sorties. Of these, the French made one large one at Metz, towards the north-east; and several large, and many small ones, at Paris. At Metz, 150,000 were inside, and 200,000 outside; at Paris, over 200,000 had arms in their hands, and 200,000 Germans kept them in. For the first eight days the Germans round Paris were short of meat; after that, there was a regular supply of everything. In Metz, the French were killing their horses for food almost from the outset. In Paris, for the last month of the siege the supply of food was of the meagrest; in eight days more there would have been absolute famine. If half a million of the non-fighting population had been transported south of the Loire before the Germans arrived, Paris would have been in better plight. The points chosen for the great sorties were not well selected. The French had no proper idea of the thinness of the investing line, which had a circuit of nearly 50 miles; nor did they ever discover where the weakest parts lay.

A plan, and description of part of the investing line of Paris, will give a good idea of the business of investment. The V. Corps had charge of the line which blocked the outlet from the peninsula on the west side of Paris. The 9th Division had the right half of the line, from St. Cloud over the heights of Garches as far as the Château de la Bergerie. This was its 1st line. The 2nd line began from the south park wall of St. Cloud, and extended with a formidable succession of redoubts, batteries and abattis past Villeneuve to the north of Vaucresson. *Plate* XVII.

The 1st line of the 10th Division was from Vacherie de St. Cucufa, by La Jonchère, Villa Metternich, to the north entrance to Bougival, and the bank of the Seine opposite Croissy.

This line was divided into two sections, to correspond to the two brigades of the division; Section I., from Cucufa Pond to La Jonchère; Section II., past Villa Metternich to the Seine.

A row of works on the plateau in front of Celle St. Cloud served as 2nd line and main position, up to Metternich Park and the ridge above Bougival.

The village of Celle St. Cloud was the cantonment for the troops of Section I. not actually on outpost duty, and Bougival for Section II.

The 3rd line of the 10th Division began at Le Butard, and passed by Bellebat, Les Gressets, St. Michel, and then followed the ridge behind the Bougival Ravine to the Seine. The Reserve was in Louveciennes.

The whole of this ground, as the map indicates, was covered with abattis, fortified houses, rifle-pits, blockhouses, redoubts, batteries, barricades and obstacles of all kinds. The greater part of the high ground between Bougival and La Bergerie, and towards Buzanval Château and the Porte de Longboyau, was covered with wood, which, however, had been cut down to a considerable extent for abattis. One colossal obstacle was the great abattis from the Cucufa Pond to the park of La Malmaison. It had an average depth of 100 yards, and had, both in front and rear, 12 rows of *trous de loup*. It was enfiladed most effectually.

Two advanced posts were thrown forward; the Porte de Longboyau in Section I., and Malmaison in Section II. At the former was a blockhouse of 40 men, at the latter only a weak detachment. The garrison of the former held its ground on the occasion of the great sortie; the latter retired to a blockhouse at the other end of the park.

The distribution of the troops was thus:—One brigade, one battery and one squadron formed the picquets, supports and local reserves for each section. One infantry regiment formed the 1st line and supports for each section, telling off one battalion for actual outpost duty. The two remaining battalions of the regiments were quartered in Bougival and La Celle St. Cloud. The outpost battalion broke up one company, 200 men, into picquets; the remaining three companies were in immediate support. Thus, 400 men acted as picquets along the front of the 10th Division, a length of 2,700 yards; and 1,200 were in immediate support, ready to turn out at a moment's notice—1,600 men to 2,700 yards. Each section had a main reserve of a brigade and two batteries; that of Section I. was quartered in Roquencourt, that of Section II. in Louveciennes.

The advanced troops were on duty for six days at a time.

The village of Rueil was a thorn in the flesh, but it could not be occupied, owing to its proximity to the guns of Mont Valérien. For this reason, also, the field battery that was at first brought into front line was withdrawn to the supports position.

In order to make the troops acquainted with the whole position, the divisions exchanged sections every month or so.

It was established as a general principle that every post, even one of observation, should hold its ground obstinately until obliged to retire by much superior force.

An officer's post, situated on the heights of La Jonchère, and supplied with the best telescopes, greatly lightened the labours of the look-out parties; but there was a great deal of thick weather, during which nothing could be seen.

The anxious time was just before daybreak. In the short winter days, anything undertaken in earnest by the French would have to begin early. A company advanced every morning some hours before sunrise from each of the sections, and advanced as far as possible to listen for movements. Just before daybreak, the local reserve paraded under arms, and remained so till the patrols reported all quiet.

The average daily number of men employed on working parties was 800, supplied partly by the local reserves and partly by the main reserves.

Much of the work was only possible at night, for Mont Valérien was only 2,500 yards from the front line, and the French advanced posts, strongly entrenched, were much nearer. Their line, facing the 10th Division, included the farm Fouilleuse, the Villa Crochard, and from there towards the railway station. Fouilleuse is barely 1,000 yards from the Buzanval Château; and Crochard about the same distance from the nearest point of the Malmaison Park. Fouilleuse has about the same elevation as La Jonchère, and rather commands the German front line in face of it.

The great importance of a skilful choice at the outset of the general trace of the investment line emerges as a most important matter. In this particular the work was almost faultless, and no great change had ever to be made later, either at Metz or Paris. Adapting of the works to the ground was usually, but not always, well done; the utmost care was taken in allotting the sections of front to separate units; a comparatively weak picquet line was found sufficient, but the first line of defence made very strong with works.

Comparatively few troops were available as a general reserve for the whole investing army—say 30,000 men out of 200,000. Counting the Reserve, 200,000 held something under 50 miles, which works out to about 4,200 to a mile, or rather less than five men to two yards; but the portion described as held by the V. Corps was more densely held than the average. At Metz, the ratio was much greater—200,000 to 22½ miles, 8,888 to a mile, five men to a yard.

The field works of the French were found to be constructed and situated with great skill.

CHAPTER XVI.

SECOND PERIOD OF THE FRANCO-GERMAN WAR.

BATTLE OF COULMIERS.

THE fortress of Metz capitulated to the Germans at the end of October, 1870. Sedan had been fought on 1st September, Paris invested during the latter half of that month. Gambetta, with incredible energy, was raising and arming forces, numerically vast, in many parts of France, and especially in the Loire region to the south and west of the capital. He succeeded in placing in the field against the Germans a force of 600,000 men and 1,400 guns.

While the bulk of these were not very formidable foes in a pitched battle, several circumstances contributed to render the task of the Germans very arduous. In spite of the bringing up from Germany of great bodies of Landwehr and Reserves, of the release of a large force by the captures of Toul and Strassburg at the end of September, the Germans were hard put to it for men; the investment of Paris swallowed up a large army, the lines of communication required guarding by very large forces now that the French nation was in arms, and it was some weeks after Metz fell that the investing force of that place was really available to help the comparatively small army covering the siege of Paris on south and west.

A great difficulty also emerged from the very nature of the struggle now beginning. The general population taking part, in the form of *Franc-tireurs*, every village and almost every homestead was a hostile post. The obtaining of information by the cavalry became almost impossible, the nature of the country south and west of Paris lending itself to irregular warfare. What Colonel Lonsdale Hale* calls the "Fog of War" became in October and November very dense, and the German commanders knew little more than that great forces were organizing behind the screen of forests and rivers.

The district of France called La Beauce is contained in the quadrilateral Chartres, Pithiviers, Beaugency, Châteaudum (*v.* inset, *Plate* XIX.). "The whole of this tract of country is a plain, intersected by a few deep ravines, forming the beds of small streams. From a military point of view such a country is well adapted for the use of cavalry and artillery, but these open plains are unfavourable to infantry, which may have to approach and attack farms and villages without being able to avail itself of cover. The employment of the three arms was considerably influenced by the nature of the country; in all the battles and engagements on the plains of La Beauce the artillery played the chief part, for it could, as a rule, open fire with effect at long ranges. More than ever had the other arms to conform to its movements. Cavalry could move unhindered in every direction, and could patrol to great distances without any anxiety

* The People's War in France in 1870–71.

as to its lines of retreat, as it could fall back in any direction it chose. Owing to the openness of the country, it was impossible for the cavalry to surprise the enemy's infantry if the latter displayed ordinary caution. The plains of La Beauce, therefore, had their drawbacks, for cavalry could not, at a distance of more than 1,000 yards, hope to make a successful attack even on raw troops, if armed, as in this case, with a good and powerful firearm. The enemy's infantry was able to hinder the movements of the cavalry by covering large spaces with a cross-fire from the numerous but widely-scattered farms.

"Our infantry had to perform the hardest and most thankless tasks in the engagements in this district. When on the offensive it had to advance without cover, and when on the defensive, it had to rely on villages and farms as its only *points d'appui.* It was impossible to make a gradual advance and at the same time to keep up a fire on the enemy's position, or to move small bodies from cover to cover. The skirmishers were forced to extend in long thin lines, while the supports could not be massed, but had to lie down or kneel in small detachments in rear of the skirmishing lines, taking advantage of every scanty cover afforded by the slight inequalities of the ground.

"The general system of tactics was also materially influenced by the nature of the ground. It was impossible to concentrate considerable forces under cover, for the purpose of striking a sudden and unexpected blow on the enemy's flank, or at any weak point in the position.

"Turning movements could be seen afar off, and parried. The general openness of the country rendered it impossible to deceive the enemy, so as to contain him at one point and fall upon him with superior forces at another. The C.-in-C.'s duties, therefore, were confined to making a correct distribution of his forces at the commencement of the battle, extending their front rather than their depth, and posting small reserves in rear of the most important points of the line, rather than forming one main body of reserve. It was seldom possible to manœuvre during a battle, and the result therefore depended almost entirely on the *morale* and numbers of the opposing forces. Modern history offers no instance of battles so completely fought out as those on the plains of La Beauce. No engagement was followed by direct pursuit from the field, as a battle terminated only with the complete exhaustion of both sides.

"The enemy was superior in numbers and the quality of his firearm; we in officers, *morale* and training."*

Beyond the open country of La Beauce, the forests imported another difficulty into the invaders' projects. Forests and villages became nests of Franc-tireurs, and the inhabitants often took a treacherous part in the fighting. The accounts of the numerous engagements, large and small, contain in many cases such words as "continued to carry on the struggle with great bitterness in the copses and farmsteads"; and then "punished the inhabitants for taking part in the fight." On many of these occasions

* This long quotation is from Von Helvig, an officer of the Bavarian General Staff. The critical reader at once sees the signs of "special pleading." In fact, this writer's comments, if not his facts, have to be accepted with caution. The battle of Coulmiers, for instance, which will shortly be described, and in which the Bavarians were beaten, can hardly be said to have been "fought out." The fact is that the Germans were getting a foretaste of "modern warfare," such as we had in South Africa. Notice for instance what Von Helvig says about cavalry in the quotation. A few pages further on, describing the engagement of Artenay, we find—"The threatening advance of our cavalry forced the hostile artillery, which had maintained its position extremely well throughout, to retire hastily."

the French fought with great vigour and courage, and sometimes with skill, but the raw state of their troops often vividly emerges, as when we read of a German battery advancing to within 600 paces of a wood full of French rifles and shelling them out of it.

The German force covering the investment of Paris on south and west was composed of the 1st Bavarian Corps (General von der Tann), the 17th and 22nd Infantry Divisions, and the 2nd, 4th and 6th Cavalry Divisions. The Bavarians and the 2nd Cavalry Division were in possession of Orléans in the beginning of November, and were linked to Paris by the Artenay road. The other portions of the covering army were at Chartres and to the north of that, and were able to give no help to the southern force in the engagement about to be described. From Chartres to Orléans is three substantial marches.

Before November the German patrols and scouts had not been much troubled by the inhabitants, but these now began to show more boldness and activity in sniping and in cutting off detachments. This, and other indications, were rightly interpreted by von der Tann as portending an early advance. Up till now, the Germans had been the attackers; now being in greatly inferior numbers, and having no distinct object of attack in the "fog," they were to experience the disadvantage of the loss of initiative.

What was known indicated that the first attack would come from between Blois and Chateaudun, *i.e.*, from south-west or west; and it was known that there were also considerable forces in the Sologne from Blois to Gien. To clear up the situation to the south-west, a reconnaissance in force was made on November 7th to Chantôme. On approaching the Forêt de Marchénoir the enemy was met in great force, advancing, and the reconnoitring force driven back rapidly with loss. This success had a great effect on the *moral* of the new French levies.

Von der Tann had now a problem before him. Were the enemy bent on first retaking Orléans, or would they separate him from the 17th and 22nd Divisions and cut him off from Paris? Should he try to hold Orléans directly, or look upon the Artenay road as the important thing?

His own available force was as follows:—

1st Division (two brigades) ...	... 5,402 infantry, 34 guns,
2nd Division (two brigades) ...	... 7,725 infantry, 36 guns,
Reserve artillery	... 16 guns, 140 infantry,
2nd Cavalry Division (two brigades)...	1,988 horses, 12 guns, 728 infantry,
5th Cavalry Brigade	... 1,069 horses,
Bavarian Cuirassier Brigade ...	... 1,098 horses, 12 guns, 688 infantry,

making a total of 14,683 infantry, 4,695 horses (including divisional squadrons), 110 guns.*

The garrison of Orléans, additional to the above, numbered 2,142 infantry, 250 horses and two guns.

* It will be noticed that these figures show a much higher ratio than usual of artillery to infantry. At three guns to every 1,000 men, this force would only have had 45 guns. The actual ratio works out to more than seven guns to 1,000 men.

While knowing he would be outnumbered, von der Tann did not know at all exactly the hostile strength. It was as follows :—

15th Army Corps.—2nd Division of two brigades and three batteries ; 3rd Division (Peytavin's) of same strength ; Reyau's Cavalry Division (three brigades) ; reserve artillery, three batteries. Total, 36,000 infantry, 3,600 cavalry, 13 batteries.

16th Army Corps of same strength, the divisions being commanded by *Admiral* Jauréguiberry and General Barry ; and there were a few battalions of Franc-tireurs. Thus General d'Aurelle de Paladines was about to attack with from 70,000 to 75,000 men and 160 guns.

To return to the problem before von der Tann. He hoped, by choosing ground that would give free scope to his artillery and cavalry, to at least repulse the French till the 22nd Division could arrive. For this purpose the holding of Orléans itself was eminently unsuitable, the town being surrounded by a broad belt of villas and garden-enclosures, where fighting would degenerate into isolated infantry struggles, and he was short of men for such a purpose. Moreover, he would lose the Artenay road.

But the total evacuation of the place was repugnant to him, for he knew it would at once be occupied by Franc-tireurs from the south ; the townspeople also had been showing an ugly spirit, and he had a large hospital there with sick and wounded. The retaking of the place, moreover, might prove very troublesome. So he determined to leave the garrison (as stated above) and to begin at once the dispatch of the trains to Artenay. The garrison was to be ready to march at a moment's notice.

The neighbourhood of Coulmiers was chosen as the battlefield. Stationed here with the whole force facing west, von der Tann hoped, by holding St. Péravy on his right, to ensure retreat to Artenay; and, by stretching his left to the Mauve, to prevent the enemy avoiding battle and slipping past to Orléans.*

Early on the morning of the 9th November, accordingly, the German strength was thus posted.

1st Division at Descures, 2nd between Rosières and Montpipeau, with a battalion forward in La Renardière and Baccon, and another in Château Préfort, the enemy having during the night begun to occupy Le Bardon. A squadron watched the road along the Loire, and on the extreme right a battalion held St. Péravy. Another battalion during the night had occupied Coulmiers and prepared it for defence. But the final distribution of the troops had to wait until the enemy declared himself.

"Coulmiers consists of stone houses and had an extensive park.† This park, at the south-west of the village, is surrounded by a thick hedge and ditch, and is the key to the defence. From Coulmiers the ground falls gently for some distance to the west, when it again rises, forming a valley, in which lies a group of houses, Carrières-les-Crottes, close to the road. Both these slopes could be swept from Coulmiers both by infantry and artillery fire. A large farmhouse, Ormeteau, lay about as far to the north of Carrières as the latter place was to the west of Coulmiers, viz., about 1,000 yards. If the enemy obtained possession of Ormeteau, Carrières would have to be abandoned,

* St. Péravy to Huisseau is 10 miles, so that von der Tann, with less than a man to the yard of front, was evidently trusting to the inferiority of the French troops, and to the expected superiority of his own artillery; but he had to acknowledge, when the day was over, the surprisingly good work done by the French artillery. As the battle progresses it will be seen that he had soon to narrow his front substantially.

† Von Helvig.

and these two places once in his hands, he could attack Coulmiers on its northern and most unprotected side, and could seriously threaten the retreat of its garrison."

When von der Tann arrived on the scene at 8 a.m., the main French advance seemed to be coming from the south-west. He accordingly ordered the 3rd Brigade to the Mauve about Préfort, the 1st towards La Renardière, the 4th at Coulmiers, the 2nd and the artillery reserve in rear at Bonneville. The 4th Cavalry Brigade and the Bavarian Cuirassier Brigade were to come down from the north to Coulmiers, but the appearance of the enemy in force opposite St. Sigismond prevented this movement.

The German advanced posts contested stubbornly the possession of Baccon, La Renardière, La Rivière, but were gradually driven back to the park of Montpipeau. During this time Carrières-les-Crottes and the other quarries were taken by the French, and the attack on Coulmiers began to look very dangerous. Finding, about noon, that Rébillard's Brigade, the extreme right of the French, was not advancing, von der Tann, seeing the state of affairs at Coulmiers and the advance of great hostile masses at Cheminiers and Champs, ordered the 3rd Brigade from Préfort into a more central position in the park of Montpipeau; but it was an hour before it could be collected, and two more before it was in a position to help either the 1st or 4th Brigades. His original plan of pushing an attack from his right south-westerly was now seen to be quite impossible, for before noon the enemy's plan became sufficiently obvious. The 2nd Brigade, from reserve at Bonneville, had to be sent north to face Jauréguiberry's Division at Cheminiers and Champs, and two of the three reserve artillery batteries to the right of Coulmiers. The 4th Brigade was extended from Vaurichard by Ormeteau to Coulmiers, five batteries were fighting on the north of the last place, and one at the south-east corner.

The Bavarian cavalry was engaged with Reyau's Cavalry Division and horse-artillery in front of St. Sigismond. The French horse advanced in great strength, as if intending to charge, but it suddenly wheeled and unmasked some batteries. After a few rounds of shell on both sides, the French cavalry retreated definitively, and took no further part in the action. It is believed that it mistook the advance of its own Franc-tireurs from the north-west (*v.* map) as the approach of the leading troops of the German 22nd Division.

Von Helvig, of the Bavarian General Staff, says:—"The improvised mitralleuse battery at Bonneville was the only available reserve. The situation could not be described as brilliant, but was not such as to make us abandon all hope. Everything depended on our being able to reinforce the 1st Brigade at La Renardière by bringing up the 3rd Brigade in time to hinder the enemy's advance in that direction. We had every reason to hope that the 2nd Brigade with its two batteries would be able to detain the enemy on the front and right flank, or even to repulse him by a vigorous attack. . . . This uncertainty and the ensuing loss of time is still more prejudicial when, as in this case, the forces acting on the offensive are four times as numerous as those acting on the defensive."

The critical period was from noon to 3 p.m., and during these three hours, as we have seen, the 3rd Brigade was engaged in assembling and marching. There is no doubt that a mistake was made in so establishing it on the Mauve that all this time was wasted. A battalion or two on the stream, and the rest of the brigade massed and ready to march at Huisseau, would have amply sufficed to prevent the French from slipping past to Orléans in any strength.

It was at 1 p.m. La Renardière and the farms round were slowly given up. The 3rd Cavalry Brigade, in this emergency, moved out to La Motte aux Taurins with two horse-batteries to fill the dangerous gap between Coulmiers and Montpipeau Park; fortunately for the 1st Brigade, which was lining the south-west edge of the park, the French did not rapidly advance beyond La Renardière, and here we probably see the effect of the comparative rawness of the French levies. The German cavalry were soon withdrawn again from a position where they were simply affording a good target to the enemy, for the hostile infantry were all among enclosures where horsemen could effect nothing against them.

Barry's (French) Division opposite Coulmiers had rather hung back till La Renardière was taken, but between 1 and 2 p.m., swarms of skirmishers advanced simultaneously and with vigour against the park, Carrières and Ormeteau. To meet this every available battalion had to get into the firing line, and a German battery joined the line itself. The devoted exertions of the five batteries north of the village enabled the infantry to hold on for some time, but the planting of two batteries at Grand Lus, and the approach of Dariès' Brigade on the south of Coulmiers, brought the park and village under a terrific cross-fire.

The three batteries on the south-east had been forced to retire,* and the park boundary was abandoned at some points. Von der Tann was beaten, and had the alternatives of concentrating and holding his ground till nightfall, or of drawing off at once and retreating north while he could. He was aware that, on the 10th, 30,000 to 40,000 of the enemy would arrive against Orléans from Gien (*i.e.*, from the south), and that disaster would be complete if, meanwhile, any part of his present opponents got round his right flank. So at 4 p.m., he ordered the evacuation of Coulmiers, and a gradual move, by échelons, of the whole force to St. Péravy, and thence by Coinces and Sougy to Artenay. This was done with deliberation and in order. But it is to be noted that during the retreat the Bavarians quite inexcusably lost touch with the French—inexcusably, because they had a large cavalry force that had not suffered. The same laxity was shown by the French on many of their numerous retreats—notably by MacMahon in his hurried retreat after Wörth. Was this due to an imperfect appreciation of the utility of keeping contact, or to something like panic? The baggage got away in good time from Orléans, but some of the train fell into the hands of the French. The hospital was abandoned to the enemy. The 15th and 16th French Corps made no pursuit worthy of the name.

The Bavarian losses are given in the Official Account at 46 officers and 720 men—about five per cent. only of the total force. It was evidently not a battle that was *fought out;* von der Tann had to consider the Artenay road. The 2nd Cavalry Division only lost 1 officer and 16 men. Now, though there was good reason for not sending it against Peytavin's Division after the evacuation of La Renardière, an opportunity seems to have been missed when Dariès' Brigade came into the open south

* It is perhaps worthy of note that, when these guns were first threatened by skirmishers, a section of the gunners, armed with captured rifles, was detached from the batteries, lay down, and kept off the hostile skirmishers for some time.

of Coulmiers, where the ground was more open, when one thinks of what Bredow's charge effected at Vionville. But the 3rd Cavalry Brigade had at this time been withdrawn behind Bonneville, and no doubt von der Tann was already thinking how useful intact cavalry would be in the retreat he foresaw.

The French had made the mistake of fighting on this front two days too soon, *i.e.*, before the troops from Gien could co-operate. With their known superiority of force, they had, even without the latter troops, a great chance of destroying the Bavarians. The detachment of Rébillard to the extreme right had, no doubt, the effect of neutralizing the 3rd Bavarian Brigade during the critical part of the fight, but it should have kept pushing forward so as to hold the latter in its place. And it seems as if their great numerical superiority should have enabled them to more completely outflank the hostile right. If a brigade had been vigorously attacking St. Péravy, where there was only a battalion and some cavalry, at the time when Coulmiers was in jeopardy, the retreat of the Bavarians would have been gravely compromised.

CHAPTER XVII.

RUSSO-TURKISH WAR, 1877–78.

By the end of the Franco-German War it had been proved beyond dispute that a great change had come over both Grand Tactics and Fighting Tactics. In Napoleon's day masses of artillery prepared the way, while the infantry would be fighting for the possession of advanced posts. Presently, the infantry columns of attack would be put in motion—great masses of men preceded by a comparatively weak swarm of skirmishers. The attack of the columns would usually be frontal, simply because this was still possible. Cavalry would also be used to aid the sledge-hammer blow.

But, by the time I speak of, there were essential changes. The cavalry dare not appear within range, except when it has to sacrifice itself to save a situation. The infantry attack, aided, as in Napoleon's case, by a great artillery, must also be helped by a flank attack. Supports and reserves play their old part, but they must be kept further back. All must succeed in getting cover, or must extend. Reconnoitring parties must go out on the flanks to give notice of turning movements. Reserves must be provided behind the flanks to resist such movements.

The idea still remained that against artillery alone cavalry had still a chance; therefore escort was necessary for the guns, even if not likely to be threatened by infantry.

Extended infantry have become the line of battle; therefore infantry training must be based on small tactical units. Only thus is there any possibility of control, and evidently a much greater individual intelligence is needed in the men. The young soldier must be taught that on him, as an individual skirmisher, depends the winning of the battle; that learning to skirmish, to make use of the inequalities of the ground, to look out for signals from his immediate commander, and to shoot coolly and deliberately, are the way to do it. Above all, that, once committed to an attack, he must never hang back or waver, that the longer he takes to dislodge the enemy the greater is his own risk of death, that if he is attacked his weapon, coolly used, can stop anything.

The mixing of units during the attack made it necessary to have practice in disorder, as a part of the proper training of both officers and men. The assiduous practice of accurate volley-firing has become of no consequence. Close order, even of small bodies, is only possible, (1) when marching to the scene of action, (2) when deploying out of range for action, (3) in movements under artillery fire, or long range infantry fire, (4) in advancing to the attack, close order for supports and reserves possible for small bodies, unless the ground is very open, (5) in the dark or in fog, for the purpose of surprise. Incidentally one may note how little of this last occurred in 1870–71.

These being the lessons from this war, one would expect that in 1877 the Russians would have learnt some of them; but Russia is usually behind the rest in such learning.

The Russians had crossed the Danube at Sistova at the end of June, 1877. The IX. Corps, about 20,000 available strength with a good deal of cavalry, passed by boat-bridge on the 10th July, turned to the right, and took Nicopolis on the 16th, capturing 7,000 prisoners and many guns. This operation had been rapidly performed and in a workmanlike manner; but the Turks were greatly outnumbered, and there had not been much scope for the display of the peculiar faults of the Russians.

General Krüdener, commanding the corps, was now ordered (18th July) to seize upon the position of Plevna, 20 miles south of Nicopolis—an important point, as being at the junction of five good roads, and being surrounded by heights which completely dominated the country to the west and south-west, from which direction Turkish movements might be expected. In fact there were already rumours that Osman Pasha was on the march from Widdin, a fortress 90 miles up the Danube. Under such circumstances one would expect Krüdener at least to discover the true situation to the south and west with cavalry. He sent six battalions, four batteries and a regiment of Cossacks by the direct road through Bryslan, three battalions, a battery and 200 horse from Bulgareni, a Cossack brigade round to the south-east of Plevna. The main column arrived at 2 p.m. on the 19th at Verbitza, and was at once checked by hostile artillery firing from above Grivitza. This column had no cavalry at all with it; its regiment was at Riben, 10 miles to the right *and rear*. Hearing the firing, the Cossacks trotted on, and got skirmishing within two and a-half miles of Plevna till nightfall. The Turks had already begun entrenching.

Plate XX.

First Assault on Plevna, 20th July, 1877.

On the night of the 19th July, therefore, the force of 6,500 men and 46 guns was scattered on a curve of 17 miles, with an enemy of unknown strength, and in an unreconnoitred position, occupying the centre. As a fact, half of Osman's Army had arrived, having made good time from Widdin. Nevertheless, Schilder-Schüldner, the Russian general, ordered an attack on all sides at daylight the next morning, but at 4 a.m., the Turks moved out a small force against the Cossack regiment near Bukova. It should be remembered that Osman's men were badly in need of rest, and that he was sparing them as much as possible; otherwise the combat of the 20th might easily have ended in complete destruction of the Russians.

The Russian attack formation was defective; a battalion of five companies deployed only one as skirmishers, the remainder advancing in two lines of company-columns, each having a quarter-company front. A regiment of three battalions advanced with two battalions in front line in this formation, and the 3rd in mass in rear. The weak skirmish line was soon overtaken by the columns, which suffered severely, being practically, closed as they were, in the very firing line.

The Turks, on their side, were too fond of their trenches; they were most lavishly supplied with ammunition and used it freely. The enemy carried 60 to 80 rounds only per man, and on this occasion all the reserve ammunition of the left wing for both guns and rifles had been left 18 miles in rear at Bulgareni, so that a vigorous pursuit by the Turks would have been most decisive.

The Turkish field artillery—steel Krupp guns—were much superior to the Russian field pieces. The Russians had eight guns to a battery, but only six in a cavalry battery.

PLEVNA JULY 30

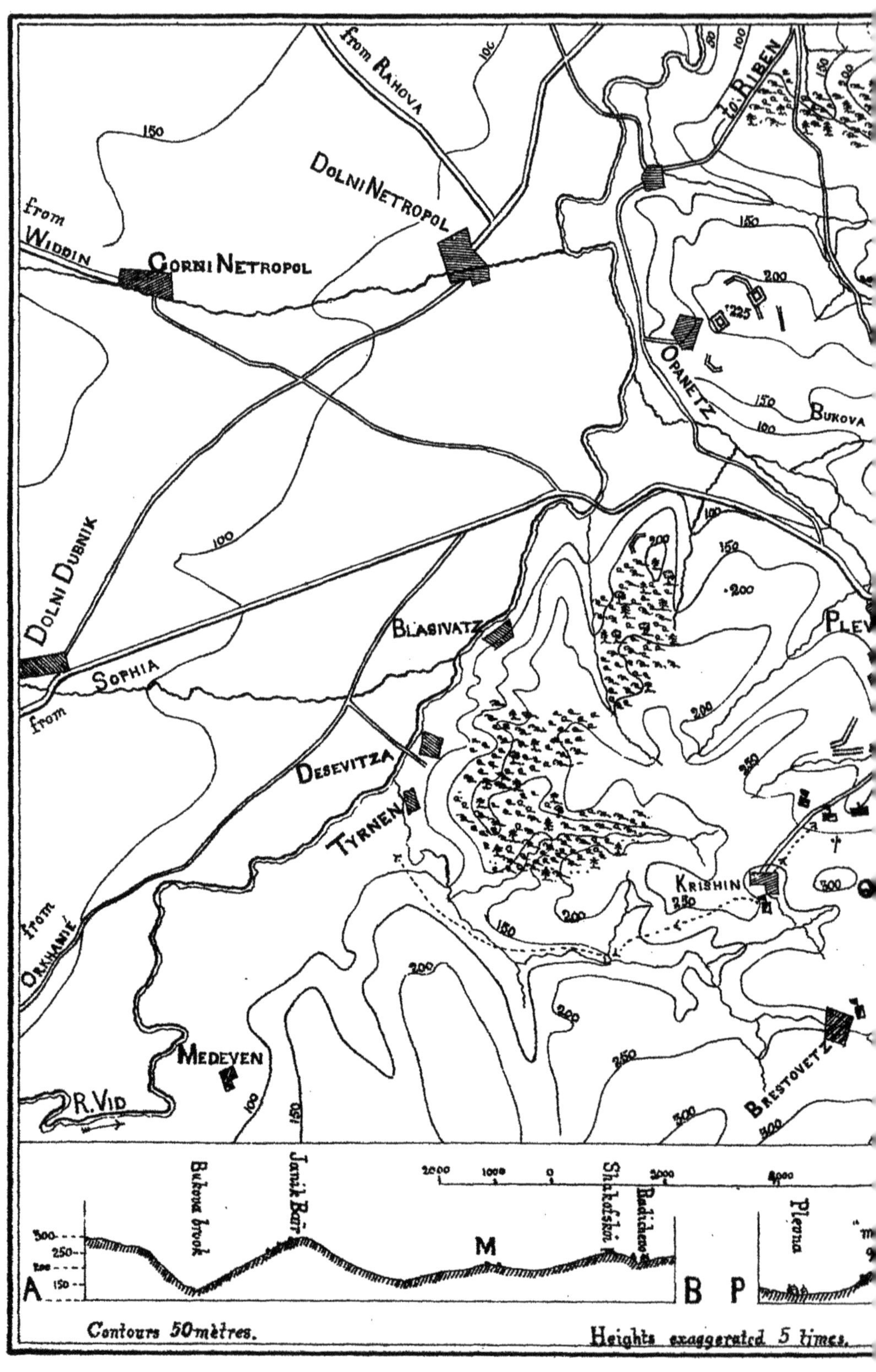

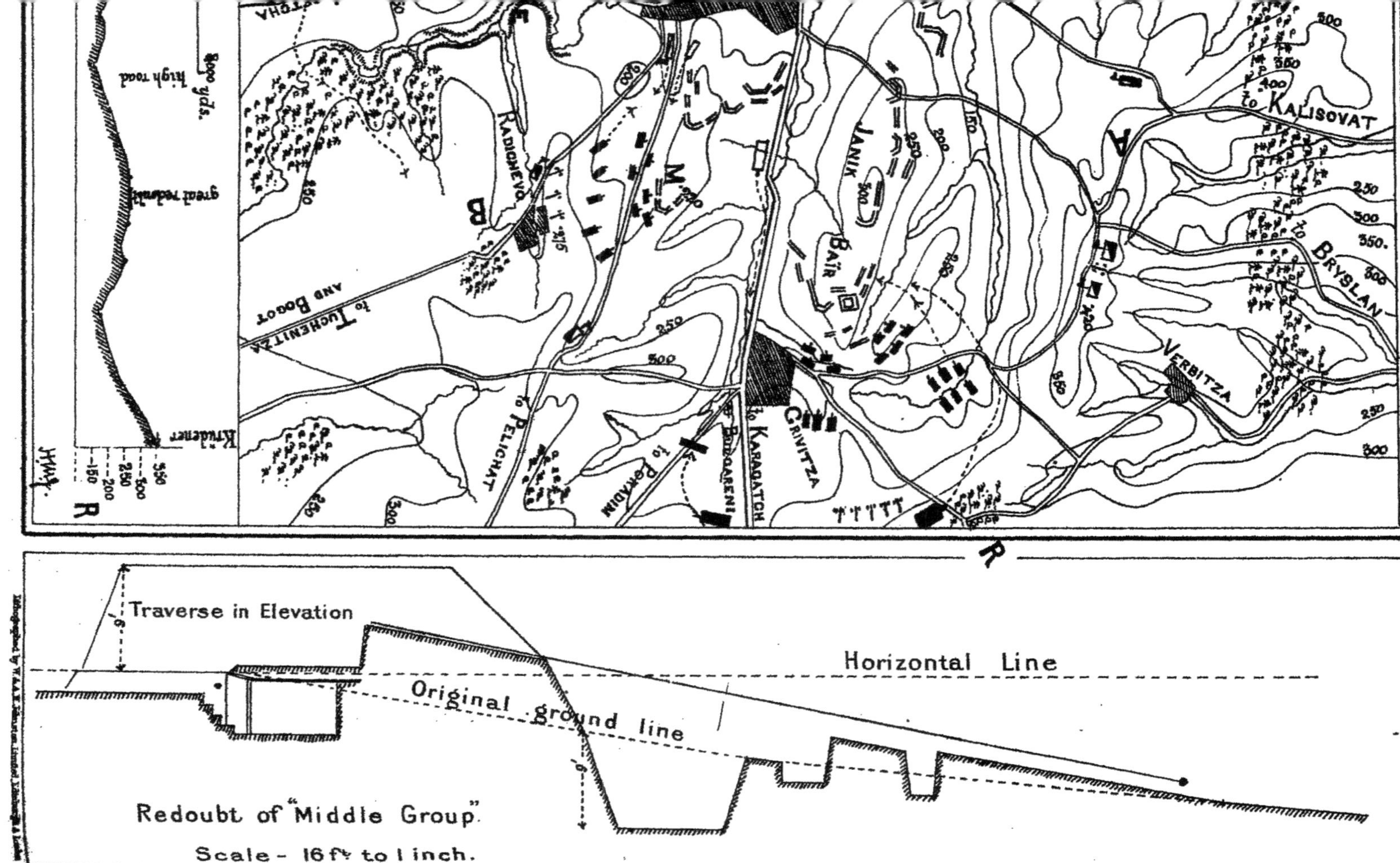

Traverse in Elevation

Horizontal Line

Original ground line

Redoubt of "Middle Group."

Scale - 16 ft to 1 inch.

Lithographed by W. & A. K. Johnston, Limited, Edinburgh & London

Proper reconnaissance would have shown that the strongest point of the Turkish position was the east end of the Janik Baïr Hill, and would have proved the absurdity of the isolated method of attacking that was put in force. Osman only deployed a small part of his total force, giving repose to the remainder; and the Russians by some skilful use of the advantages of the ground, gained, with their right attack, the west end of the Janik Baïr above Bukova; and the three battalions from the east gained the Radichevo ridge. Both these forces penetrated almost to the town, but were met by such a hail of bullets from hedges and ditches there that they were checked entirely and retired to cover. The centre attack, at 9 a.m., had failed, and the two flanks were not in sight and not in communication except by a detour of seven miles. It is said that the Turks in the centre were at this moment about to retreat, when Osman threatened to fire upon them with his reserve guns from his position east of Plevna, if they left their trenches. Before noon, when the fight was becoming stationary, the Turkish reserves formed in the town and sallied out on all three sides in great strength, driving back the enemy headlong and inflicting great loss; but the pursuit was hardly carried beyond the original front line. The Russians lost two-thirds of their officers and over 30 per cent. of their men; the Turkish loss was heavy.

In addition to the faults pointed out, it may be noted that the artillery preparation of the centre attack was far too meagre, and that no reserves whatever were kept in hand.

As soon as this defeat became known to headquarters, Krüdener was ordered to proceed with the whole IX. Corps, a brigade of the XI. Corps, and a division of the IV. Corps, and an extra cavalry brigade, the three last bodies being under the command of Prince Shakofskoi.

The Turks meanwhile strengthened the Grivitza redoubt and the lines of trenches between it and Bukova, and began a group of redoubts just east of the town.

By July 30th the fortifying of the position had proceeded to the extent shown in the plan. Some of the works were very incomplete, especially the advanced positions marked M, where there was barely cover for a man kneeling; but wherever the work was complete, the cover was most thorough, the siting of the trenches and redoubts was done with astonishing skill, and in many cases two or three tiers of rifle fire were provided.

By this date, 30th July, Osman had his whole force in the place, probably about 40,000 men, had retaken Loftcha from the Russians, and was still in a position to receive reinforcements and supplies from west and south-west, the enemy only occasionally sending across the Vid a small cavalry force, which was easily disposed of by the Turks. The Russians, since the 20th, were only occasionally seen hovering far out to the north-east while collecting their strength, as above noted. It is worth stating that Krüdener, knowing that Osman was now in great strength, did not care about the job before him, but a peremptory order from headquarters on July 28th left him no alternative. This time, however, a careful reconnaissance had been made.

SECOND ASSAULT ON PLEVNA.

The battle began at 8 a.m. by some shells fired from the Grivitza redoubt. Krüdener was moving from the following positions:—10 squadrons and 6 guns from Bryslan, 9 battalions and 40 guns from Koioulovtsi,* 9 battalions and 40 guns from Tristenik,*

30th July.

* East of "R" in Map.

6 battalions and 24 guns from Karagatch, 12 battalions, 8 squadrons and 54 guns from Poradim, 12 squadrons and 12 guns from Bogot,—the total strength being only 30,000 men, as the losses at Nicopolis and on July 20th had not been made good. The guns were 80 9-prs., the rest being of smaller weight.

Krüdener was making the same error as was made on the 20th, as he was going to attack an enemy, who must be at least equal to his own numbers, over far too wide a stretch of front. His orders were :—

(1). Cavalry on the extreme right, to move forward at 6 a.m., observe the enemy and protect the right flank.

(2). Right flank (Wilhelminof) to attack the position north of the high road (Janik Baïr Hill), 31st Division in front, 5th Division in reserve.

(3). Left flank (Shakofskoi) to leave Poradim at 5 a.m., and attack between Radichevo and Grivitza with two brigades.

(4). Cavalry on extreme left (Skobeleff) to leave Bogot at 5 a.m., and block the Loftcha road about Brestovetz.

(5). Main reserves (a brigade), under the C.-in-C., at Karagatch (ten miles from Grivitza), to be under arms early.

(6). Krüdener's headquarters one and a-half miles east of Grivitza.

The key of the position was supposed to be the Grivitza redoubt, and against this 18 battalions and 80 guns were to deploy, but room was only found for 40 guns. This main attack deployed rapidly on the heights to the east of the redoubt, and there was very soon no doubt as to where the main blow would fall. This in itself was no small advantage to Osman, whose internal communications were of the quickest and most commodious; according to a Turkish account, the Pasha commanding the north half of the defence was soon informed that the hostile movement on the Opanetz side was a mere demonstration, whereupon he brought troops along towards the right.

From 8.30 a.m. till 2.30 p.m., the centre attack was carried on solely by artillery, at 3,000 yards' range; at the outset the shooting was wild, owing to the fog which did not disperse for an hour or so.

The left wing reached the Radichevo ridge without opposition, and here they were within easy range of the Turkish "middle group" of redoubts and epaulements. The fight here was also an artillery duel till 2.30 p.m., the material damage done to each side being fairly equal.

At that hour both infantry attacks commenced. Shakofskoi's six battalions descended the slopes to the Pelichat road; as they came out of the brush at the bottom they were met by a murderous fire from several tiers of trenches. It will be seen that the point at the salient, where the redoubt was hardly begun, is a weak point. The battalions pressed on in spite of terrible losses, took the point mentioned, and carried several further trenches by 5 p.m. The battalions were now simply holding on, in a line stretching east and west. The Turks had retired in good enough order to save the guns from the captured works.

The right attack advanced its infantry in two columns against the Grivitza position, one of five battalions with six in support, coming from the north, the other of four battalions with three in support, from the east. The leading troops of the north attack carried the first trench, 1,000 yards north-east of the redoubt, were checked at the second trench, rushed it when the next battalion in rear arrived, drove the Turks across the little ravine, and approached the redoubt itself. A major and three companies reached the parapet, but were there destroyed. The regiment that made this attempt lost 29 officers and 1,006 men.

The next regiment, two battalions, then tried the same line and failed in the same gallant manner, and shortly afterwards the reserve of six battalions made an attempt from a little more to the right, but failed with heavy loss.

These attacks had been made piecemeal.

The east attack advanced just north of Grivitza village. The first battalions were stopped dead 400 yards from the redoubt; the reserve regiment of three battalions coming up on their left was checked abreast of them, by the fire of the trenches south of the redoubt. A terrific exchange of musketry fire went on in these positions till 6 p.m.

Of the reserve of six battalions, three had already been sent to help Shakofskoi, and one to the right of the north attack, where the Turks were beginning to advance.

There was no fresh power now available, beyond some cavalry and two battalions of infantry. Nevertheless Krüdener gave the order for a fresh attempt at all points. The assault was made with the utmost desperation; a general was killed within 100 paces of the redoubt, but the work could not be won.

That retirement was *ordered* by Krüdener speaks volumes for the passiveness of Turkish defence. A regiment from Sistova arrived in the nick of time to cover the retreat; "a wild and disorderly conflict in the vicinity of the redoubt continued most of the night."

Shakofskoi had practically his whole force now embroiled; he had possession of half the works on the high ground he was now occupying, but he was hard pressed by fresh enemies from the town on his left, and he was gradually being cut off from Krüdener by the advance of strong hostile reserves along the high road to Grivitza. The promised regiment from Krüdener's main reserve, marching south to reinforce Shakofskoi, crossed the head of these fresh Turkish troops, deployed to the right, got into a sharp engagement, and never reached Shakofskoi. The general gradually withdrew to the ridge from which he had started his attack in the morning, held it during the night, and marched back to Poradim next morning. The enemy had made a determined counter-attack after 5 p.m., deploying considerable bodies of horse on the flanks, and had been repulsed with some difficulty.

During all this time Skobeleff had been doing good work on the extreme left. Under cover of dense fog, and of the orchards and vineyards, he had worked forward unmolested to Krishin, and gone forward in person to within one-third of a mile of the town with 200 horse and 4 guns. From here he could see the mass of the Turkish reserves, 20,000 strong, between Plevna and Grivitza. As soon as he heard Shakofskoi's guns opening at 10 a.m., he displayed his four and opened also; at once six Turkish guns, and then many more, replied, and Skobeleff had to get back to Krishin.

Skobeleff's object was to assist Shakofskoi, and also be in a good position to harass the Turks if they were forced to retreat out of Plevna. So he established himself at Krishin, sent a squadron to find fords across the Vid, pushed out detachments towards Loftcha to guard against interference from the south, detached a squadron to the right to keep communication with Shakofskoi, and moved forward again with a battalion, four squadrons and four guns towards Plevna. From 10 a.m. to 4 p.m. he maintained himself close to the town against a far superior force of infantry and guns, who, however, did not attack with any great vigour. By that time it was clear there was no danger from the south, and he had just been informed by his right squadron that Shakofskoi was on the point of making a decisive attack. He accordingly brought up his full strength and rushed the Turks back to the edge of the town. But Shakofskoi came to a standstill, Skobeleff could do no more, and at dark retreat was perforce ordered. Cossacks, mounted or dismounted, covered the retreat. It will be seen that Skobeleff's tactics were excellent on this occasion; he kept himself in communication with the important attack on his right, and made his own chief attempt simultaneously with Shakofskoi's.

The Russian losses were 169 officers and 7,136 men.

Comments. These two fights, the 1st and 2nd at Plevna, show sufficiently wherein the Russians had failed to learn the lessons afforded by the War of Secession, and by 1870–71.

In the first combat there was practically no artillery preparation; in the second the fog greatly hindered its efficacy at the outset, and no advantage was taken of the obscurity to bring forward the infantry firing-line early and obtain possession of the ravine north-east of the great redoubt. This would have given two important advantages; the Turks would have been compelled to line their parapets from the first with rifles on account of the infantry menace, thus exposing themselves to the artillery fire, instead of lying hidden and safe—in every account we read that on neither side of Grivitza was there any infantry firing till 2.30 p.m., and when the time came for the infantry attack, the first firing-line at least would have had a shorter distance to go. Skobeleff's bold and skilful advance demonstrates that the fog *could* be taken advantage of.

The two chief attacks were frontal, were made in masses before the fire of skirmishers had effected anything worth speaking of, and the artillery, though it damaged Turkish guns and portions of parapets, had not in the least shaken the infantry defenders, for the reason above stated. The north and east sections of the Grivitza redoubt attack were made each on a narrow front, directed solely at the redoubt; this exposed each of them, as it got near, to a most deadly slanting fire from the flanking trenches. The north section certainly attacked the trenches with its last reserve and took them, but with frightful loss; and being the last reserve there available, there was no sufficient "go" left in them when they turned upon the redoubt. All accounts say that the attack of the east section, made against the redoubt, was altogether repulsed by the fire of the flanking trenches on that side, which were not captured at all. The redoubt, in fact, acted as a sort of lure; the Russian commanders failed to see that 2,000 rifles in trenches are a more important enemy than 500 behind the parapets, though more massive, of a redoubt.

Shakofskoi made a better bid for success by attacking on a broader front; thus Turkish trench-men were not able to get in a slanting fire on the columns. By his better method, Shakofskoi got some trenches and a redoubt, while he still had the five battalions of the 2nd line to resist the 1st and 2nd counter-attacks made by the enemy. To the end of the fight in fact, Shakofskoi's troops "possessed still a certain cohesion, which rendered tactical action and leading possible," while here the Turks were using some of their reserves; but Krüdener's attack used up every available regiment without accomplishing anything, the Turks meanwhile having to bring up no reserves to repulse him.

Krüdener's main reserve, as stated, was only one-sixth of the entire force; half of it was used up pretty early on the extreme right of the north attack, while the other half was unexpectedly embroiled with the Turkish reserve, as the latter was advancing along the high road. An attacker's advantage lies in his being able to choose the point that shall be his chief objective; it will often happen, as here, that he cannot tell, until he is completely committed to the battle, at which of two points it will be possible to break in; when that possibility becomes reasonably plain, the time has arrived for sending in a powerful reserve at the indicated point.

There was no proper co-operation of the three arms; the preparation, as shown, was left exclusively to the artillery; when the infantry advanced Krüdener's guns gave no help, probably because they were directly behind the advancing battalions; only a few of Shakofskoi's helped. Now the Franco-German War had clearly enough shown that at this crisis of the attack you must contrive, if you are to win, to have every gun and every rifle joining in, except the rifles of the last reserve perhaps. If necessary, the guns must shift from their original positions; above all things they must continue firing.

The cavalry, except of Skobeleff's little force, did nothing; there were 30 squadrons of them, therefore probably not less than 4,000 men. If reconnaissance had shown that the ground was not going to allow of their action, surely 2,000 of them could have been temporarily dismounted to hold ground, at least. Most of them were trained to dismounted work; half at least of the squadrons were armed with much the same rifle as the infantry. The topography was such that numerous ravines afforded shelter for the horses, especially on the north side. If these 2,000 dismounted cavalry had simply feigned attack with their horses in the ravines behind them, Krüdener would have had a further reserve of 12 battalions to his hand; if he had concealed these on the high road east of Grivitza and stayed near them himself, seeing for himself how both Wilhelminoff and Shakofskoi were faring, he would by 4 p.m. have known that the latter's attack was the more promising and would have been able to reinforce it substantially, just at the moment when Skobeleff was doing his best to help on the other flank by drawing off the Turkish reserves issuing to counter-attack against Shakofskoi. But Krüdener paid attention to the Janik Baïr attack alone, and thus lost the grip that a C.-in-C. should keep of the whole situation.

But the excellence of Skobeleff's tactics is beyond dispute. He used all his arms, kept a careful lookout for his safety while pushing forward to rashness, kept an eye on what his right flank neighbour was doing, and timed his own supreme effort to help that neighbour.

Development of tactics must surely result from every important campaign; somebody will learn it, and those who do not will be the sufferers; the Russians had learnt nothing of tactics from the immediately previous wars; the Turks had learnt nothing of strategy from them. The proceedings of such belligerents may not seem worthy of much study, but war is eminently a practical art, as Alison says; and the proceedings, however crude, of any belligerents armed with the latest weapons will surely afford lessons from the mere results.

When the War of Secession began, it was a case of two halves of a nation learning to fight—having to teach themselves the art of war—but from the very beginning there were valuable lessons to be learnt, while towards the end both belligerents were probably ahead of Europe in their practical military skill. The very nature, however, of the war—

a sort of family quarrel—and the distance of the theatre from Europe, caused most European soldiers to consider it a war of no account. The tough Turkish soldiers won battles, but their leaders had learnt nothing of the effectiveness of immediate and unrelenting pursuit. Their best general allowed his fine army to be shut in, having learnt nothing from the successes Lee achieved by keeping his freedom of manœuvring, and the collapse of Bazaine and MacMahon that was due to letting themselves be surrounded.

In the map the places to which roads lead are stated; the distance of the places from the edges are as follows:—Rahova, 40 miles; Riben, 4; Kalisovat, 3; Bryslan, 5; Karagatch, 10; Bulgareni, 18: Poradim, 7; Pelichat, 5; Tutchenitza, ½; Bogot, 2½; Loftcha, 15; Orkhanié, 51; Sophia, 90. The road that leaves the map at "R" communicates with Koioulovtski and Tristenik, distant respectively from the edge, 4 miles and 4½ miles.

On the 31st August, 1877, Osman had about 50,000 men in Plevna, and had been working hard at the entrenchments. The enemy had not yet attained this strength.

They had the 4th Roumanian Division observing from the Vid to the Bryslan road, the IX. Corps from that road to the Bulgareni road, the IV. carrying on the line to the Loftcha road, the total strength being about 45,000, the troops in the IV. Corps section numbering between 15,000 and 20,000.

Take Plevna as centre, and draw a semi-circle of six miles radius; the curve, 20 miles long, represents the Russian line, the two Russian corps having about eight miles each.

Battle of Pelichat, August 31st.

This was plainly to offer a chance to an active enemy in the centre, and Osman tried it on August 31st, against the centre of the IV. Corps. Advancing on the Pelichat road with much cavalry in front, he found a lunette close to the road and 1,500 yards in front of the village, an entrenched battery the same distance to its left front, Sgalevitza village to its right, one and a-quarter miles off, covered on both flanks by lunettes and trenches, and a wood midway between the villages held by two battalions. One and a-half miles in rear of Poradim was a reserve of three regiments and some guns.

The Turkish cavalry, having pushed back the Russian outposts, spread off to the flanks, and the infantry took the Pelichat lunette after a short, but fierce, struggle. Three batteries near Pelichat then pounded the Turks, and the lunette was retaken by infantry.

The 2nd line of Turks was not close enough to prevent this; it proceeded to extend to its right, and tried again for the lunette, but without success. Some irregular horse showed their uselessness in pitched battle by circling round into the village, looting and fire-raising, but without affecting the issue of the fight.

Meantime Turkish batteries got into position against Sgalevitza. By 10 a.m. the Russian C.-in-C. was assured that Osman was making this one attack only, so he ordered down a brigade of the IX. Corps to come on the left flank of the Turks, nine other battalions to strengthen the reserve at Poradim; but the brigade, marching badly in the great heat, was too late to join in.

By noon the Turks had 50 guns in line, and a great show of infantry was preparing to make three distinct, but simultaneous, attacks.

The Pelichat lunette was the objective of the right attack; it was repulsed to a great extent by gun-fire, but no small effect was produced by the timely appearance on the Turkish extreme right of six Russian squadrons and a horse battery, demonstrating again the uselessness of Osman's numerous cavalry, of which he had come forth with no less than 2,500.

The centre attack was directed against the wood; two battalions and a battery from the reserve at Poradim, coming up in the nick of time, helped to repulse this attack. The left attack was the most powerful and vigorous; it reached the trenches of Sgalevitza, being well assisted by artillery fire to the last moment. It was only beaten back by using the supports in an active counter-attack. In all the cases, the Turkish guns prevented pursuit. A feeble effort was made by Turkish horse to turn the right of Sgalevitza, but it was easily frustrated by a single battery.

At 3 p.m., the Turks tried again, but in rather a half-hearted fashion, preparing however for it with well-directed artillery and infantry fire. Two Russian regiments from Pelichat and one from Sgalevitza then attempted to attack in turn, but the Turkish gun-fire soon drove them in. At 4.30 Osman ordered retreat, all the Turkish guns firing steadily to prevent pursuit. Ten Russian battalions and some squadrons slowly followed for three miles, but could not get to close quarters.

The Russians lost 30 officers and 945 men.

Osman called it a "reconnaissance in force." The attack he made was purely frontal, and against the entrenched part of the IX. Corps' position. In making a sortie, which this practically was, you have no time to spare in outflanking operations. To do so widely on the right (*i.e.*, north) of Sgalevitza would have brought the attacker dangerously near the IX. Corps; and while there was no such danger on the other side of Pelichat, such a movement might, if unsuccessful, have rendered the regaining of Plevna a difficult operation.

We have seen how well the Turks used their artillery.

Recapture of Loftcha by the Russians, September 3rd.

A few days later orders were issued for the retaking of Loftcha from the Turks, who had several thousand men there and had entrenched according to their fashion. The Russians, too, were taking to the spade, for on arriving within range they fortified their right wing. Skobeleff, with the Caucasian Brigade and a few infantry battalions, was to make the left attack. The Turks had batteries on both sides of the Osma, and the infantry of the right attack suffered badly from these before they could fire a shot.

From 6 a.m. till 2 p.m., September 3rd, the artillery preparation went on from the mouth of 60 guns, the enemy, with far fewer, keeping it up manfully. Then Skobeleff captured a hill that completely dominated the town and the entrenchments on the hither bank, crossed and occupied the town. The key of the position had been correctly recognized. The right attack crossed on the other flank, and combined attacks were made on the entrenchments on the far side. These were successful, and the cavalry did severe execution on the flying enemy. The Russians lost 39 officers and 1,477 men, the Turks over 2,000 in the fight, and many in the retreat.

Here again the "preparation" was left entirely to the artillery, and the infantry of Skobeleff were made to advance 1,000 yards in the open against an enemy far from shattered. On the right the infantry was so placed, straight in front of the hostile trenches across the river, that it was compelled to attack prematurely—*i.e.*, without waiting for Skobeleff's capture of the "key."

Now the Osma was fordable at many places, and the Russians easily outnumbered the Turks; therefore outflanking from the beginning was clearly indicated.

We saw how, in 1870, German corps commanders intelligently supported each other's operations, often without orders. During the attack on Loftcha the IV. Corps, not far distant, was informed of the march of a hostile column from Plevna towards the battle, but the IV. Corps' commander did nothing.

The powerful Cossack brigade was well directed, by joining in, when the issue was no longer doubtful, on the enemy's *left* flank and rear; it cut him off from all chance of rallying upon the column from Plevna. But it is possible that the Turkish commander would in any case have retreated south-westwards as he did, in order to continue covering the Teteven and Jablonitza Passes through the Balkans. The column from Plevna also manœuvred next day to its right, *i.e.*, in the same direction, probably not with any intention of turning the Russian left—for it did not fight—but simply to warn the enemy off from an immediate march on the passes, which were not yet properly held.

Skobeleff's order, given to his troops the evening before, is worthy of note:—

"In the first part of the action the preponderating *rôle* belongs to the artillery. The order of attack will be communicated to the chiefs of batteries, who are recommended not to scatter their fire. When the infantry moves forward to the attack, the artillery will support it with all its efforts. Special vigilance is then necessary; the fire will be accelerated if the enemy should unmask any reserves, and pushed to the utmost limit if the attacking column meets any unforseen obstacles. When the distance permits, shrapnel will be used against the enemy's trenches and troops." (It seems that only the 4-prs. fired shrapnel). "The infantry must avoid disorder in the struggle, and make a careful distinction between the forward movement and the attack. Do not forget the necessity of aiding your comrades at any sacrifice. Do not waste your cartridges. Remember that the nature of the country makes it very difficult to supply ammunition. I mention once more to the infantry the necessity of silence and order in fighting. I call the attention of all the soldiers to the fact that in an intrepid attack the losses are a minimum, and that a retreat, especially a retreat in disorder, results in great losses, and in shame.

"This order to be read to every company."

Even here, one of the chief Russian faults seems to appear—a want of insistence on the importance of the *special* preparation by rifle fire.

THE BOMBARDMENT.

The Russians now brought up more Roumanian divisions against Plevna, and determined to try the effect of a prolonged bombardment. By September 8th, three of these divisions, with Russian cavalry, held the northern sector from the Vid to the Bulgareni road, two being forward towards the Janik Baïr, and one in reserve about Verbitza. The Turkish entrenchments had been greatly strengthened and multiplied on the Janik Baïr. Krüdener's IX. Corps took a

v. Plate XX.

small sector between the Bulgareni and Pelichat roads; he was to have 6 batteries and 20 siege guns in front line, along the ridge between Grivitza and the Pelichat road. The IV. Corps was to construct emplacements and trenches on the Radichevo Ridge, putting six batteries in front line. In rear of him at Tuchenitza stood a division, a brigade and 28 guns. The main reserve of 9 battalions, 30 guns and 2 cavalry regiments was posted behind IX. Corps. The Turks, noting the weakness of the advanced portion (M) of the "middle group" of works, which Shakofskoi had carried in the second assault on July 30th, had added immensely to the strength of the entrenchments here, putting also a redoubt and trenches about the 200-contour on the Radichevo-Plevna road. *Plate* XX. shows a section through a face of M as completed. It will be noted that the whole "middle group" position was commanded both from east and south, but Osman considered he had not enough troops to extend his line to the Radichevo Ridge and the "300" ridge to the east of M.

It is worthy of note that the Turks did almost nothing in the way of entanglements or any such obstacles.

Only the lightest baggage followed the Russian troops to these forward positions; numerous dressing stations were prepared, and light country carts requisitioned to carry the wounded. In fact, preparations were made for a decisive struggle.

On the night of September 6th, in profound silence and in perfect order, the materials (gabions, fascines, platforms, etc.) for the batteries were carried to within 2,000 to 2,500 yards of the Turkish works, each corps having the services of a Sapper Company. By 6 a.m. on the 7th, the emplacements were made and armed—a smart piece of work. Two 95-foot field observatories, made of ladders, were also completed.

The cannonade continued the whole day, the infantry ready for battle as close up as they could get cover. The Turkish guns replied briskly.

Next day, the 8th, the Roumanian divisions arrived, and the Russian cavalry on the north being thus relieved, crossed the Vid and moved up behind Plevna to Dolni Dubnik.

The Russians at Tuchenitza moved towards Krishin.

During the night the guns had been advanced at most parts nearer to the Turkish works, and ten light batteries were added at the shorter range. Nearly 200 guns were firing from the Russian side, chiefly at the Grivitza redoubt and the redoubt at M.

During the night of the 8th a slow fire was kept up, with the idea of preventing the Turks from mending their parapets.

The whole of the 9th was similarly spent in cannonading; by the end of the day the Russians had lost 300 men.

On the 10th the bombardment was continued; but a change in the weather to "much rain" interfered with the bringing up of ammunition, which was naturally running short. Except in the case of *chaussées,* the roads of Bulgaria soon became useless under such circumstances for heavy traffic.

It was plain that the Turks had not yet been thoroughly shaken, though their return fire slackened greatly at times; but the gradual giving out of the Russian gun-carriages, and the ammunition difficulty, determined the attacker to make his assault next day.

Skobeleff again commanded the advanced troops on the extreme left. He found on the 8th that the enemy had now a good redoubt (unnoticed before) on the height immediately north of Krishin, and several more on that tongue of land that stretches towards Plevna in a north-east direction. His 20 guns opened from behind Brestovetz on the Krishin redoubt. After some hours of firing, an advance was made on both sides of the high road to take the "second knoll" Q, two battalions advancing through the orchards in first line, with one 500 yards in rear, and two further back as reserve. Presently the "third knoll," W, was also captured, but a counter-attack of Turkish Reserves soon drove the battalions back to "Q." Even this portion of the "Green Hills" was not destined to be retained, for Skobeleff now heard that the general assault would not take place for some days, held that "Q" was dangerously advanced, and retreated to the "first knoll," K, having lost 900 men. There was something faulty in these proceedings.

Early next morning, having his right on "K," centre at Brestovetz, guns in its rear, his left in trenches to the west, and Prince Imeretinski behind with a reserve of 7 battalions and 26 guns, he was attacked twice by the enemy, who got within 60 paces of the hastily-constructed trenches, but were repulsed, chiefly by gun-fire.

On the 9th orders came for Skobeleff to retake the "third knoll," and reinforcements were sent to him. At daybreak on the 10th the "second knoll" was rushed without trouble, but the lack of entrenching tools was then apparent. Skobeleff wisely postponed the attempt on the "third knoll" till the 11th, when the general assault was to take place. During the night he had 16 guns taken across the ravine to the point "E" from which they could sweep the "third knoll."

The Third Assault.

11th September.

Elaborate arrangements were made for the great assault, some of which miscarried. All the guns were to open heavily at daybreak, there was to be a simultaneous pause at 9 till 11 a.m., to make the Turks mass their infantry in preparation for attack; at 11 a.m. the fire was to be resumed, and to stop again suddenly at 1 p.m.; then again to be resumed at 2.30, and to be continued by all the guns whose fire should not be masked by the infantry advance at 3 p.m.

Now Skobeleff wished to take the "third knoll" early, and make his attack on the Krishin redoubts at 3 p.m. The firing for this purpose, and some stir in the Turkish trenches of the "middle group," drew the advanced regiments of the centre, or Radichevo, attack to engage prematurely at 11 a.m.; these lost half their men and nearly all their officers, having taken some trenches and been driven out again.

Meanwhile Skobeleff, who had 10 battalions and 34 guns, had taken the "third knoll," and was facing, north-westwards, a line of three strong redoubts joined by breastworks, with tiers of rifle trenches on the slopes below the works. At 2 p.m., while waiting for the appointed hour, 3 p.m., he was hotly attacked and had to bring up his second line. Repulsing this attack with difficulty, he advanced across the ravine at 3 p.m., but progress up the opposite slope against the eastern and middle redoubts was stopped half-way up. Bringing up the whole of his supports, a slight advance was effected; but the Radichevo attack having failed, the western portion of the "middle group" of Turkish works turned their guns on to Skobeleff's right flank. That general had now to determine whether to throw in his last reserve, or to throw up the job. He chose the former, and the centre redoubt was captured by 4.30 p.m. Thousands of Russians in disorder sought the shelter of the work, but it was commanded from the Krishin redoubt and threatened with a counter-attack from the left. Two events combined to make this counter-attack ineffectual; Captain Kouropatkin collected 300 men of all kinds and led them out against the enemy; General Leontieff, with a large cavalry force, had been ordered across the Vid to cut the Turkish communications; passing south of Krishin at the critical moment, he perceived the situation, unlimbered his horse batteries, dismounted some squadrons, and occupied the village, which was within 600 yards of the redoubt from which the Turkish counter-attack was issuing. By 5.30 p.m. the east redoubt (*i.e.*, the one nearest to Plevna) was in the hands of the Russians.

The Radichevo attack had utterly failed, with a loss of 110 officers and 5,200 men.

Attack on the "Middle Group."

Grivitza Attack.

The right, or Grivitza, attack was made against the Great Redoubt, by the Roumanians in two columns from the north and north-east, and by part of the IX. Corps from the left of Grivitza village. The attacks were meant to be simultaneous, and were made in the same formations as on July 30th. The left wing of the Russian attack from the south lapped round the rear of the work, and got in by the weak gorge; the Roumanians worked in over the ten-feet-high parapet of the front. The Roumanians lost 56 officers and 2,500 men—eloquent testimony of the bravery with which they received their baptism of fire; the Russian Brigade lost about half these numbers. The Roumanian column from the north had an unpleasant surprise; they found that what they had supposed to be a mere subsidiary shelter-trench, 400 yards to the north of the great redoubt, had been transformed into a massive work similar to the other. This ignorance of what they were going to attack was a common feature of the Allies' proceedings, and does not show their staffs in a brilliant light.

During the night that followed, the Russians in all the captured works were subjected to repeated attacks, and worked for dear life at adapting the intrenchments for their own use; but the lack of a proper supply of tools and of trained Sappers was painfully apparent.

Another difficulty that cropped up was the rifle-ammunition supply; part of the troops were armed with the Berdan and part with the Krenka rifle, whose cartridges are not interchangeable.

12th September.

On the 12th Skobeleff was refused reinforcements, but ordered to hold what he had gained. He was attacked again and again by great numbers, Osman having been enabled, by the cessation of attack elsewhere, to bring all his reserves against the Russian left. At the 5th assault the Turks retook the centre redoubt opposite "W"; and Skobeleff, seeing that his men in the eastern redoubt were being exposed to unavailing slaughter, retired them, held on to the "first knoll" during the 13th, and retired at 7 p.m. by order to Bogot.

The Radichevo attack had cost 110 officers and 5,249 men; Skobeleff's attack, 160 officers and 5,600 men (one account says 8,000); Grivitza (Russian) attack, 22 officers and 1,305 men; total, 292 officers and 12,154 men. The Roumanian losses have been already given.

Comments.

The chief comment to be made on the "grand tactics" of these days is the inadequacy of the general reserve. It was pretty obvious that the Turkish resistance would be very tenacious, but as the assault progressed, the weaker points would appear. Thus Skobeleff was right when he asked for reinforcements, for the complete capture of the whole Krishin-Plevna position, along with the entrenched camp of the hostile reserves behind it, would have knocked the bottom out of the Turkish defence, and Skobeleff would have done it had he had 5,000 fresh men to his hand. The Krishin position dominated the town, and in that town Osman had his vast stores of ammunition and food. It should have been recognized early that the "middle group" was too strong to be taken. An attacker should not fritter away his whole strength by assaulting equally everywhere, especially when, as in this case, his numbers are not substantially greater than the defender's.

Strategically, it may be said that it should have been recognized after the assault of July 30th that Osman should be completely invested before any further attempt was made. As is well known, this course was now adopted; and herein Osman committed the unpardonable sin. He said that he remained because he hoped to be assaulted again, and was confident of repulsing any such attack with much bloodshed to the enemy; and repulsed assaults, even when the victor makes no pursuit, are not to be accepted with impunity. Such losses as 14,000 or 15,000 men in a day shake the moral of the best troops, unless something obvious has been gained by the sacrifice. This is what Osman hoped for.

The natural suitability of the Turkish position for entrenched defence is remarkable, and the ability with which the works were sited is noteworthy. For this very reason the system of keeping a great intact reserve for the chief assault is indicated. The Grivitza redoubt, though taken, did not let the Russians into the heart of the defence; the "middle group" had obviously been enormously strengthened, and approach from due east to its weakest point, the salient "M," had been rendered difficult by trenches and batteries on the reverse of Janik Baïr, that could fire up the little ravines between "M" and Grivitza. Attack at one of the extreme flanks of the semicircle of the Turkish position seems indicated. Its left flank, above Bukova, would, if gained, have brought the attackers directly above Plevna and within artillery range; but the supply of heavy ammunition at that part of the front would have been almost impossible, owing to the badness of the roads. There remains the other flank, where Skobeleff fought so well.

This general's fighting tactics, by night and by day, showed that he had learnt something from 1870; but even he confined himself too much to frontal attacks. Though the direction he took, to the "third knoll," was against the inner flank of the immediate enemy, he soon found that he was himself outflanked by the nearer works of the "middle group." At St. Privat the Prussian Guards had acted in a similar manner, and could make no headway till the Saxon Corps deployed beyond them against the extreme of the French flank. It is worthy of consideration whether Skobeleff would not have done better by directing his attack so as to envelope on three sides the Krishin redoubt first, having previously entrenched the "first knoll" to prevent a Turkish advance along the high road.

A careful student of this battle, who had also fought in 1870, says:—

"Skobeleff's method of conducting the action exhibits the traits discussed in connection with the events of July 30th, on a larger scale; careful reconnaissance of the enemy's positions; *general* preparation of the attack by concentrated artillery fire, careful formation of the infantry in the fighting line, special reserves in rear of the flanks, and a general reserve in two lines; *special* preparation of the attack by brisk fire of strong, well-covered skirmish lines, increased wherever possible by that of some batteries, or at least some guns, brought into the front line; advance of the infantry by stages, occupation of the new line reached, and *renewed* preparation by infantry and artillery fire from that point; great care for the safety of the flanks, wise husbanding of the Reserve; wherever practicable, new reserves were formed from troops previously engaged; notwithstanding the unfavourable nature of the ground for cavalry, the latter was at the right place at the decisive moment, was properly employed, and took a decisive part in the action; lastly, the indefatigable efforts in collecting scattered men, which alone made it possible again and again to lead formed and half-fresh detachments into the action." To this we may add a very important matter—that this general understood the necessity of bringing up supports just before the firing-line is beginning to waver, and the futility of reinforcing *after* the firing-line has begun to retreat.

The centre or Radichevo attack was a complete muddle from the beginning. Either no report of the premature attack was made to the C.-in-C., or the latter shirked his duty, for the local commander, not knowing what to do, held back the rest of his troops till 3 p.m.; and thus his command did not bring its force to bear in combination. The effect of its disastrous losses seems to have been to deprive the supreme command of all hope of eventual success, and to have had the curious mental effect on the C.-in-C. of shutting him out from a large and general view of the *ensemble* of the operation he was engaged in. It is a thing that has often happened, as witness General Buller at Colenso when Long's guns were abandoned; it is a thing that has caused many authorities to opine that the C.-in.-C. should not see the fighting, but sit in a telephone office far in rear.

The employment of the Russian cavalry invites comment. A large force of horse was sent across the Vid when the bombardment of September 8th began, the objects being two—to draw away troops from the eastern defences, and to cut off the retreat of the Plevna defenders, if these should abandon the place. The first object was effected, for between the 9th and 13th considerable bodies of Turkish infantry appeared on the west bank of the Vid, which shows how much Osman had in hand in his defence of his works. Large parties of Turks were also seen working at entrenchments on the right bank.

THE INVESTMENT.

On the 9th the Russian cavalry commander, Kryloff, heard of a substantial body of Turks, some 10,000, being about Gorni Dubnik, six miles from Dolni Dubnik on the Sophia road. Now, if this were true, General Kryloff had become for the moment the commander of a force covering a siege, *i.e.*, his business was (1) to verify the information, (2) to meet this body of enemies as far as possible from the locality of the siege. Thus Napoleon, to cover the siege of Mantua in 1796, marched to meet the Austrian relieving army in the Tyrol; the Germans, to cover the investment of Metz from MacMahon, moved far to meet the latter. Instead of acting thus, Kryloff took up an absurd defensive position near the Vid, thus exposing himself to be attacked simultaneously by the reinforcement from outside and the garrison of Plevna from inside. The cavalry force was from 6,000 to 7,000 strong, but Kryloff had been feebly told "to save his troops as much as possible in the execution of his task." All he effected was to delay the arrival at Plevna of the reinforcements and valuable convoy for 48 hours, the unwieldy column being not very mobile; he did not stop to fight, but took up another position from which he "observed" the column entering Plevna. And he had 6,000 good cavalry with him.

The cavalry investment, in fact, proved ineffectual; and Osman continued to obtain supplies and ammunition and news until the complete and final investment was established. It was Chefket Pasha who had successfully brought in the 10,000 troops and the convoy. He now fortified and held Dolni Dubnik, Gorni Dubnik and Telis, stages on the Sophia road, and got in again with a convoy of 2,000 wagons and 5,000 troops on October 8th; this time most of the great Russian cavalry force is unaccounted for, and the cavalry on the spot were easily disposed of by the 5,000 troops sent by Osman to meet the convoy.

If this sort of thing had been allowed to go on, it would have been a case of another Sebastopol; but on October 24th the Russian Guard and other reinforcements had arrived, and it became possible to complete a real investment of Plevna, after disposing of the fortified *étappen* stations above mentioned. The one at Gorni Dubnik was only taken after incurring a loss almost equal to the whole strength of the garrison; the tactical leading of the attacks on the place hardly deserves the name, the troops going straight at it from 1,000 yards without any *special* infantry preparation.

The eventual investment line round Plevna extended 44 miles, and the total investing army numbered 75,000 Russians, with 400 field guns and some siege pieces, 25,000 Roumanians, with nearly 100 guns—total 100,000 men and 500 guns. The Turks had 50,000 men and 100 guns. Thus the investing force had rather less than three men to two yards, a thinner line than the Germans used at Metz and Paris.

Attempt to Break Out.

Osman attempted to break out to the westward on December 10th, with 35,000 men and 60 guns. On this side of the Vid the investing force numbered 25,000 to 30,000 in infantry alone, and a lot of entrenching had been done. The operation of crossing the Vid by two bridges and some fords was done with remarkable speed and order, and compares very favourably with Bazaine's operations at Metz under similar circumstances. A force attempting thus to break out has no hostile flanks available to strike at, and has no time to spare. A rapid concentration, secretly effected, and a headlong dash are the only chance, and these Osman provided. For a moment it looked as if it might succeed, but a check occurred and the chance was gone. Meantime the works on the eastern side were rather shamefully abandoned by their defenders, and all was over.

General Comments.

To say that Plevna and other scènes of this war opened the eyes of military Europe to the value of entrenching on the battle-field is to make a true statement, while at the same time the lesson might have been learnt from the War of Secession. In 1866 and in 1870 comparatively little entrenching was done, except in an investment line; but in America and in Turkey the great utility of digging oneself under cover from day to day was soon recognized. The Russians had not learnt the lesson beforehand, so their supply of tools was totally inadequate; the Turks learnt it from the outset, owing to their numerical inferiority and, perhaps, natural aptitude. The French, outnumbered though they were, did not turn naturally to this aid, in spite of the lessons they might have taken to heart from Virginia. The Germans, in their steady triumphant progress, rather scorned the digging business.

In 1877 the Turks developed a fire-tactics of great simplicity, based on the principle that, as soon as there was any chance of your bullets reaching the enemy, you should fire as rapidly as you can. The Turks fought mostly from stationary positions, and the ammunition supply was thus easy enough; but also, when they made one of their few sustained attacks, they managed to keep the men supplied in a marvellous way.

The Turks, in fact, never wavered in their lively appreciation of two important things:—deliver as much fire as you can, and get underground on all occasions. In the latter their skill was extraordinary; defilade seemed to come natural to them; the trenches always had a good field of fire, and forward works were seen into from those in rear, and seldom afforded cover to the attacker when taken. It was only natural that the Turkish rifle-fire system should give rise to much wordy warfare, the echoes of which are still to be heard; but on the whole the idea of unlimited unaimed fire did not eventually gain many adherents. Germany, in particular, adhered to the principle of holding fire till you have a reasonable chance of hitting. No doubt, Turkish fire did some damage at long ranges and when unaimed, but the Russian attack methods, as has been shown, lent themselves to rendering unlimited shooting particularly effective against them.

The war showed that an army must in the future have an abundant supply of tools. In the advancing force, these must be carried by the men, whenever there was any chance of having to use them at short notice; the whole of these might be lost in a day, for men will drop them in emergency; and therefore a great surplus must be carried in wagons.

The Russian cavalry have by some people been excused for their ineffectual work as "screen" and as "advanced cavalry," by statements that they were outnumbered, in the first part of the war at least, by the Turkish horse. It is true the latter were nominally very numerous, but they were nearly all "irregular"—in fact, very irregular; and all accounts from observers on their side agree that for the most part these squadrons were not amenable to orders, and were therefore of hardly any value. The lesson is that the traditional "swarms of irregular horse" are not to be depended on as a substantial aid—in fact, no arm requires to be more systematically trained and more thoroughly disciplined than cavalry, while it is also true that a rapidly-raised horse, if composed of good riders, intelligent, and fired with a true patriotism, may be of very great use. This was shown in America; but the Bashi-bazouks and Tcherkesses of 1877–78 were mostly thieves.

CHAPTER XVIII.

"SMALL WARS."

THE British nation has had, during the last century, more fighting to do against savage tribes and irregular armies than against civilized troops. The peculiarity of such wars lies mainly in the fact that the difficulties of the campaign are due more to the geography and topography and climate than to the fighting capacity of the enemy, though, to be sure, the latter has sometimes been of the finest quality.

A few typical instances of such combats will accordingly be described, before proceeding to the Boer War and the struggle in Manchuria.

INDIAN MUTINY.

In the Indian Mutiny of 1857–58, our difficulties were due to the terrific heat of the season when the struggle began, to the great preponderance of force on the part of the mutineers, to so many of them being soldiers trained by ourselves in our own methods of fighting, to the great supplies of artillery and ammunition in the enemy's hands, to the fact that they seized at the outset upon the strongly-fortified city of Delhi, and to the painful lack of organization in the British Indian Army of that period. Audacity and dash saved the British cause in India—a judicious putting aside of the "rules of war," conjoined in many cases with fine tactical skill.

Thus, for instance, the strong bastioned fortress of Delhi, manned by 30,000 to 40,000 native troops, wielding a powerful artillery and disposing of unlimited ammunition and supplies, was besieged by a British force of one-quarter the hostile numbers—besieged, but far from invested, for the circuit of the fortress was such that only the north face could be attacked by the small British Army, and the enemy was able to issue at will on every other side, to threaten the flanks and rear of our lines, and to receive supplies and reinforcements at any time. Yet this great place was taken in four months, by a wonderful mixture of skill and rashness.

On May 11th, 1857, the first body of mutineers entered Delhi, having escaped from the military station of Meerut, 40 miles distant. In a few days there were no Europeans left alive in the city, and the enemy quickly set to work to prepare for its defence.

To deal with this situation there were a few British regiments about Simla, 180 miles to the north, about 1,000 faithful native troops and the Meerut garrison. These concentrated at Alipur, 14 miles north of Delhi, on June 7th—3,000 British, 1,500 native and 22 field guns. A spirited action was fought successfully next day at Badli-ki-Serai, six miles further forward, against advanced troops of the enemy; and General Barnard, wishing to occupy the chosen "ridge" before the enemy had again time to post himself in strength, pushed on at once and drove him off the ridge back to the city.

Here we became rather the besieged than the besiegers for some time, for it was

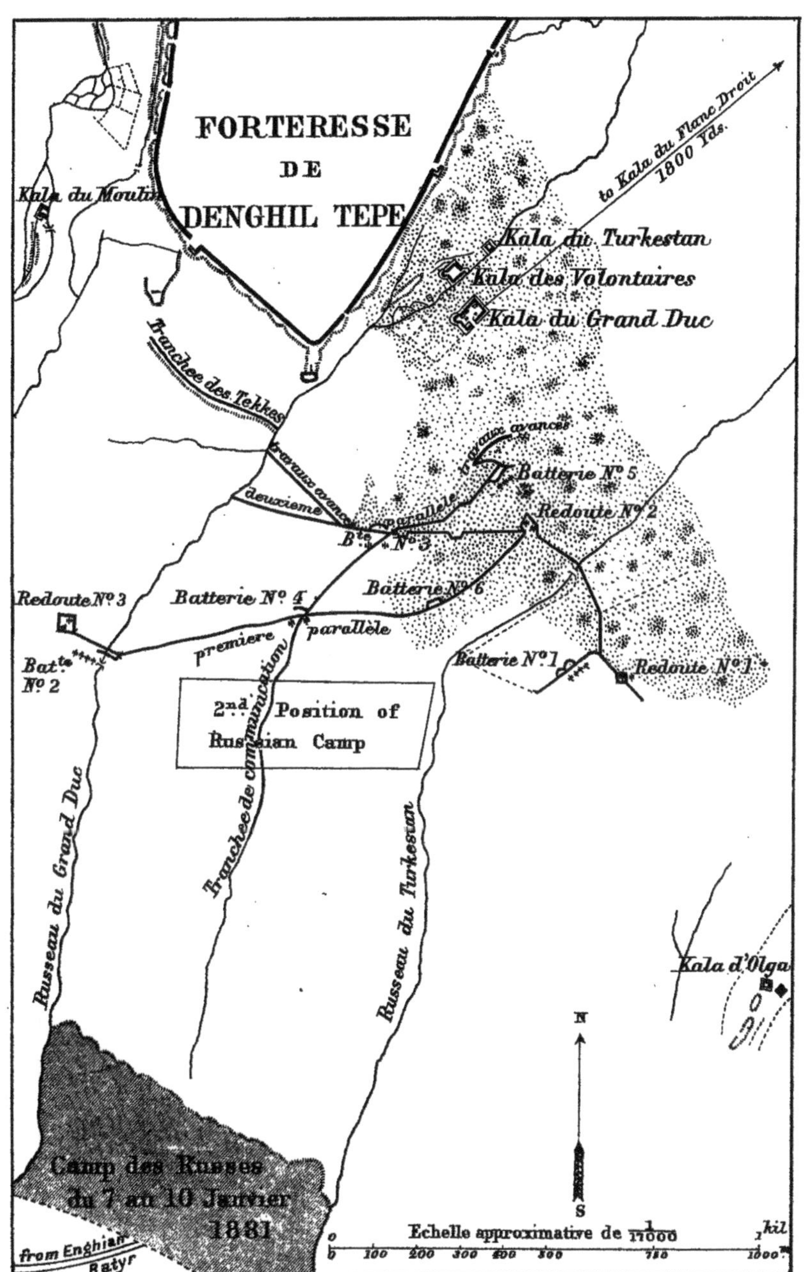
FORTERESSE
DE
DENGHIL TEPE
Kala du Moulin
to Kala du Flanc Droit 1800 Yds.
Kala du Turkestan
Kala des Volontaires
Kala du Grand Duc
Tranchee des Tekkes
travaux avances
Batterie Nº 5
deuxieme parallèle
Redoute Nº 2
Bte. Nº 3
Redoute Nº 3
Batterie Nº 4
Batterie Nº 6
premiere parallèle
Batte. Nº 2
Batterie Nº 1
Redoute Nº 1
2nd. Position of Russian Camp
Tranchee de communication
Russeau du Grand Duc
Russeau du Turkestan
Kala d'Olga
N
S
Camp des Russes du 7 au 10 Janvier 1881
from Enghian Batyr
Echelle approximative de $\frac{1}{11000}$
0 100 200 300 400 500 750 1000m
1 kil

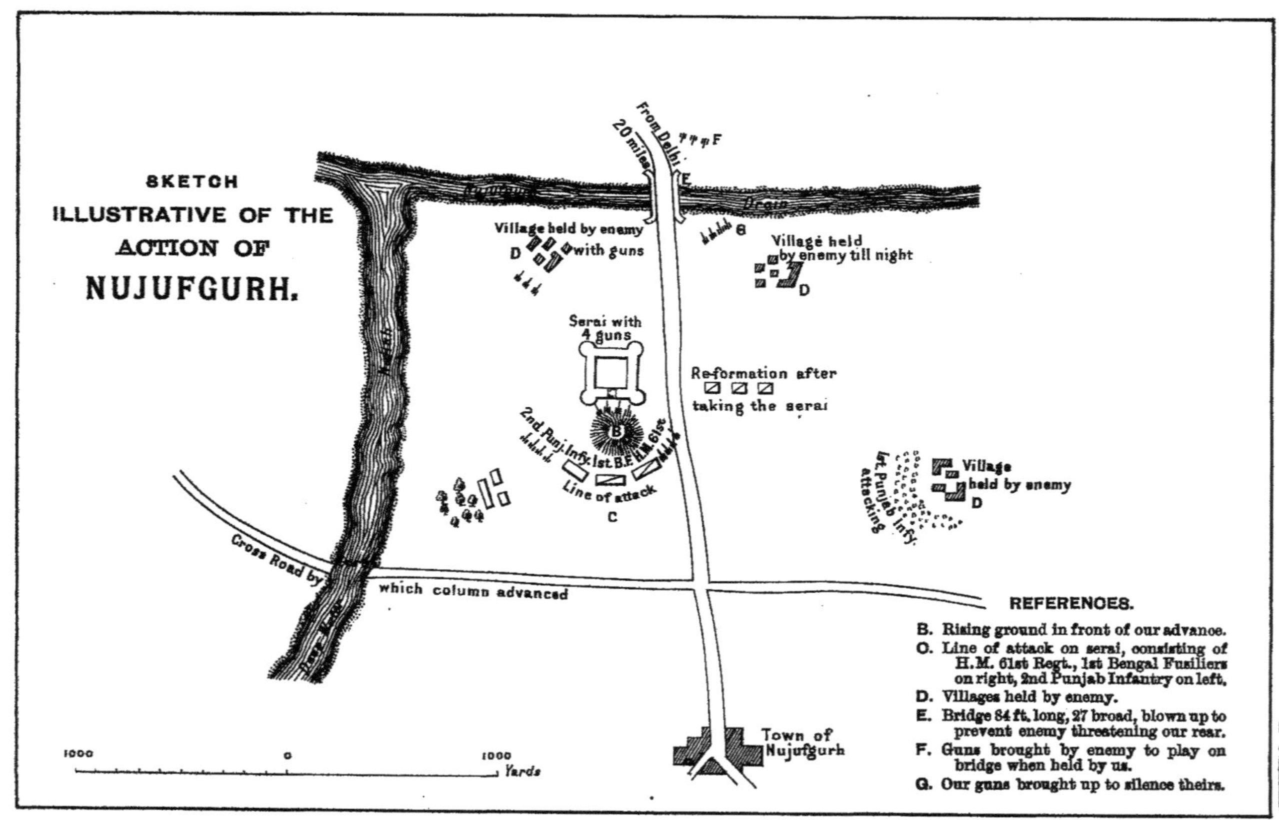

Lithographed by W. & A. K. Johnston Limited Edinburgh & London.

many weeks before the siege-train arrived. The mutineers heard of its approach, and it was in connection with an attempt of theirs to cut it off that there occurred the combat of Najafgarh, which I am about to describe.

ACTION OF NAJAFGARH, 24TH AUGUST, 1857.

On 7th August the famous Brigadier-General John Nicholson arrived from the Punjab with his column—one and a-half battalions of British infantry, one of native infantry, a battery of horse artillery and two squadrons of native horse. On his way he had saved the arsenal at Phillour, the only place where a siege-train existed, and it was following him at a long interval. On the 24th we found that a large body of the mutineers with 18 guns had left Delhi with the avowed object of intercepting the siege-train, which was now within a week's march of the ridge. Nicholson was sent out westwards next morning to deal with them, leading a force of 2,000 infantry and cavalry, and 16 horsed guns. It was the rainy season, the roads were quagmires, and the line of march was through swampy country. At noon word came to Nicholson, who was resting his men, that the enemy were 12 miles ahead at Najafgarh, and he pushed on to attack them before nightfall. Arriving about sunset at a flooded branch of the canal, he found the enemy posted on the other side, obliquely to our line of advance, their right unturnable on the main canal, in their centre a strongly-built serai, on their left a village. Behind the right, which had nine guns, was the canal bridge, and on the other side of the Delhi road another village was held; the serai had four guns. The water at the ford was breast-high, the passage of it slow, and under the enemy's fire. There was some loss in crossing, and the mutineers then retired to their main line. Nicholson, with no time for a proper reconnaissance, nevertheless saw at once that the enemy's right was safe, and that an attempt to turn his left would risk the line of retreat; so he chose the centre as the point of attack. Leaving 100 men of each battalion as reserve, to cover the ford, and to watch the enemy's right-hand village, he led the 61st, the 1st Bengal Fusiliers and the 2nd P.N.I. behind the slight rise (B in plan), with four guns on the right and ten on the left flank, and some squadrons in support. The 1st P.N.I. was sent towards the village that formed the hostile left. Nicholson addressed his three regiments:—"Hold your fire till you are within 30 yards, fire a volley, and charge with the bayonet." A few minutes of rapid artillery fire was all there was time for, and the line rose to advance, sinking ankle-deep in the swamp between the serai and the rise. The mutineers fought well, and fired grape rapidly, but the serai was taken. The enemy, seeing the British re-form and move towards the bridge, made an attempt to save the guns of their right flank, but these stuck in the soft ground and were abandoned.

v. Plate XXI.

Meantime, their left flank village was cleared, and the 1st P.N.I. moved on to the village near the canal. The quick advance of the British main body towards the bridge had cut off the defenders of this village, who now had assailants on three sides of them. They resisted with desperation, repulsed the 1st P.N.I. and killed its commander, and the 61st had to be sent to help. The place was not actually taken at all, but was evacuated in the dark.

In this episode there is a lesson, and one that has been utilized in later times. The "desperate" resistance was due to the impossibility of escape. The lesson is that in dealing with uncivilized troops, it is often advisable not to prevent all chance of escape to a force posted in a strong position. Our force in these wars is generally small, and there must be no waste of life; impress the posted enemy that you have the whip-hand of him, as Nicholson did in taking the serai, but encourage him to quit such a post as this village by giving him an apparent road to safety. Then turn the guns on him, and send the cavalry to catch him in the open.

It was now dark, and the troops bivouacked in the mud. Next morning the bridge was destroyed, and Nicholson marched back to Delhi. The mutineers made no further attempts to interfere with the march of the siege-train.

On the day of the final assault and capture of Delhi, there occurred a smart piece of tactics that is worthy of note. The attacking force was in four columns, with a reserve column. Three of them assaulted at those portions of the enceinte that had been breached, towards the east end of the north face of the fortifications. The 4th column was to try to force the Lahore Gate, near the north end of the west face. To do this, it had first to clear some suburbs on that side, in effecting which it was assailed by great numbers who drove it back; this counter-attack became very menacing and dangerous, for the reserve column had been already thrown into the main assault, it looked doubtful whether the 4th column would be able to prevent the counter-attack from penetrating right to the British position, and there were few troops there now. So a few squadrons of British and native horse were ordered to take post at a point, close to the walls of the city, from which they could threaten with an instant charge any hostile troops venturing to cross an open strip to the British lines. The mere posting of the cavalry there stopped the forward move of the counter-attack, but the squadrons suffered severely in their two hours' parade under the fire of the ramparts.

I shall now shift the scene to the Russian operations in Central Asia, and comment upon the capture by Skobeleff of the oasis of Geok Tepe in 1880.

Russians in Central Asia—1880.

In reading the accounts of such campaigns, one is struck by the fact that the preparation required is generally a much more complex and delicate affair than the execution.

The expedition of the previous year had failed.

Skobeleff started from Tchikichlar on the Caspian, about 260 miles from his objective. The first half of the route was up the valley of the Atrek; after that there were great waterless stretches; whole marches had to be made between water and water. The "struggle against nature" was going to be very severe. The enemy, Tekké Turcomans, were numerous and brave, and Skobeleff's force was not large enough to guard such a line of communications. So the expedition had to become a flying column, carrying all its requirements with it, or close in rear.

Skobeleff arrived at Tchikichlar on May 19th, and found between that place and the most advanced post 4,570 effective infantry and gunners, and 1,786 horses. He set to work to despatch forward supplies and transport, and issued most urgent instructions on the care of transport animals. Camels were chiefly relied upon, and these were bought in thousands; at the end of the expedition only about three per cent. of these had survived.

The important strategic point of Khodja Kala was reached by the advanced guard on June 11th, and the place was fortified. From here reconnoitring parties were sent towards the Oasis, and discovered from prisoners that the inhabitants were collecting in the central citadel, were fortifying it strongly and had determined on a stout resistance.

Further preparations had now to be made, and reinforcements obtained. By September Skobeleff had 4,150 infantry, 1,000 cavalry and 1,200 gunners with about 60 guns, but the further move against the main objective was not made until December, on the 12th of which month the position of Enghian Batyr, six miles from the citadel, was occupied; and again there were delays while supplies were being brought forward. *v. Plate* XXI. Skobeleff knew now that it would take him every fighting man he had to capture Denghil Tepe,* and that he would probably have to besiege it by regular process; if he once committed himself to the work he would have no men to spare to escort convoys, so he must have, at the front itself, enough of everything for two or three months. The enemy now numbered 30,000 warriors, of whom 10,000 were horsemen; they had some 5,000 rifles among them, and the great mud fortress had one 6-pr. gun.

The enceinte measured 1,400 yards on the east and west faces, 850 yards on the north, and 680 yards on the south. The huge parapet was 14 feet high, 35 feet thick, at the base and half that thickness at the top. Along the top were two clay walls, four and a-half feet high, for musketry, the exterior one being notched for firearms at every yard. The ditch varied from 7 to 14 feet in depth, with a width of 15 feet for the most part. The escarp and counterscarp were almost vertical. The terreplein was covered with thousands of tents and mud huts, and was cut up with numerous excavations to give cover. A great mound, 70 feet high, stood near the north-west corner. There was an ample water-supply from wells within the enclosure; a branch of the river was led through the place, but the Russians cut this off. None of the streams shown were formidable obstacles. The ground on the south was cut up with frequent irrigation channels, which the Russians found very useful for cover. The terreplein could be seen into from various points, notably from the top of the Kala du Flanc Droit, 1,600 yards east of the enceinte, whence a post of observation, using a heliograph, gave useful notice of massings of Tekkés preparatory to their frequent sorties. For Skobeleff had chosen the south-east corner as the point of attack, and had soon found it necessary to have resort to parallels and sapping. When the digging began, the Tekkés responded by pushing forward trenches as shown on the plan; they also occupied the Kala du Moulin, and the three Kalas† on the east side.

The labour on the siege-works came very heavy on the small Russian force, who had also to resist numerous sorties made generally at night and with great vigour. About the time the works had reached the condition shown in the plan, a night attack on them was made by 4,000 Tekkés, who charged without firing towards the Russian right. Fortunately for the besiegers the enemy had chosen the moment when the guards of the trenches were on the point of being relieved, so that twice the usual number of rifles were immediately available. The Tekkés rushed simultaneously the Mortar Battery No. 5 and Redoubt No. 2. The single gun in the redoubt had barely time to fire one round of grape; the gunners in the battery had a hand-to-hand fight. In a few minutes the Tekkés had possession of the redoubt, and of Nos. 5 and 6 Batteries. Spreading over the ground between the parallels, they threw themselves on to Batteries 3 and 1, while fresh masses marched out between Redoubt No. 1 and the Kala d' Olga, thus completely

* Tepe = high ground.
† The Kalas were redoubts of mud walls, loopholed.

turning the Russian siege-works, and threatening the communications with the camp. The defenders of Battery No. 1 and Redoubt No. 1. managed to hold their own, but the situation was extremely critical till reinforcements arrived from the camp. Then the Tekkés were gradually driven back to their fortress. The Russians had lost 2 guns, 6 officers and 121 men *hors de combat*—mostly killed, for the Tekkés fought with the sword and made sure that a man who was down was dead.

After this sortie, Skobeleff moved his camp close up to the 1st Parallel, so as to have his whole force ready at hand for the next one; but it interfered seriously with the men's repose, for rifle-shots from the enemy reached them. He also gave a very notable instruction for the guard of the trenches. As soon as a night-sortie was on hand, the rifle-men were to get out of the trenches and line up some yards in rear in the open. The Tekkés did no firing on these occasions, and the parapet of the trenches was no sort of an obstacle; its summit, in fact, gave a good standing ground for lunging at the man in the trench with spear and sword. Nowadays we should keep him off with barbed wire. At the commencement of the campaign Skobeleff had warned his officers:—"Keep in mind, as a fundamental principle, that in Central Asia close order is the thing." Skobeleff, anxious to annul the effect of the sortie on both parties, next day turned all his guns on the three Kalas, made breaches in them, and captured them all, in broad daylight, with a loss of 61; while the infantry attack was being made, every gun turned its attention to the adjoining portions of the enceinte, and there was no interference from that quarter.

The Kalas were at once organized for defence, joined to the siege-works by covered ways, and strongly garrisoned. The top of the highest of them afforded a point from which the whole interior of the enceinte was visible. Towards evening of the day following the capture, great masses of Tekkés were seen preparing for another night attack, numerous cavalry hovered about in rear of the camp, and the preparations seemed to foreshadow an attack at all points; for the enemy moved out against the Kala du Flanc Droit, and were also in force about the Kala du Moulin, as well as in the ditch of the south-east corner. On this occassion Russian guns, firing in the dark, did some damage to their own side. The chief effort was made against Redoubt No. 3, which was taken by the enemy, who, when expelled again by Colonel Kouropatkin, succeeded in carrying off a gun with them. The final repulse at all points was only effected after a loss of 150 officers and men.

It was now January 12th, losses and sickness were increasing, and provisions and munitions were diminishing. Skobeleff began accordingly a systematic breaching operation near the south-east angle; a large gap was effected, but during the night the Tekkés had completely rebuilt the parapet; so a sap was pushed forward from the Kala des Volontaires to within 20 yards of the great ditch, a shaft was sunk to below its bottom level, and a gallery mined forward to under the parapet. A ton of powder, all that was available, was put in, and preparations made for the final assault. The Tekkés had heard the underground work going on, but, knowing nothing of mines, thought the enemy were simply engaged in the absurd attempt to debouch into the terreplein through a hole in the ground.

The preparations were complete on January 23rd, and orders issued for the assault next morning. Three columns were to operate—Kouropatkin was to lead the 1st, 12 companies and some guns and Sappers, against the breach the mine would make, and he was to be ready at 7 a.m. at the "Grand Duc." The 2nd column, eight companies, was

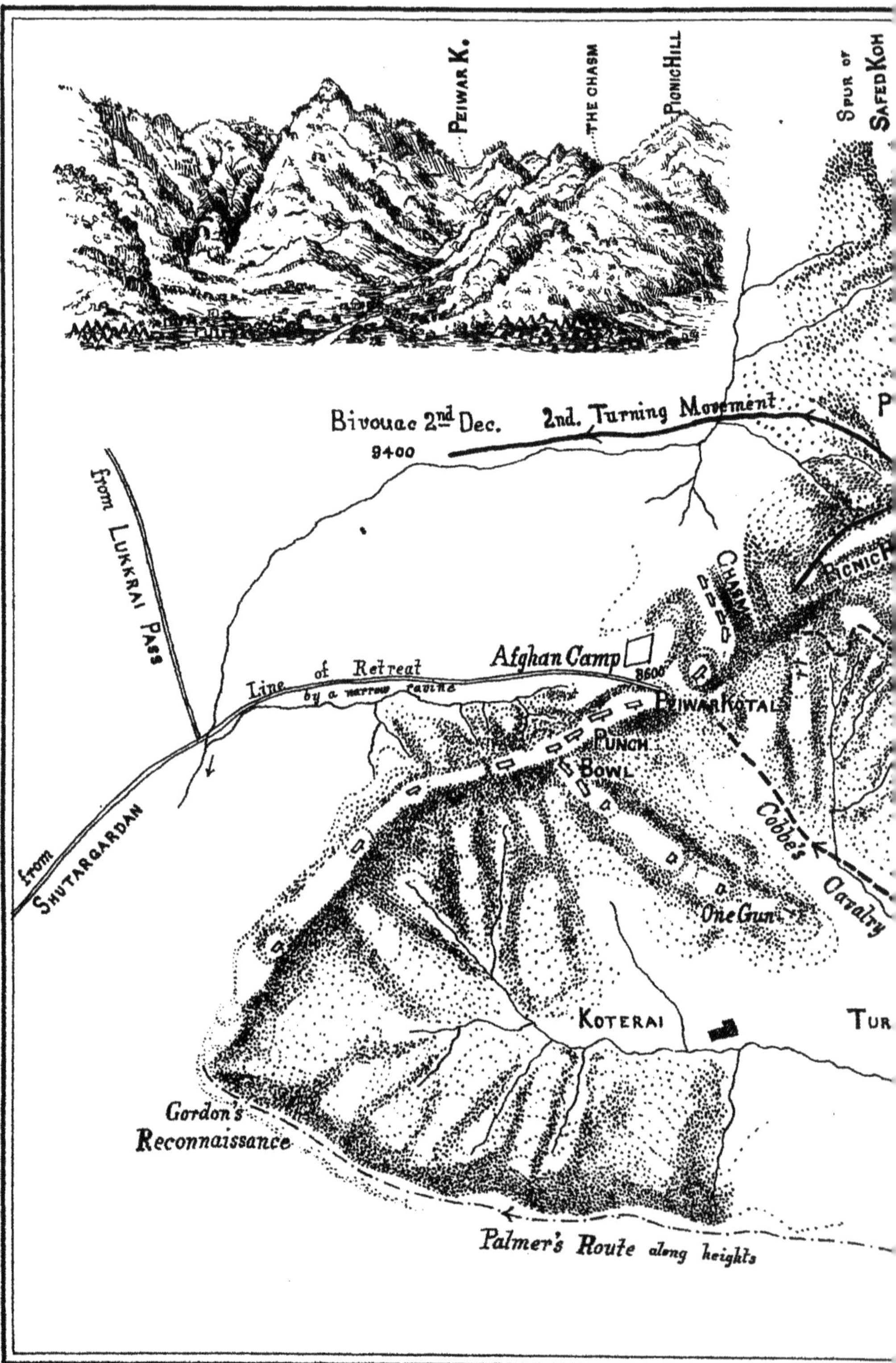

Peiwar K.
The Chasm
Picnic Hill
Spur of
Safed Koh
Bivouac 2nd Dec.
9400
2nd. Turning Movement
from Lukkrai Pass
Chasm
Afghan Camp
8600
Line of Retreat
by a narrow ravine
Peiwar Kotal
Punch
Bowl
from
Shutargardan
Cobbe's
Cavalry
One Gun
Koterai
Gordon's
Reconnaissance
Palmer's Route along heights

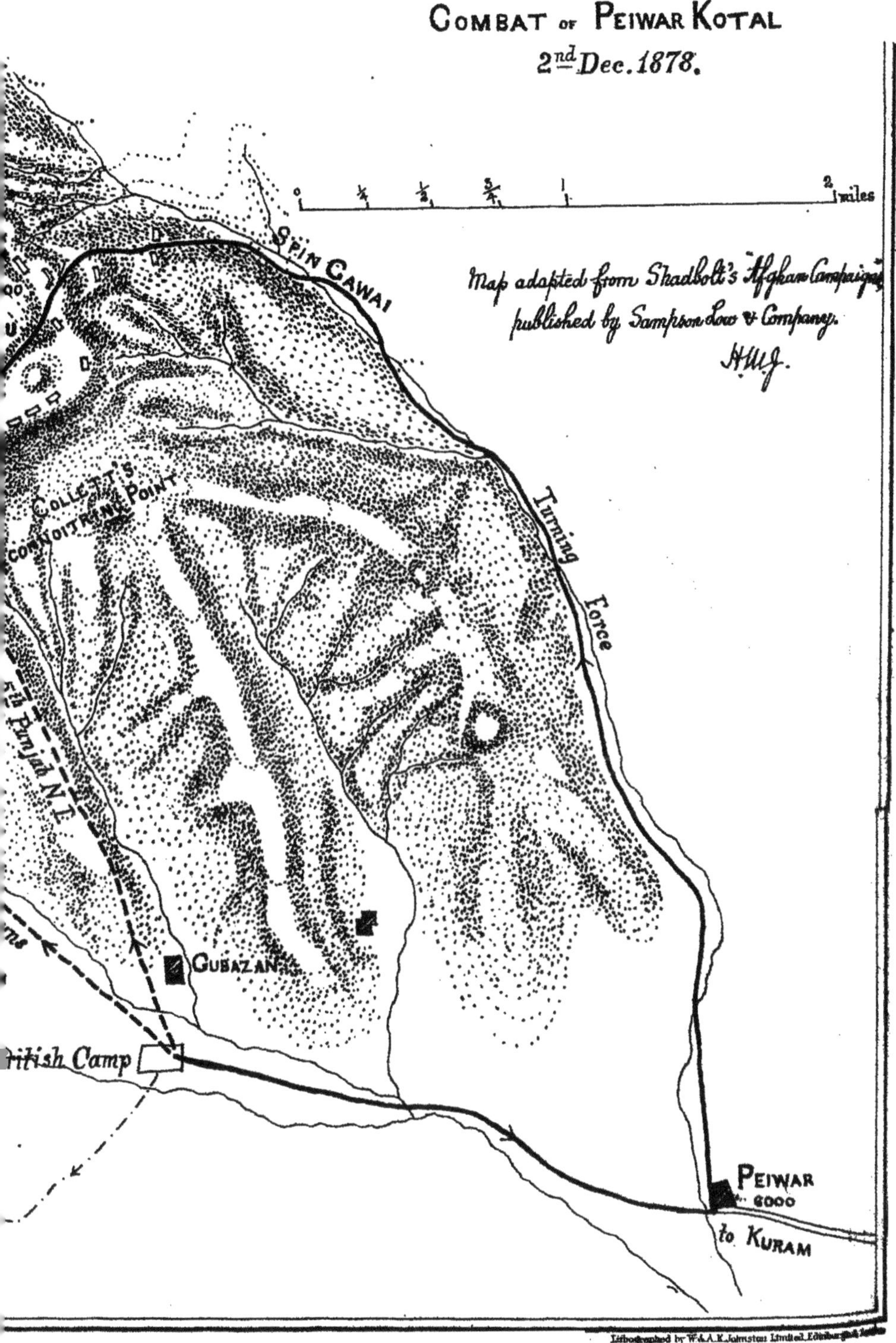

Lithographed by W. & A. K. Johnston Limited, Edinburgh & London

to take the new partial breach that the guns were to make at the angle. The 3rd column, five companies, guns, etc., was to capture the Kala du Moulin and some trenches about it, and then direct its possible of fire on the rear of the Tekkés defending the breaches. When these were taken, the 3rd column was to escalade into the place. Skobeleff kept a reserve in hand of 21 companies and 24 guns. The operation was to begin with half an hour of violent fire from all the guns, and the effect was to be signalled from the top of the Volontaires. Immediately the mine was sprung, all the guns were to fire twice, and then the assault was to be made. If the mine failed, all the guns were to be brought into a single battery at No. 5, and to direct their fire at a single point of the east face, till a breach was complete.

While the guns were doing their preliminary bombardment, the 3rd column was set in motion, and quickly took the Moulin. The Tekkés, thinking the artillery fire against the corner to be a mere demonstration, massed opposite the 3rd column; the mine, being sprung, opened a gap 140 feet long, and filled up the ditch with débris. The Tekkés were naturally thunderstruck, and many fled; but some brave thousands held their ground. The assault commenced, and the guns turned their attention to searching out the interior of the great enceinte. A very lively hand-to-hand struggle ensued, and some skilful tactical work had to be put in by Kouropatkin before a firm lodgment was effected. Sappers followed with gabions and fascines to entrench the points as they were gained. Finally, there was a keen fight for the great mound in the north-west corner and for a Kala near it. Then the last of the defenders evacuated the place by the north in full flight. Infantry pursued for six or seven miles, and cavalry for ten. The Russians had lost over 300 killed and wounded; the enemy lost in thousands.

PEIWAR KOTAL, 2ND DECEMBER, 1878.

This was an attack successfully carried out by General Roberts in Afghanistan.

The army of invasion had been constituted into three attacking columns—the Peshawar Valley Field Force, under General Sir Sam Browne, based on Peshawar and numbering 10,000 men and 30 guns; the Kuram Valley Field Force, under General Roberts, based on Kohat and numbering 5,500 men and 24 guns; the Kandahar Field Force, under General D. Stewart, based on Quetta and numbering over 13,000 with a powerful artillery.

The Kuram Valley leads to the Shutargardan Pass, one of the few practicable avenues to Kabul from the east. The route is from Kohat to the frontier post of Thal, on the Kuram River, 55 miles of fairly easy going; thence, up the river bank, the distance to Kuram Fort (8,000 feet above the sea) is about 40 miles, and the "going" was arduous. From Kuram the route leaves the main river, crosses several tributaries and spurs, and reaches the Peiwar Kotal in about 15 miles of difficult marching. The pass is at a point on a ridge jutting out southwards from the Safed-Koh and at right-angles to the latter. The Safed-Koh is the great range along the south side of the Kabul River. Where the Peiwar ridge leaves it, the Safed-Koh rises to 15,000 feet; between this point and the Peiwar Pass is the Spin Gawai, very difficult, but affording a means of turning the Peiwar, if it should be held in force by the enemy.

Plate XXII.

Arriving at Kuram on November 24th, General Roberts made a reconnaissance in force against the front of the position. It was found that the Peiwar was held in great strength, and that entrenchments and obstacles abounded. Therefore the general resolved to halt and rest his men for a few days, which would give time also for a more extended reconnaissance, which might result in the discovery of a means of turning the position. The reconnaissances then undertaken showed that the enemy held a mile or more to the proper right of the pass, and about two miles to the left, and that they had defences and troops at the Spin Gawai; but, without force, it was not possible to ascertain whether, the Spin Gawai being surprised, the advance thence along the ridge to the Peiwar would be difficult or the reverse. The east side of the ridge was very precipitous, and the whole covered with dense pine forests.

The plan was explained to commanding officers on December 1st, the camp having been moved forward as far as possible (*v:* plan) so as to deceive the enemy. The general's intention was to throw the bulk of his troops at the Spin Gawai by a secret night march, the remainder to do their best in front as soon as the flank movement should begin to take effect.

General Roberts himself commanded the turning force, which consisted of 72nd Highlanders, four native regiments, a mountain battery and four field guns on elephants, together with a company of Sappers. The remainder, under General Cobbe, numbered about 1,000, besides Major Palmer's native levies, who proved untrustworthy.

Silently, and leaving camp standing, the turning force marched at 10 p.m. on the 1st December. The march was short, but difficult; it had been intended that there should be a three hours' rest just before touch was gained with the defenders of the Spin Gawai; but as that point was reached two unknown men in the 29th N.I., the leading regiment, fired their rifles. Fearing this might mean treachery—a warning to the Afghans—General Roberts halted the column, retired the 29th N.I. to the rear of the column, and pushed on. The enemy had not, after all, been aroused by the shots, and the leading battalion, Goorkhas, reached the foot of the pass at 6 a.m. They reached the first barricade unobserved, extended round it, and carried it without much difficulty; the defenders, having fired one volley, were bayoneted at their posts. The Afghans were now completely aroused, and two more stockades at intervals of 300 or 400 yards had to be taken by the Highlanders and Goorkhas. The Mountain Battery, fighting well in the front line, lost its commander, Captain Kelso. Bodies of Afghans tried a move against our right flank, but the 29th N.I. retrieved its reputation by repulsing them in gallant style. By 6.30 the top of the pass was in our hands, and our men were ready for the rolling-up movement towards the Peiwar. An hour later a point was reached whence visual signalling was opened up with General Cobbe, and he was instructed to begin his frontal attack.

General Roberts now led the advance for the clearing of the forest-clad ridge, the field guns having been brought forward by the elephants. At every step the resistance became more and more obstinate, as each body of Afghans fell back upon the next in support. When more than half-way to the Peiwar, a precipitous-sided gully was reached, on the opposite side of which the enemy were posted in great strength. The 29th N.I. and the 2nd P.I., now in front, made two attempts on this position without success, the Afghans each time counter-attacking with great spirit. One of these advances was only finally disposed of by bringing up the Highlanders and Goorkhas again into front line; but even then, no perceptible advance could be made.

While this was going on, Cobbe had opened fire at 6.15 with artillery upon a hostile gun that was seen to command the approach to the pass, infantry moving to a position under cover, ahead of the artillery. The 5th P.I. felt their way forward along the spur to the right front, when the signals from General Roberts were made out; by 11 a.m., after a determined duel with the opposing artillery, that of the Afghans having an enormous superiority of position, two of the latter's guns on the left of the pass were observed to retire. Our infantry had been steadily stealing forward, and were by noon 1,400 yards from the summit. The 5th P.I. had made the most progress and were in touch with General Roberts's men, and joined them in attacking the gully. By 1 p.m., some companies of the frontal attack had gained the summit of a steep ridge, whence they could fire, at 800 yards range, at the Afghan main batteries in the pass. The latter, however, held on with great gallantry.

By 1 p.m. General Roberts recognized that further direct advance on the Peiwar was barred, and he began a second turning movement; but his men had been marching all night and fighting since 6 a.m., so an hour's halt was ordered and a meal partaken of. About this time two of Cobbe's guns, shown in the map near the end of the route of the 5th P.N.I., found a point from which they could fire into the Afghan camp.

The second turning movement was continued at 2 o'clock. The 2nd P.I. held the ground opposite the gully, the 29th retired to guard the Spin Gawai and the field hospital that had been established there, the Highlanders, three native regiments and the mountain and field guns moved so as to strike the Afghans' line of retreat.*

The effect on the enemy was at once apparent; and the commander of the frontal attack, finding resistance slacken, pushed on and was in the main position by 2.30 p.m.

The final break and flight was effected with the greatest precipitation; they left their dead behind, 18 guns, and vast quantities of stores and ammunition, which seem to have been there accumulated with the expectation of holding the position throughout the winter. Our native cavalry captured some more guns during the pursuit.

Major Palmer's native levies could not be induced to run any risks, and had no hand in the result.

We had 2 officers and 19 men killed, 72 wounded. The enemy's force amounted to 4,500 regulars, and a large number of tribesmen. The portion of our turning force that actually fought numbered 2,263, in the flank movement, and about 800 in the frontal movement. The Afghans had muzzle-loading rifled Enfields, an excellent weapon; their artillerymen were well trained, the weapons being mountain guns and some 12-pr. howitzers. Our force had breech-loading Sniders, mountain guns and muzzle-loading field guns. The Afghans, fighting behind barricades made of huge tree-trunks, were not so much handicapped by the muzzle loading of their rifles as they would have been in the open.

It was now seen that the Afghan position had been one of enormous natural strength, both in front and to their left of the pass, and that their preparations had been most skilfully made. But for the second turning movement, the position could not have been taken *without the certainty of severe loss.*

In these three cases that have been selected out of "small wars," the conditions as to tactics were entirely different from each other.

At the time of the Indian Mutiny, both sides were armed with muzzle-loaders,

* The route is shown in the Map from "Two guns" to "Bivouac 2nd December" and "Camp 3rd December."

the enemy was composed of troops trained and drilled by ourselves in our own mode of fighting in line, with skirmishers to begin the fight. At Najafgarh, as has been pointed out, night was approaching, the enemy's right flank was unturnable, his left flank was too far off from our ford, and, moreover, the turning of it would have no effect in threatening his retreat. Therefore, Nicholson brought his chief attack against the serai in the centre, which covered the only bridge the enemy had in rear of him. The latter erred in squandering his force by holding the distant left-flank village.

In Skobeleff's operations, close order was the indicated formation for receiving attack, and for attacking, simply because the enemy's fire was less to be feared than the sweeping rush of his sword-men and spearmen.

In both these cases, then, the fight is in close order, the ground being flat and open; but at the Peiwar Kotal, the problem was entirely different. If the civilized troops now had breech-loading rifles and rifled artillery, so had a considerable number of the enemy, and his shooting capacity was about equal to our own. Extended order was therefore indicated, and the formidable difficulties of the ground made a surprise flanking movement almost necessary. It is worth while to observe the position on the map of the British camp, where it stood for some days before the attack was made. Being well forward from the village of Peiwar, where the routes to the Spin Gawai and the Peiwar Kotal diverge, its position would tend to relieve the Afghans of some of their anxiety for the former avenue to their left flank. And, in fact, the actual taking of the Spin Gawai was effected with little trouble, doubtless because some of its original defenders had been withdrawn when the advancing of the British camp to the position shown had seemed to them to lessen the probability of attack by the Spin Gawai.

An uncivilized army has usually been found easy to surprise, the minds of such people, unaccustomed to abstract thought, being ill-adapted to the mental effort of imagining the unseen. On the field of battle, on the other hand, the leaders often show a remarkable tactical ability.

A purely defensive attitude employed against a powerful Afghan force 18 months later at Maiwand led to disaster. The Afghan army was executing a march across the front of a British brigade, about 2,500 strong. The latter marched out on to a flat, stony plain, near enough to the enemy to compel him to halt for battle, and then halted, allowed him to develop from line of march to line of battle without molestation, and was quickly enveloped by the greatly superior numbers of the enemy. The Afghan artillery was numerous and good, and was able to bring a concentric fire upon the brigade, posted in the open without a vestige of cover. But nothing but purely passive defence was attempted. Our gunners, however, plied Ayub Khan's Herati regiments so effectively that they twice recoiled, and later information shows that they would have fled at that moment if we had attacked. Passive defence under such circumstances is the very worst thing; an enemy of this kind is more affected by evidence of activity on your part than by anything else. Your counter-attack may be a last effort of desperation, but to a savage foe—and especially to an Asiatic foe—it looks like a proof of the little damage he has been able to inflict on you.

In these cases, more than in others, is Napoleon's *dictum* true—"In war the moral is to the physical as three to one."

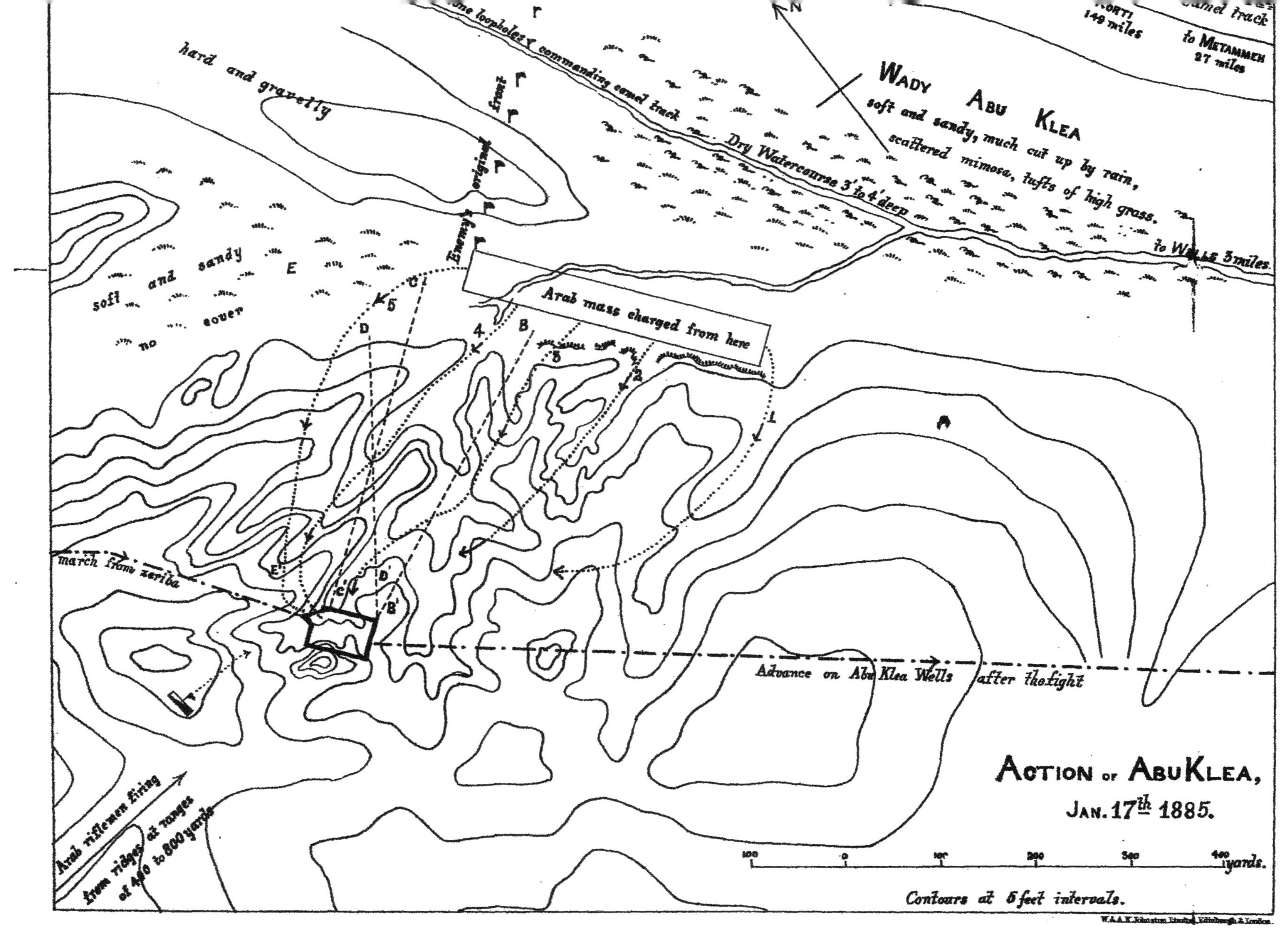
Action of Abu Klea,
Jan. 17th 1885.
Wady Abu Klea
soft and sandy, much cut up by rain, scattered mimosa, tufts of high grass.
to Metammeh 27 miles
Korti 149 miles
camel track
some loopholes commanding camel track
Dry Watercourse 3' to 4' deep
to Wells 3 miles.
hard and gravelly
Enemy's original front
soft and sandy E no cover
Arab mass charged from here
march from zeriba
Advance on Abu Klea Wells after the fight
Arab riflemen firing from ridges at ranges of 400 to 800 yards
Contours at 5 feet intervals.
yards.
W. & A. K. Johnston Limited, Edinburgh & London.

PL. XXIII.

right face of square
massed transport
left face of square
Arab mass before the charge
from

Section on B'B, shewing the good field of fire, front of left face; the troops about B' had a still better field against charges 1 & 2.

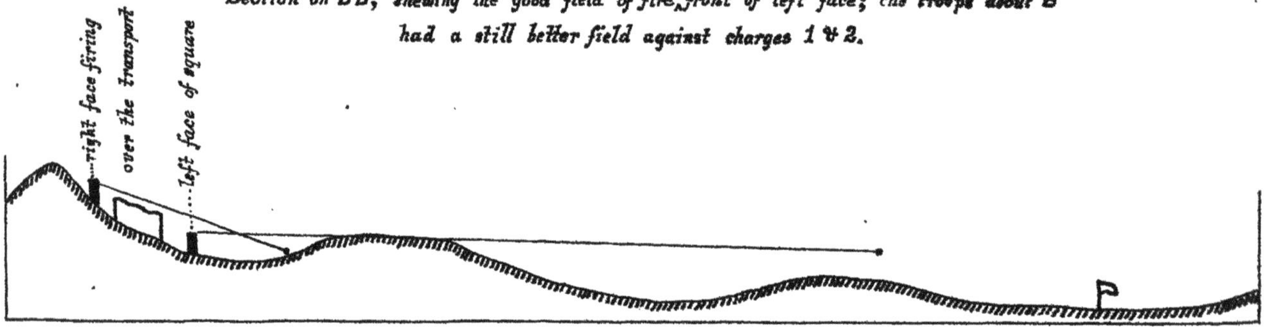

right face firing over the transport
left face of square

Section on C'C, shewing bad field of fire from rear part of left face.

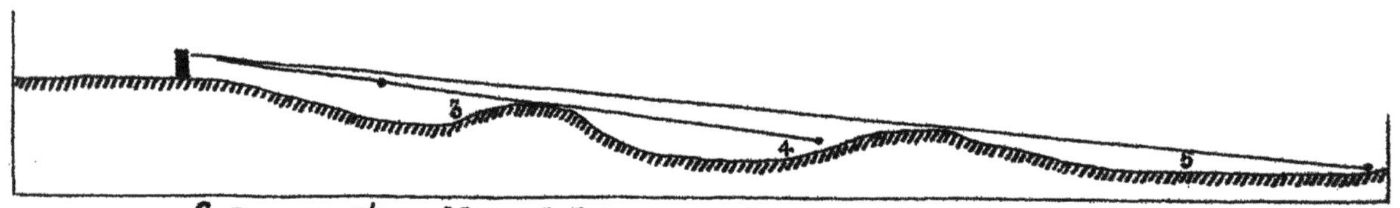

Section on D'D. Mounted Infantry at left front corner taking in flank charges 3, 4 & 5.

CHAPTER XIX.

DESERT WARFARE.

BATTLE OF ABU KLEA, JANUARY 17TH, 1885.

THE tardily planned expedition for the relief of General Gordon, besieged at Khartoum, was commanded by Lord Wolseley. It reached Korti, the point at which the Nile was to be left and the desert march commenced, on December 16th, 1884. This march was to reach the Nile again at Metemmeh, about 150 miles from Korti. The caravan track afforded a well-defined route, on which there were known to be two or three points where wells existed, but of a precarious quantity and quality. At Jakdul, more than half-way across, a good supply of water was expected.

Plate XXIII.

The familiar difficulty of insufficient transport prevented the despatch in a single body of the whole force with which Lord Wolseley had planned to occupy Metemmeh; so detachments were sent forward by stages. The desert march was a surprise to the Arabs, and Jakdul was reached without opposition by convoys that established a depôt there. By January 8th, 1885, all was ready for the start of the Desert Column from Korti; it was commanded by Sir H. Stewart, and amounted to 98 officers, 1,509 men, 2,228 camels and 155 horses.

The following is an extract from the orders given to Sir H. Stewart:—

"Having reached Jakdul and established 150 Sussex Regiment there, you will proceed to Metemmeh with—

1 squadron 19th Hussars,
3 guns R.A. on camels (7-prs.),
Guards' Camel Regiment,
Heavy " "
M.I. " "
250 men Sussex Regiment,
Naval Brigade,
Detachment R.E.,
A section Field Hospital and half Bearer Company."

Whether the column was going to reach Metemmeh easily or only by hard fighting would depend on whether Khartoum had fallen or not. Nothing was known of this, except that Gordon was in great straits.

On January 14th, Stewart marched from Jakdul with substantially the force mentioned above, viz., about 1,800 combatants of all ranks and 1,118 transport animals.

The wells of Abu Klea were now the immediate objective, 50 miles forward and 20 from Metemmeh. Starting in the afternoon, 10 miles were covered; at 5 a.m. on the 15th the column moved again and reached the half-way hill at 9.30; and marched again in the afternoon for some hours. At daybreak on the 16th the advance was resumed, the squadron being sent far ahead to reconnoitre. The march was continued till 11.30, and about noon the cavalry reported definitely that the enemy was in strength on the route to Abu Klea. In the afternoon the advance was continued in close formation, the three camel corps in front, baggage next, and the Sussex in rear. Descending into the narrow valley leading to the wells, a halt was made three and a-half miles from them, masses of the enemy being descried in front. Here the force bivouacked, sending out a picquet to a hill on the right front whence the hostile flags and main body were visible.

Stewart's report now speaks as follows :—"During the night a continuous light fire at long ranges" (mostly from the right) "was kept up by the enemy, doing little damage. Upon the 17th it was plain that the enemy was in force. During the night they had constructed works on our right flank, from which a distant but well-aimed fire was maintained. . . . Under the circumstances no particular hurry to advance was made, in the hope that our apparent dilatoriness might induce the enemy to push home. The camp having been suitably strengthened (zeriba) to admit of its being held by a small garrison—viz., 40 M.I. and 125 Sussex—and the enemy still hesitating to attack, an advance was made to seize the Abu Klea wells."

The time had now come for the square formation ; it was made up according to the diagram, *Fig.* 1, *Plate* XXIV.

As the square moved out of the hollow in which it was formed, a heavy fire was opened upon it, causing several casualties. As it advanced, the cavalry and a few of the M.I. skirmished ahead for the purpose of keeping down this fire, a square in motion being perfectly defenceless. Shortly the guns were taken off the camels, put together, and dragged by the gunners. Some hostile horse and spearmen, who were trying to work round to the zeriba, had to be attended to by the Hussars.

The advance had commenced along the bottom of the valley, but, the view becoming too restricted, it edged off to the right and then moved parallel to the valley along higher ground, which, however, being undulating and rough, was difficult for the camels and guns.

Plate XXIII. When within about 1,500 yards of the line of flags which marked out the hostile front, the square halted and the guns fired some shells, whereupon several hundred Arabs were seen to rise and retreat. The square then advanced on a line that would bring it about 500 yards beyond the proper left of the flags ; "when nearly abreast of them the enemy delivered his attack in a singularly well-organized charge of 5,000 men, horsemen in front, commencing with a wheel to the left. A withering fire was at once brought to bear upon the enemy" by the M.I. in the square. "The rear portion of this (left) face,* taking a moment or two to close up, was not in such a favourable position, and I regret to say that the square was penetrated at this point by the sheer weight of the enemy's numbers."

What had happened was this. The skirmishers, retiring at speed before the charge, did so directly towards the square, and masked the fire of the square till the charge was within 200 yards or less. Thus was nullified to some extent most of the effect of the good tactics that had directed the square to pass the hostile flank at 500 yards' distance.

* "Heavy Camel Regiment, 225 men," on diagram.

The fire of the left front angle (M.I.) was so effective that the Arabs swerved rapidly to their right and brought their greatest weight to bear upon the left rear angle. At this point the skirmishers were followed so closely by the enemy that one of them was speared just outside the square. Along this part of the charge, also, the ground was such as to screen a good part of the movement (*v.* Section 2, *Plate* XXIII.). At the critical moment the Naval Brigade's Gardner gun jammed, and the sailors, standing manfully to the hand-to-hand struggle, lost half their men.

The heavy camel men at the left rear angle had edged up the left face to join touch with the M.I.; this created a gap with the Naval Brigade, and some Sussex men rushed across from the right face to endeavour to fill the gap. But the heavy camel men were pushed back into the centre of the square and carried these Sussex men with them.

A great piece of good luck supervened at this crisis of the struggle. "The camels, which up to this time had been a source of weakness to the square, now became a source of strength. The spearmen by weight of numbers forced back the rear face of the square on to the camels; these formed a living traverse that broke the rush and gave time for the right face and front face to take advantage of finding themselves on higher ground, and to fire over the heads of those engaged in a hand-to-hand struggle on to the mass of the enemy behind." This fortuitous position of the square was the good fortune referred to.

v. Section 2.

A desperate conflict ensued in the centre of the square, but *the fire on the supporting masses* caused them to waver and fall back; within five minutes those Arabs who were at actual grips with our men were disposed of.

During this *mêlée* Arab cavalry had charged the right rear corner, but had been repulsed by fire.

The enemy had had enough of it and withdrew sullenly across the valley. The Hussars went forward to find the wells, and occupied them after a few shots.

Our losses were 9 officers killed and 9 wounded, 65 men killed and 85 wounded—being about 10 per cent. of the force engaged; 1,100 bodies of the enemy counted on the ground round the square speaks volumes for the determined nature of the attack.

This fight is a model for desert tactics, except in the faulty action of the skirmishers in retiring directly on to the face of the square that was about to be charged. The excuse was great, for the charge was delivered at lightning speed.

This kind of fighting is of an extremely anxious nature for the commander; the desert gap from water to water must be bridged over quickly; not only must the enemy be repulsed, but he must be so beaten as to be driven quite off the field; that is, the moral effect of his losses on him must be sufficient to achieve this, for you cannot break up to pursue a foe of this quality who still outnumbers you by four or five to one.

Another point to note in connection with this sort of work is that all the defenders of the square must be staunch men; any part of it may be the chief point of attack. It was fortunate that Lord Wolseley constituted his whole force of British troops, and rejected the offer of Egyptian regiments. A year before this fight Baker had ventured to the relief of Tokar from Suakin, with 3,500 Egyptians and blacks. The result was a butchery in which 112 officers and 2,250 were killed and wounded, Baker and his staff making ineffectual efforts to form the square, and eventually cutting their own way through to safety, sword in hand.

El Teb and Tamai.

A month later Sir G. Graham, V.C., R.E., was provided with a British force of 2,850 infantry, 750 mounted troops, 150 Marines, 8 7-pr. guns.

In his first encounter with the Arabs at El Teb on 29th February, 1884, they stood on the defensive behind entrenchments, having mounted the Krupp guns they had captured from Baker. They were thus abandoning the advantage of their extreme tactical mobility. Graham turned the left flank of their line and proceeded to roll it up. They were 6,000 strong, and fought each point with reckless bravery, but the divided masses were taken one by one; 2,100 bodies were buried in the captured position. The British loss was 189 killed and wounded.

The force was, a fortnight later, in zeriba eight miles from Suakin. The task now was to clear the Arabs out of Tamai, seven miles further on across a gravelly plain beset with clumps of mimosa bushes. The advance was made in solid column, and the bivouac zeriba was only 1,400 yards from the enemy's position.

Next morning two squares were formed, and these moved slowly forward. At El Teb the enemy's position had been manifest; here there were hidden ravines, and no such attack as the last could be safely initiated. This change of tactics is worthy of special notice.

As the 2nd square neared a ravine, a great mass of Arabs suddenly sprang up, charged with great fury, and without actually breaking the square forced it back in some confusion. The machine guns were temporarily lost, and four 7-prs. were left unprotected; but the gunners, turning the guns this way and that, and using inverted shrapnel when the small allowance of case-shot was expended, held their own until the battle was restored.

The other square was vigorously attacked, but suffered no sort of reverse.

On this occasion the enemy numbered no less than 12,000; their loss was 2,000, or about 16 per cent. General Graham had 221 casualties, or about 10 per cent. of those engaged. The cavalry did good work in supporting the 2nd square when it was rallying.

Not only was the advance in square good under the circumstances, but the formation of two squares was eminently suited to the occasion. In the rough ground, studded with bushes and intersected by water-channels, a single square comprising the whole force would have been unwieldy, and might have shown gaps at the critical moment. When that moment would come, no one could tell, the ravines completely hiding the hostile masses till they rose. Two squares in échelon would not both arrive at the danger point simultaneously, and the rear one would flank the other before it was itself attacked. If the danger point were known, as at Abu Klea, and it were necessary to have two squares on account of numbers, then the échelon advance would of course be so arranged that the leading square should not mask the fire of the other when the rush came.

Combat of Tofrik, 22nd March, 1885.

A less successful combat, in which practically the whole of the transport was lost, was Sir John McNeill's fight with the Arabs at Tofrik, a point about six miles out from Suakin.

Sir Gerald Graham had been given a force about 13,000 strong, with the double mission of (1) destroying the power of Osman Digna, (2) of guarding the construction of a narrow-gauge railway from Suakin to Berber. The latter work was only pushed a few miles, and then the project was abandoned by the Government of the day.

Tamai had been again occupied by Osman Digna, so Graham sent Sir John McNeill with the following force in that direction :—

1 Squadron 5th Lancers.
Berkshire Regiment.
Battalion of Royal Marines.
1 Field Company R.E., with a Telegraph Section.
Naval Detachment with four Gardner guns.
Ammunition Column.

These were the British troops.

There was in addition the following Indian Contingent :—

15th Sikhs.
17th Bengal N.I.
28th Bombay N.I.
1 Company Madras Sappers & Miners.

The total force contained about 4,000 combatants.

Graham's orders for the force were that it was to march eight miles towards Tamai, make three zeribas, of which the centre one was to be capable of holding 2,000 camels, and the flanking ones to be garrisoned by a battalion each. This place was thus to become an advanced depôt for water and stores for an advance in force on Tamai. The Indian Contingent was to escort back the "empty" camels, and the R.E. were to reel out their telegraph line and keep up communication with headquarters at Suakin.

The force moved off from Suakin in two squares at 7 a.m. on March 22nd, and encountered difficulties from the formidable nature of the mimosa scrub and from the careless loading of many of the camels, which caused such frequent halts that it was determined to end the march at Tofrik, six miles out. This was telegraphed to Suakin and approved by Sir Gerald Graham.

The diagram *Fig.* 2, *Plate* XXIV., shows the clearing in which the zeribas were to be made, and the position of affairs when the rush of 5,000 Arabs took place. The surrounding bush was dense and rose to eight feet in height, so that the cavalry vedettes, less than half a mile out, were lost to sight. Small picquets were put out along the edge of the bush, and the work of making the zeribas was commenced.

The important thing was to get the flanking zeribas completed. About one-third of the force was told off into cutting and building parties; the British brigade stood in square to the east of where the great central zeriba was to be, as a rallying point in case of need; the camels were watered and fed near them, led into the centre and there unloaded, being afterwards "parked" near the British square. The Indian troops were in line as shown, and were to remain thus till the flanking zeribas should be ready for occupation.

The Marines' zeriba was nearly ready by 1.30 p.m., and was then garrisoned by that battalion, with the addition of two Gardner guns. The Berkshire zeriba was not yet ready, and its colonel represented that his men had had no food since 4 a.m. A little delay therefore occurred in manning this zeriba; half the battalion received its rations and marched into the zeriba, the other half remaining where shown, near the camel "park."

The former half piled arms in the zeriba, consumed its rations, and went out to cut the bush still required for the completion of its enclosure. The 15th Sikhs were in a long line facing west, the 28th N.I. and two companies 17th N.I. in a long line facing north, the other six companies covering the central zeriba on the south.

At 2.30 p.m. the unloaded camels had begun to file out for the return march, and were in a confused mass between the left rear of the six Indian companies and the half-battalion of the Berkshire, who were having their dinner. In a moment the cavalry came galloping in from west and south, with the Arabs on their heels. The Berkshire working party got back to their arms, the Indian picquets rejoined their lines with speed, and fire was rapidly opened. The Lancers rode in round the Berkshire corner, and foolishly forced their way through the right half of the six companies of the 17th N.I. This rendered the regiment unsteady. Sir John McNeill instantly perceived this, and ordered some of the Berkshires from their inner face to support. But it was too late; the 17th fired a single volley and broke, some climbing in through gaps into the Berkshire zeriba, the rest bolting for the Marines' zeriba.

The Arabs now stampeded the baggage animals, driving them on to the half-battalion of the Berkshires in the open, who were in square, masking their fire; they swarmed over the half-finished fence of the zeriba and over the half-built sandbag parapet that was being prepared at the outer salient for a Gardner gun. The naval men in charge of these guns stood to them manfully and suffered severely, as did the Berkshires, who were soon engaged in a confused hand-to-hand struggle all over the interior of the zeriba; 112 Arabs were killed inside the zeriba before it was cleared.

The 15th Sikhs and 28th N.I. stood firm, as also did the two companies of the 17th next the Marines' zeriba. The Official Account says, "There never was a doubt as to the result of the attack on these regiments. The Sikhs bore the brunt, and hundreds of dead Arabs were afterwards counted in front of their position."

The Berkshire half-battalion in the open were having their dinner when the rush came; it formed rallying square, and repulsed two determined attacks, when the baggage animals had cleared from the front, killing 200 of the enemy. The Marines and Berkshires then cleared the area traced out for the central zeriba.

The whole affair lasted 20 minutes; in that time 1,500 Arabs were killed; the British brigade lost 135 killed and wounded, the Indian 149, and nearly 200 followers disappeared or were killed; 500 camels were missing.

Before the attack, a squadron of Hussars had left for Suakin; at about two miles it met a squadron of Indian cavalry, and simultaneously heard the firing at the zeriba. The Hussar major took command of the two, hastened back, and met half-way camel drivers, some native infantry, a few British soldiers and a lot of transport, fleeing for Suakin, closely pursued by Arabs who were cutting them down as they ran. The squadrons promptly formed line to the right and cut into the pursuit, checking it; then every available man dismounted and fired volleys. The pursuit seemed stopped, and the cavalry followed, again firing volleys; but a part of the enemy was observed to be working round to again get at the fugitives. This was stopped by a troop; the two squadrons had despatched a great number of the enemy.

Continuing their advance, they were able to shepherd a mass of the enemy past the east side of the zeribas, where the Arabs suffered so severely from the enfilading fire that they dispersed, and the combat was over.

Comments.

Though the affair was eventually a success, the Arabs having been defeated, there are several comments possible on Sir John McNeill's methods.

The clearing was so small that a proper lookout, and a proper field of fire, were not attainable. To send out men on foot far into the bush to act as picquets would simply have been to expose them to certain massacre. The cavalry vedettes even could see hardly better than a man on the ground, but the bush was not absolutely continuous, and the clear patches could have been seen well from a captive balloon. There was one with the Suakin Field Force, but the point which was to have been reached by this expedition was in fairly open ground, and so the balloon was not sent out. So this matter may be dismissed; but the reflection occurs whether in such work, when for any reason a balloon is not available, it would not be wise to carry the materials, all ready prepared, for a light field observatory, sending along with it a detachment of R.E. who have already been exercised in putting it up. A well-drilled detachment would hoist an observer 60 or 80 feet in a couple of hours or less.

When the fighting began, the following points came under notice. The Sikhs and the 28th in a single line stood their ground, though assailed by a numerous foe who did not fear death; the half-battalion of the Berkshires in the open, though it had only time to hurriedly form a rallying square, and was further hampered by having the camels stampeded on to it, held firm; the Marines in a completed zeriba did likewise. But the half-finished zeriba of the Berkshires was entered at the first rush, and the 17th N.I., their ranks broken by the fleeing Lancers, gave way immediately. The case of the latter may be at once dismissed as an accident, for the reason stated; but the Berkshire zeriba involves a fundamental question.

We saw that Skobeleff, on the way to Geok Tepe, included among his instructions to his officers, "In Central Asia, close order is to be the rule." His enemies there had just the same characteristics in fighting as these Soudan Arabs, extreme bravery and an ardent desire to get to close quarters, great individual speed and a high average of physical strength.

It seems, then, that it should have been an instruction to Sir John McNeill's officers—and should be a standing instruction to all officers under such circumstances—that a work, zeriba or other, that is incomplete, either from having gaps in it or from not being yet a high enough or broad enough obstacle, should not be trusted to; that is, it should not be lined as if it were complete. Men, lining such an obstacle, do not get literally shoulder to shoulder; there is a natural tendency, under the severe experience of a rush of fanatics, to edge behind the best parts of the obstacle, and there is usually no time for the officers to correct such faults. If this half-battalion had formed a rallying square right in the centre of their zeriba, they would probably have suffered less. As Skobeleff said, "In Central Asia, the order of fighting is close order." If there is time, during the period when an expedition is being assembled and prepared, the troops should be practised in the special kind of work that is before them; sudden alarms given during the practice are particularly educative.

The great lesson to be learned from all such actions is that the superior weapon alone does not ensure victory against such foes. Their number is usually so great, their rush so rapid, their contempt of death so admirable, that no care in adapting one's tactics to the occasion is thrown away; and the fate of Hicks Pasha and of Baker Pasha's force shows that one's men must be of as staunch a sort as the enemy. He is usually least formidable when he is deluded by the possession of a gun or two to stand on the defensive. In the attack he often shows remarkable tactical skill, but on the defensive his tactical ideas do not often go beyond a rather passive defence. Under these circumstances he has no "grand-tactical" mobility to speak of; each body may be depended upon to defend its own point to the death when its turn comes; but you can choose your point, and know you will not have the whole lot to fight at once.

That you must know, also, your enemy's peculiarities is a necessary corollary to what has been said above. The Italians once thought that they had, in the Abyssinians, an enemy that they could play with, but the frightful disaster of Adowa taught them the truth in a cruelly drastic manner. The great night-march was to be a surprise to the ignorant foe, but he was ready for them; a few more days' delay would have compelled the Abyssinians either to attack the Italians in a chosen position, or to evacuate the whole district. This was because they had "eaten up" the countryside, and an Abyssinian army had most of the disadvantages in this respect of a European army, without the latter's elaborate and expensive supply arrangements. But the Italians did not know, or did not heed this, and plunged into the disastrous attack.

CHAPTER XX.

THE BOER WAR.

HAVING interpolated two chapters on Tactics as exemplified in "Small Wars," I now turn to that great struggle in S. Africa, which tried so severely our strength, both as a nation and as an empire.

Before entering upon a description of some of its battles and of the tactical developments that came about, the reader should understand that this book is an attempt at a History of Tactics, and not at all a book in which the main aim is to argue out the correctness or incorrectness of the tactical ideas employed. Neither is it a mere chronicle of what soldiers did. Praise or blame is hardly to be avoided, if one is to give an intelligent and reasonably interesting account of what happened; but the writer has no mission in this chapter to plunge into controversy such as is naturally raging in the military magazines. His aim is to give as clear a picture as possible of the new things that were seen, and of the new things that were done in consequence.

With this purpose in view, some battles and combats of the war are chosen for brief description, each to be followed by some comments, and the whole to be again followed by some more general remarks bearing on the whole effect of this war upon tactical ideas.

BATTLE OF BELMONT, 23RD NOVEMBER, 1899.

At the date of this action the redoubtable Cronje was still 300 miles away, besieging Mafeking, the Boers who were collecting to bar Methuen's progress to Kimberley were estimated at under 5,000, and Methuen had over 8,000 men available at Orange River Station, not counting the Highland Brigade, some distance down the railway, and the 12th Lancers still at Cape Town.

Methuen's plan was to fight rather than manœuvre; to all suggestions that he should go round the Belmont position, he replied, "I intend to put the fear of God into these people." But he underestimated the deadly effect of the Mauser over the open veld, and the difficulty of reaping the fruits of victory when you have only a squadron or so of horse, and the enemy are all mounted men.

The Boers, on their side, divined Methuen most perfectly, unhesitatingly assumed that he would keep straight along the railway all the time, and entrenched themselves on the tangle of kopjes east of Belmont Station in perfect confidence. *Plate* XXV.

The British bivouac on the 22nd November at Thomas's Farm was about two miles from the forward kopje nearest the railway. That evening our cavalry came under rifle fire, and were also saluted by a gun on the kopje mentioned.* To a slight extent this revealed the Boer position, but the absence of good maps, and the extreme difficulty of close reconnaissance in such country, brought it about that the plan for the next morning's attack could only be very partially carried out.

* This led our men to christen the kopje "Gun Hill"; two hills to the rear were similarly named "Table Mountain" and "Mont Blanc."

The 18th Battery, R.F.A., moved out and answered the gun, till the Boers withdrew their two Krupps and their "pom-poms" behind the kopjes; this gave us the advantage in the morning of having no artillery fire to deal with.

The available force was the Guards' Brigade of 4 battalions, the IXth Brigade of 3½ battalions, the 9th Lancers and a few irregular cavalry, 210 horses, 3 companies M.I., making a total of 900 mounted men, 2 batteries R.F.A., 2 companies R.E.; while 400 bluejackets and Marines with 4 12-prs. were arriving. The transport, supposed to be cut down to a minimum, still stretched over five miles of road.

The real topography of the Boer position was as shown in the plan, but Methuen's maps divided "Gun Hill" into two parts, of which the more easterly joined on to "Mont Blanc." The ganger's hut was to be the point where the Guards were to rendezvous in the dark, for Methuen did not like the prospect of advancing against the kopjes in broad daylight. This hut had been estimated by a reconnoitring officer as 800 yards from "Gun Hill"; its real distance was 1,800 yards.

The plan was to send the IXth Brigade against the west face of "Table Mountain," the Scots Guards and Grenadiers against the west and south of "Gun Hill," the two Coldstream battalions to pass round the left of the Scots, and, as soon as the latter were in possession of "Gun Hill," to push on along the high ground that was supposed to afford a way to "Mont Blanc." The IXth Brigade, having secured "Table Mountain," were to make a wide sweep by the north end of "Mont Blanc" and thus assist in its capture.

At 2 a.m. the march commenced, the IXth Brigade leading up the railway. The Guards arrived at 3.20 at the hut, and deployed, Scots and Grenadiers in front line, men at five paces interval. The underestimation of the distance brought it about that, when the 1st line was still 350 yards from the hill it was plainly visible, and was received by a blast of fire. The whole brigade—1st line, supports and reserves—was completely deployed, and the attack was carried out in the most approved fashion. Advancing with long rushes, alternating with short pauses, they reached the comparatively "dead" ground at the foot of the steep heights, and formed up there for the assault. As they climbed, the reserve companies endeavoured to keep down the magazine-fire that was being poured down from the top; but the assault, from the very circumstances of the case, had been unprepared by artillery. The two battalions, however, gained the crest after considerable loss, in time to see the burghers disappearing on their horses out of range, and then hastily joining their comrades on the 2nd line of defence. The 18th Battery arrived just at this time on the right of the Coldstreams, along with the naval men and their guns, and began pounding at the east kopjes.

The IXth Brigade were also rather late in arriving at their objective; the 75th Battery supported their attack on "Table Mountain." The 5th Fusiliers, in 1st line, went straight at the south-west corner of the hill, but were checked by a steady fire from skilfully placed sangars; the Northamptons worked out to the right of them, into the bay between "Gun Hill" and "Table Mountain," got on to a sort of spur joining the two, enfiladed the sangars, and eased the pressure; within ten minutes of the Guards' success at "Gun Hill" the IXth Brigade had possession of the west crest of "Table Mountain," but a considerable check now ensued on account of the fire of Boers who still held on to the north and east edges of the plateau, and the IXth Brigade were not able to at once sweep on to "Mont Blanc," as the plan had intended.

The Scots and Grenadiers, too, were to have been supported on their left by the two Coldstream battalions—in fact, the whole attack was to sweep from west to east; but these battalions moved to the right. The battle had become a "colonels' battle," each battalion doing what seemed immediately obvious. The 1st Coldstreams, coming under fire from the south end of "Mont Blanc," turned to the east; the field and naval guns bombarded the point from 1,300 yards, and it was taken without trouble, the Grenadiers coming down from "Gun Hill" to support. Soon after this the Boers were driven entirely off "Table Mountain," when the Yorkshires of the IXth Brigade had worked past its south side. These troops, with the 2nd Coldstreams, now cleared the ground between "Table Mountain" and "Mont Blanc," and the whole swept the latter clear, and captured the Boer laager. It was only 7.30 a.m.

More than half the mounted troops had spent the time covering the 75th Battery and demonstrating against a thin line of Boers with a "pom-pom," to the west of the railway. These withdrew with great skill at the last moment, and our cavalry could manage no pursuit on this side. On the right flank, the irregular horse, the remaining squadron of 9th Lancers and one company M.I. covered the naval guns and the 18th Battery, attempted a pursuit, and were ambushed.

Now, Methuen had made up his mind to a sledge-hammer attack from west and south, the Boer line of retreat being north-east. The very smallness of his mounted force should have suggested to him that it should be kept together, that the bulk of it should have rested out of range to the south-east, and careful signalling arrangements made for it. Instead of this, the force was halved, was kept moving about, and horses and men were tired before the time for pursuit came. The battery on the left flank might quite well have been covered by a single company of infantry, it being very unlikely that the Boers would operate strongly on that side. Pursuit by gun-fire was attempted, but unsuccessfully; the naval men made desperate efforts to drag their guns on to a small kopje on their right, from which the line of retreat was visible; but the brigade were exhausted by their efforts during the previous day and night. Want of supreme control of the action is apparent; as soon as the final attack on "Mont Blanc" was started, Methuen should have been planning for pursuit. The naval guns might easily have been spared, or the 18th Battery; and one or the other should have at once been directed to attend to the line of retreat, along with, as has been suggested, the bulk of the massed cavalry.

The casualties amounted to 25 officers and 270 men. The enemy lost about 100.

His leaders had not yet learnt that the modern rifle is most effective when firing along the flat; his men were all on the edge of steep-fronted kopjes, and were only able to meet the near attack by a plunging fire. The edge of the high ground afforded a good mark for our guns to train on, and the defenders had to crane over and expose themselves when dealing with the close attack.

BATTLE OF ENSLIN, 25TH NOVEMBER, 1899.

Again at Enslin, two days later, the Boers disposed themselves in the same manner; *Plate* XXV. again Lord Methuen proceeded on the same method of putting "the fear of God into these people." Again he had his cavalry halved on both flanks, and again one-half was nearly cut off, and the pursuit stopped. At this fight the Boer guns replied manfully to the great preponderance of pieces against them. Our attack was from the south and south-east, and the great mobility of the enemy enabling a body of his men to hover far out on our right flank, troops had to be ordered out to the east after the fight began, and to some extent the original plan was thereby disarranged.

There was a commanding kopje at the south-east corner, and this was to be the chief point of attack. Front line against it was composed of 245 naval men and a company of North Lancashires, with the Yorkshire on the right rear to pour a covering fire on the crest, along with a field battery. But the enemy quickly saw the plan, and hastened rapidly in twos and threes, for the most part unseen, into the lower boulder-strewn ridges in the centre of the position. In such ground they afforded a very bad mark to our guns, and were able to bring an almost level enfilading fire on to the naval attack. This was a first-rate tactical performance on the part of the Boers; if these reinforcements had mounted the high crest to increase the volume of fire from there, they would simply have been playing into the hands of our gunners, who were battering that edge with shrapnel. Nearly half our front line were down before the "dead ground" was reached, three out of four being hit from their left. When the Yorkshires arrived, the climb began, the Boers holding on to the last, but having taken the precaution some time before to send away their transport. The centre was cleared about the same time by the rest of the IXth Brigade.

The first attacking line had started at only four paces interval, and had rather converged on nearing the objective; this, but still more the flanking fire from their left, was the cause of the frightful losses it sustained. The Yorkshires, fresh from the Tirah, kept at eight paces, and were also too far round to the right to be seriously troubled by the flanking fire; hence they suffered much less. It seems that the whole attack on the high corner should have been made from due east.

De la Rey was the Boer leader who saw that the comparative ease with which the kopjes had been stormed was due to their steep faces and the dead ground; and by the strength of his influence and character, he was able to induce the Boers, who were about to be strongly reinforced, to make a completely new departure at Modder River. For the first time, the flat-trajectory was going to be allowed its full chance on the defensive.

Modder River Battle.

Misleading maps, defective scouting and a determination to go straight for the enemy, regardless of mere strategical considerations, are the salient points of Methuen's next move. The Boer headquarters at Jacobsdal, where they had much of their supplies, was well to the south-east of Modder River Bridge; still Methuen made no attempt upon it. The one idea was to stick to the railway and charge straight forward. This time, on the level plain, with the Boers lining the bush-covered south bank, our guns had no distinct target to aim at, and there was no friendly dead ground to help the preparation of the final assault. The sleet of lead, driving horizontally from the unseen rifles in the bushes, brought the advance to a dead halt, more than 1,000 yards from the Boer firing line. Fortunately for us, the enemy had not had the nerve to withhold his fire till we were nearer; practically, Methuen's whole force was at once involved, and distraction ensued. The C.-in-C. careered about the field, and could not be found when wanted. On our extreme left, west of the bridge, progress was gradually made, through the help of a kraal and a small donga; the river was crossed, and a small village occupied. Advance from it towards the Boer right centre was checked, to a considerable extent, by the fire of our own guns. The 75th Battery had pluckily advanced to within 1,700 yards in the open, but it is doubtful whether the consequent flattening of the trajectory of its shells may not have been rather a disadvantage than otherwise, dealing, as we were, with riflemen having their weapons on the ground and their bodies close up to the vertical scarp of the river bed. It was a case in which the modern field howitzer would have been very useful.

The gradual collecting of about 1,000 men by Pole-Carew about the village beyond the Boer right settled the contest, and during the night the Boers withdrew.

Magersfontein

The majority of the enemy were downcast over these three defeats, and Cronje, who was now in command, had most of his men for some days towards Jacobsdal, quite out of Methuen's way to Kimberley; but the British leader was apparently not aware of this, and decided to await reinforcements and rest his men and horses. Meantime, Cronje plucked up heart again, and, with greatly increased numbers, entrenched himself at Magersfontein and Spytfontein, right across the road to Kimberley. This time again the Boers had their most effective firing-trench on the flat, along the foot of the heights, so well made that we never guessed at its existence till the level blast of magazine-fire from it created such havoc in the Highland Brigade.

For the third time Methuen was indulging in a night-attack. The plan was sound, and quite likely of success, if the existence of the low trench had been discovered by reconnaissance; but an awkward belt of thick scrub being encountered just before dawn, within a few hundred yards of the foot of the heights, *and* the said trench not being suspected, the brigade passed through it before deploying, alarmed the Boers, and was caught manœuvring in the rapidly growing light.

The extraordinary extension of the Boer front in this fight is worthy of notice. Their left was down on the Modder River, their right on the west of the railway, the shape of the whole being a flattened ∽, Magersfontein kopje in the re-entrant. The great double curve must have measured over ten miles, and this was held by less than 7,000 men, with five Krupps on the main ridge and some "pom-poms" on the left flank.

Methuen had been strongly reinforced, and had now some howitzers and a 4·7 naval gun. These and three field batteries had bombarded Magersfontein kopje for one and a-half hours, late on the afternoon previous to the attack. This warned the Boers that the usual attack would come next dawn, and told them where it would come. The material effect was only the wounding of three Boers, for the enemy, seeing no attack advancing, were able to withdraw from the trenches in safety, and the trench on the level was not touched by the bombardment.

The reconnaissance had been of the most sketchy description, and what was done by the cavalry in that way only accentuates the demand that many authorities make for a corps of specially trained men for the purpose. Real scouts, for instance, two or three of them, would have reached that belt of bushes during darkness, would have lain there concealed for a whole day, and would next evening have brought back invaluable information. A very risky service?—Certainly, but what of that? A battery, or a couple of guns, and a threat by infantry would have watched the gap between the bushes and the enemy during the day, and would have dealt with any attempt by suspicious Boers to search the bushes. The balloon did useful work during the fight.

The comments of the German Official Account on this battle of Magersfontein are so apposite, and throw so good a light on the state of British tactical training at this time, that I quote portions almost *verbatim*.

"Lord Methuen rightly decided to continue the action, notwithstanding the mishap to the Highlanders, and it was only the method which he adopted that was wrong. . . . The idea of approaching the enemy under cover of darkness was, considering the ground was devoid of cover, an excellent one, and an attack was bound to be all the more effective if it should partake of the nature of a surprise. But when the bombardment ceased on the evening before, the Boers had time to recover from its impression . . . they could no longer be surprised." It is to be noted that the Germans did almost no night-advances in 1870.

The account then points out that no success was achieved in locating the Boer guns, or in doing any personal damage. This matter will be referred to in a more general way in a later chapter.

The next comment is on the small amount of reconnaissance done by the cavalry, during the interval between Modder River battle and this one. The "Account" allows for the difficulty cavalry would have in scouting in such open country, beset also with wire fences, but does not make sufficient allowance on this score. "If no other method had been possible," the German says, "the mounted troops ought to have endeavoured to gain information by dismounting and advancing carbine in hand. We may, indeed, assume that this was quite feasible had advantage been taken of the strips of bush."

The writer has not seen his own men going down before the fire of a modern magazine-rifle, on open ground and in a brilliantly clear atmosphere. His troopers would have to dismount more than a mile from the enemy, and even then most of their horses would be *hors de combat* before the crawling scouts got back to them. The remarks are an echo of 1870, when the German cavalry had an easy time. Here the cavalry scouts, their "advance, carbine in hand," seen the whole time, would not be likely to get home again safely to tell what they had seen. More guile than this was required; I have suggested *real scouts*, who would enter the most forward bushes in the dark, and spend the day there, motionless, watching, and discovering the location of the masked trenches. While lying there, a demonstration of attack would be made, just sufficient to compel the enemy to line those deadly low-ground trenches on which he laid greatest stress. A pair of skilled scouts among the bushes would easily, by prearranged signal, and without betraying their position to the enemy, succeed in letting the demonstrators know when they had seen all there was to be seen.

"The omission to despatch strong detachments round the Boer flanks partly accounts for" the bad information; "purely frontal reconnaissance will always give poor results." This is quite true; but in this case, when the attack was to be frontal, the really important thing was to find out what was in front.

"The Black Watch had been pushed forward to cover the guns on the afternoon before the battle, and the battalion thus became acquainted with the ground to be traversed, but no patrols were sent as far as the enemy, nor do any attempts appear to have been made to employ flag-signallers to announce the effect produced by the guns."

The clause about "patrols" is absurd; the matter of signalling the effects of gun-fire will be dealt with more generally later.

"The battalion should have been advanced to within about 900 yards of the hostile position, and have remained there as the pivot on which to form up the troops destined

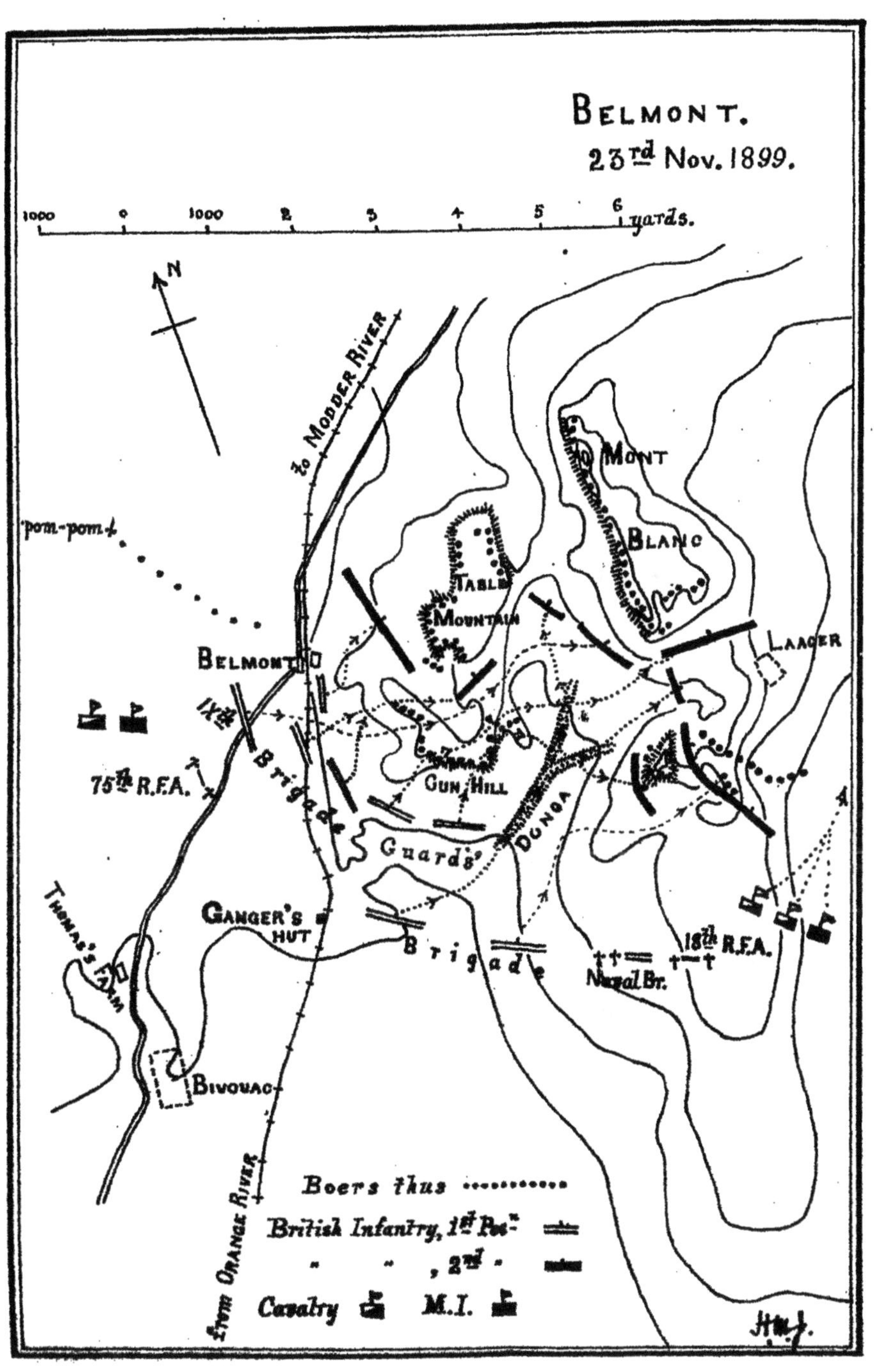

BELMONT.
23rd Nov. 1899.
1000 0 1000 2 3 4 5 6 yards.
N
to MODDER RIVER
pom-pom
MONT
BLANC
TABLE
MOUNTAIN
LAAGER
BELMONT
IX Brigade
75th R.F.A.
GUN HILL
DONGA
Guards' Brigade
GANGER'S HUT
THOMAS'S FARM
Naval Br.
18th R.F.A.
BIVOUAC
from ORANGE RIVER
Boers thus
British Infantry, 1st Posn
" " , 2nd "
Cavalry
M.I.

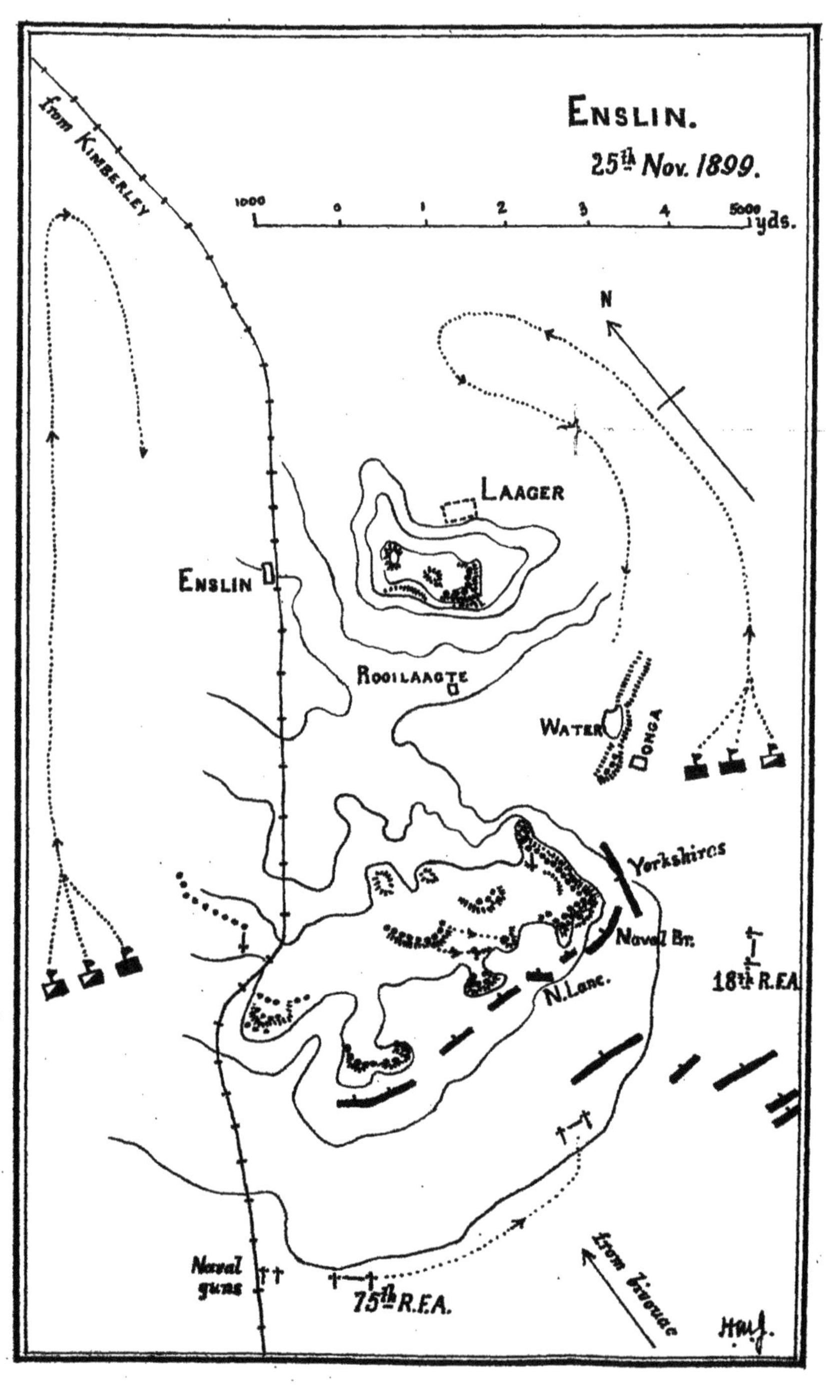
ENSLIN.
25th Nov. 1899.
1000
0
1
2
3
4
5000
yds.
N
from KIMBERLEY
LAAGER
ENSLIN
ROOILAAGTE
WATER
DONGA
Yorkshires
Naval Br.
18th R.F.A.
N. Lanc.
Naval guns
75th R.F.A.
from bivouac
HWJ.

MAGERSFONTEIN.

DEC. 11th 1899.

MARCH AND DEPLOYMENT PRESCRIBED FOR THE HIGHLAND BRIGADE.

800x 800x 800x

2 companies 2 companies 2 companies

SEAFORTHS B.W. ARGYLLS

reserve

H.L.I.

in quarter column

BLACK WATCH
SEAFORTHS
ARGYLLS
H.L.I.

Each of the 3 leading battalions to have 2 Cos. in front line at 5 paces' interval; 2 Cos. in support, 230 yards in rear, similarly extended; 4 Cos. in reserve, 230 yards in rear, extended. As soon as formation is complete, the whole to lie down, with bayonets fixed, and await the order to charge. The H.L.I. to act according to circumstances.
In the confused deployment that actually took place, the Seaforths found their way to the right.

H.M.J.

W. & A. K. Johnston, Limited, Edinburgh & London.

to carry out the night-attack." This would have been "to give away the whole show," more than the bombardment had already done.

"Military history shows that when troops destined to carry out a night-attack throw themselves down as soon as fired at, it is only by bringing up reinforcements that they can be induced again to advance. Night-attacks require formation in depth just as do actions fought by day, although of a different kind, and the British regulations also affirmed this principle; but the Colonial campaigns" (*i.e.*, our 'small wars') "had caused the necessity for such a formation to be lost sight of. Lord Methuen . . . kept the Guards, who were intended to support the Highland Brigade, nearly two and a-half miles away from what was to be the decisive point."

This is a complete misconception on the part of the German writer. The Highlanders were to be their own supports, the Guards to be the C.-in-C.'s reserve; he expected the former to succeed, and the Guards were to be used for the next stage, which would quite probably be due eastwards. Whether the Highland Brigade by itself was likely to be adequate, is another question. *Plate* XXV.

"Three thousand men," (should be nearly 4,000) "to be supported by 3,850 others in *reserve*, were to carry out the actual attack on the enemy, who was 6,000, or perhaps 7,000, strong." But only about one-third of the enemy was opposite the Highlanders. "Pole-Carew, with 1,900 troops, was to make a supplemental attack along the railway towards Kimberley" (*i.e.*, on our extreme left), "and 1,450 men were employed in minor and unimportant tasks on the field of battle, or else were not in action at all."

A mistake in Methuen's orders in relation to Pole-Carew is merely mentioned, but not animadverted upon. This force was given the foolish instruction "not to become seriously engaged." Such an order takes the heart out of a subordinate general; the C.-in-C. should explain to him the general idea, and leave the rest to the subordinate. Pole-Carew could have done great things that day, but for this hampering form of order. To this matter also I shall return in a more general way later; it is intimately connected with the great question of a uniform system of tactical training of commanders.

"If the night-attack was to succeed, it was bound to be carried out rapidly and in such a formation that deployment would have been the very simplest matter when collision with the Boers should occur." It may be supposed that the critic means "before" when he writes "when." He would have preferred a "line of columns." These would have had to keep closely side by side, and it is more than doubtful if anything would have been gained. In the method planned, the Black Watch and the H.L.I. would have been in position simultaneously; A and B Companies of the former would have advanced 460 yards, C and D about 250 yards, and the remaining four companies would have been extending. The two intermediate battalions would have followed this extension to left and right, and then repeated the operation. *Plate* XXV.*

"The cavalry and mounted infantry were employed solely as infantry. . . . Even if their dismounted action were justifiable until the arrival of the Guards, it was no longer so when a sufficient force of infantry had been brought up." This refers to the dismounting of a lot of men on our right flank, when the Boers, finding that Pole-Carew's force did not mean business, were able to gallop round reinforcements to extend their left.

* Note that the diagram purports to give the *intended* deployment.

The great difficulty of getting men extricated from action in open country is here not sufficiently allowed for by the German critic. The real mistake here was in not knowing in time that the Boer left was so extended. It was supposed at first that the direction of the mounted troops would take them outside the Boer flank.

Want of initiative seems to have been shown in the Guards' attack, when they did come up on the right rear of the Highlanders; half-hearted attacks in the midst of a battle are not of much use; it is possible that the commander was *waiting to be told* to push hard. At 7 a.m., when he began, the Boer left was still very weak.

The attempt to retrieve the Highland failure by trying to push in between them and the Boer left shows a lack of *coup d'œil;* to go into a corner like that is to place yourself where the hostile converging fire is strongest, where you get it on both flanks. Granted that we thought at first that the east end of the Magersfontein ridge would be the Boer outer flank—as soon as it was seen that it was not so, the direction of that east end of the heights should have been avoided. Pole-Carew should have been told to push with all his might, if he was not doing so, a battery or two and the mounted troops should have had the job of keeping the Boer left from advancing any nearer the Highland right flank, the mounted troops being enjoined to sacrifice themselves in a charge if necessary, and the Guards should have reinforced the Highland left before the second break of the latter brigade took place.

The British gunners did first-rate work; field and horse artillery got within 1,300 yards of the enemy's trenches and kept down his fire by continuous work.

The failure of the night-attack had somehow lost us the initiative; from that moment Methuen was practically on the defensive, and the way in which 9,000 men, with an artillery much superior to the enemy, got surrounded on two sides by an enemy of considerably smaller numbers, is certainly something very remarkable. Was it his mobility alone that did it? Half of his left wing was in the open, with no trenches worth speaking of.

The fact is that it has come about that, on open ground, an enemy in position where he can cross his fire, seeing from both faces the whole of the *terrain*, can safely extend to an amount unthought of previously. A thousand men, occupying a deep re-entrant angle of 2,000 yards to each arm, may be, if they can see all the ground and the enemy is foolish enough to attack deep into the angle, as strong or stronger than if he were massed along one arm only of the angle. This, of course, has always been the case, but previously for far shorter distances; for such short distances, in fact, that a glance shows at once the absurdity of attempting any such attack. But the great distances of these days, without any corresponding increase of power of vision, produce illusion, which must be corrected by an increase of trained imagination.

Second Phase of the Boer War.

Insufficiency of numbers, lack of mounted troops, and, as some assert, a deficiency of strategical ability, brought our arms to a standstill after Magersfontein. The general political situation, as well as the military situation in South Africa, demanded that the Empire should put forth its strength, and that the best available commanders should be appointed to direct the operations of the great army now assembling from all parts of the world. The choice fell on Lord Roberts and Lord Kitchener, the latter to be Chief of the Staff.

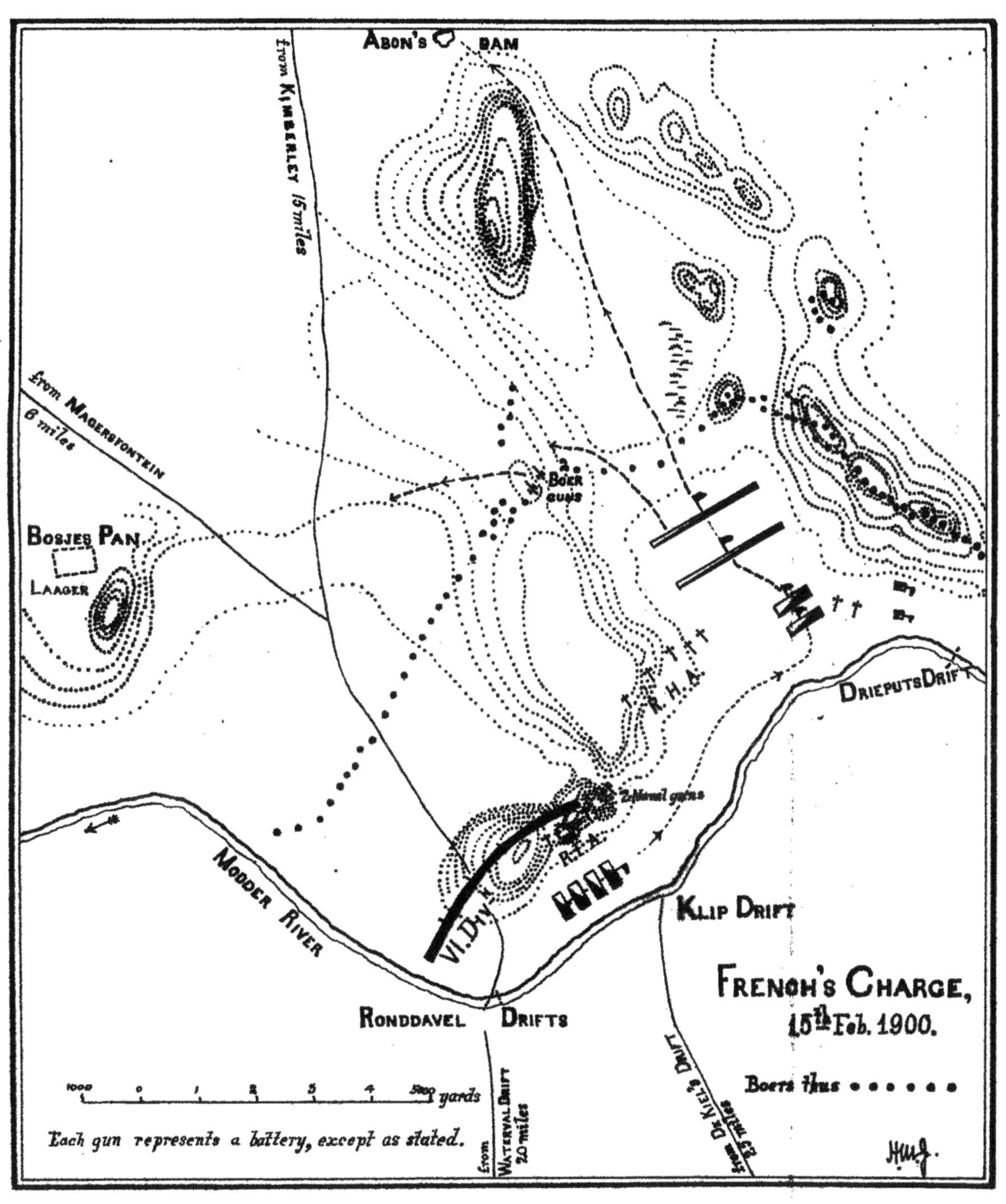

ABON'S DAM
from KIMBERLEY 15 miles
from MAGERSFONTEIN
6 miles
BOER GUNS
BOSJES PAN
LAAGER
R.H.A.
DRIEPUTS DRIFT
2 Naval guns
R.F.A.
VI. Div.
MODDER RIVER
KLIP DRIFT
FRENCH'S CHARGE,
15th Feb. 1900.
RONDDAVEL DRIFTS
Boers thus
1000 0 1 2 3 4 5000 yards
Each gun represents a battery, except as stated.
from WATERVAL DRIFT 20 miles
from DE KIEL'S DRIFT 2.5 miles

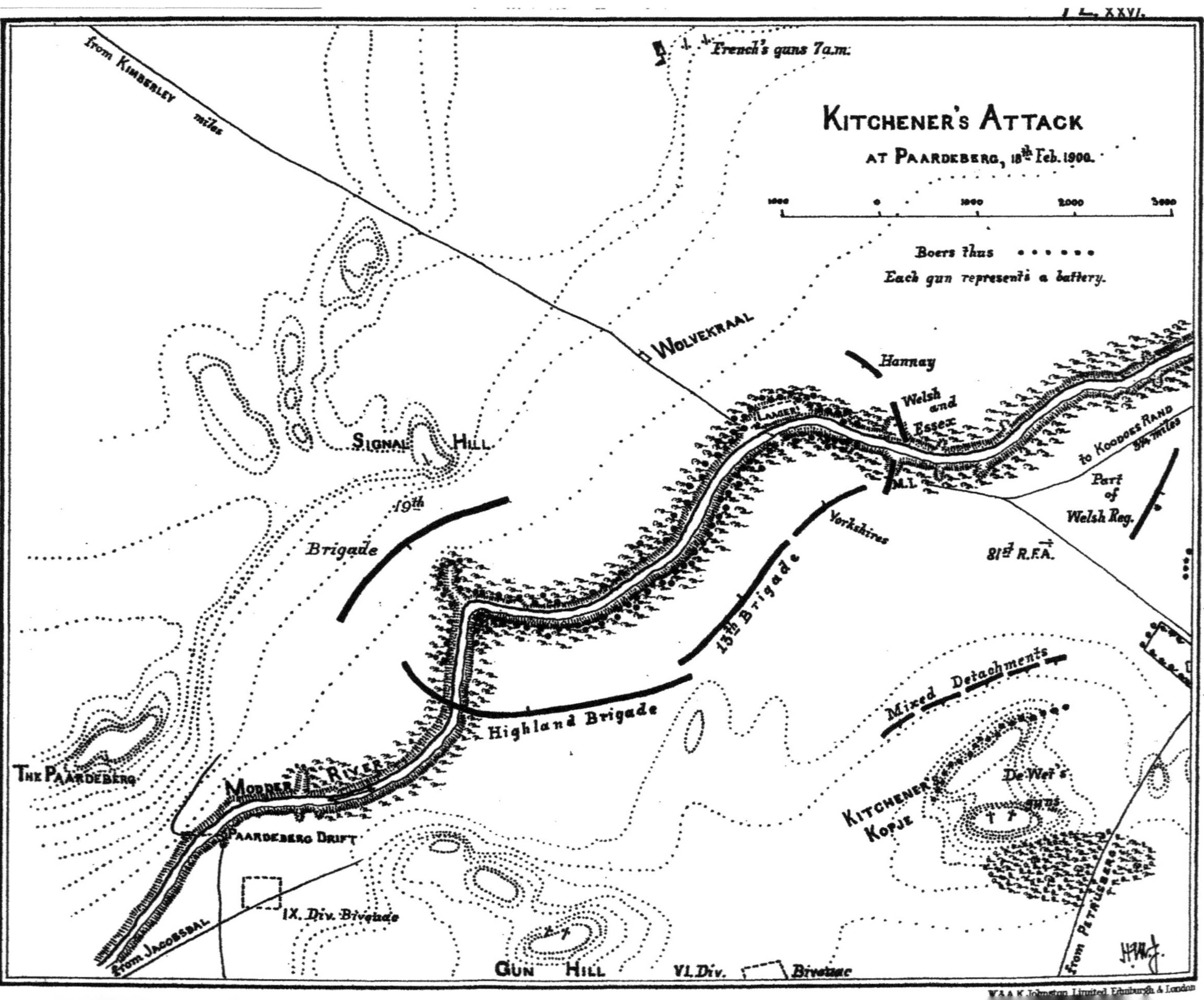

Pl. XXVI.
Kitchener's Attack
at Paardeberg, 18th Feb. 1900.
1000
0
1000
2000
3000
Boers thus
Each gun represents a battery.
French's guns 7 a.m.
from Kimberley
miles
Wolvekraal
Signal Hill
19th Brigade
Hannay
Welsh and Essex
Laager
M.I.
to Koodoes Rand
Part of Welsh Reg.
Yorkshires
13th Brigade
81st R.F.A.
Highland Brigade
Mixed Detachments
De Wet's guns
Kitchener's Kopje
The Paardeberg
Modder River
Paardeberg Drift
IX. Div. Bivouac
from Jacobsdal
Gun Hill
VI. Div.
Bivouac
from Petrusberg
W. & A. K. Johnston, Limited, Edinburgh & London

The mighty army now available, and the fresh spirit infused by these two distinguished and tried men, quickly altered the complexion of affairs, and a steady offensive became the order of the day.

The new C.-in-C. elected to operate on the western scene of conflict, and he organized and concentrated, in rear of Methuen, a force, whose first object was to be the clearing, by a strategic move, of Methuen's road to Kimberley, and whose first objective was to be Bloemfontein, the capital of the Orange Free State.

The strategy by which the first object was achieved, and its *pros* and *cons*, do not come under our subject; but in the course of the operations that set Cronje flying from Magersfontein there occurred a remarkable tactical event, which has been much debated, and which claims our attention.

I refer to General French's

Cavalry Charge at Klip Drift, 15th February, 1900.

French's Division consisted of three cavalry brigades and a mounted infantry brigade. It led the great flank march, aiming at a point on the Modder River, 14 miles in a straight line above Magersfontein, where two adjacent fords, Ronddavel and Klip Drifts, afforded alternative crossing places. These having been secured without much difficulty, the cavalry halted till troops of the VI. Division should arrive to hold them.

This march, made on February 13th, 1900, proved very trying to the horses, especially those of the gun-teams; 500 animals were put *hors de combat*. The unexpected difficulties in getting the transport of the main body across the drifts of the Riet River, 25 miles to the south, had delayed the leading infantry division (VI., Kelly Kenny's) to such an extent that the division did not begin to arrive until 1 a.m. on the 15th.

Meantime, fugitives from Ronddavel Drift had told Cronje that French was there on the Modder with a great force of cavalry and many guns; but the Boer leader was slow to believe that the British were about to leave their railway for anything more than a feint. He sent, however, Commandants Froneman and De Beer, with 800 men, to drive away the "presumptuous cavalry." These arrived before dark on the 13th, but missed the chance of attacking French's exhausted troops, and contented themselves with taking up defensive positions along the kopjes to the north of the two drifts.

At a late hour on the 14th, Cronje learned that the VI. Division was following French, and he now guessed that something serious was intended. About the time the division was arriving at Ronddavel, Cronje was engaged in shifting his head laager from Brown's Drift, only six miles down stream from Ronddavel, to a more hidden position at Bosjes Pan, three miles to the north-east; but for the moment he remained in his trenches at Magersfontein.

French's orders were to go right into Kimberley; a great moral effect was hoped to be thereby produced, and there was always the chance that the enemy, outmanœuvred, might make a combined rush on Kimberley as a last desperate move. But there was now an unknown number of a redoubtable enemy, barring the way among the kopjes; for all French knew, the main body might now be in his front, 36 hours having elapsed. *Plate* XXVI. Reconnaissances during the 14th had shown that the kopjes running from Drieputs Drift towards Abon's Dam were occupied for two or three miles from the river, while the kopjes running north-east from Brown's Drift were also held. Between the north ends of these two lines was a low, irregular ridge, two and a-half miles from the river; it looked like rideable ground, and was the direct way to Abon's Dam. French determined to charge through this, trusting to speed.

By 8.30 a.m. the cavalry had been relieved of the outpost duty on the line of kopjes covering the drifts, and was soon formed up behind them in column of brigade masses, guns on the left flank, in the following order, 3rd, 2nd, 1st, 4th Brigades. The ammunition column of the 2nd Brigade was alone to join in the dash, the others having to give up their horses to remount the gun-teams.

French now explained his intentions to his brigadiers, and discussed them with Lord Kitchener, who promised to follow next day with every available man and gun. The leading brigade moved up stream for about two miles, followed at about a mile by the 2nd Brigade, with which rode French. "The cavalry scouts soon succeeded in ascertaining the strength and extent of the enemy's position, because the Boers, contrary to their usual custom, opened fire on them at long range. . . . In consequence of the reports sent in, French ordered his seven batteries of horse-artillery, which were soon afterwards joined by two batteries of the VI. Division and two naval 12-prs., to come into action.* . . . The artillery opened fire at 2,200 yards, spreading it along the entire Boer position, and soon succeeded in silencing the three hostile guns. At the same time the VI. Division began to push forward against the Boers in their front."

The horse-batteries were to continue firing till the charge was well under way, and then join the rear brigades. When the wheel towards Abon's Dam occurred, the 3rd Brigade led the 2nd by 800 yards. Gordon opened out the 3rd Brigade into double line, five yards intervals between the files, and the whole division thundered towards the col at a gallop. The Boers opened a heavy fire from both flanks, and some hurried towards the col; before it was quite realized what was happening, the cavalry were over the col, and sweeping on towards Abon's Dam. The speed of the charge, the open order, the cloud of dust, and the effective support of the batteries, rendered the fire unusually ineffective. About 20 Boers were speared by the Lancers; one of the leading regiments swerved to the left in the hope of cutting off the Boer guns, but these had been withdrawn the moment the charge began. By 11.50 a.m. the division was assembled at Abon's Dam, where there was water, and the road to Kimberley was open. The ride has cost only 2 killed and 17 wounded.

The *Times*' historian of the war says :—"The charge marks an epoch in the history of cavalry. With other cavalry charges it has little in common, save the 'cavalry spirit'—the quick insight that prompted it, the instantaneous decision that launched it against the enemy, the reckless, dare-devil confidence that carried it through. . . . It was not a sudden charge upon an enemy already fully engaged with other troops or retiring in disorder. It was no instance of shock tactics, for there was nothing opposed to it on which a shock could take effect. Nor was its success in any way connected with the particular weapon carried by the cavalry. . . . It was only possible by the very tactical conditions that made these self-same frontal attacks so futile and costly when attempted by infantry . . . a position which, with infantry, would have required the best part of a day, and have probably involved heavy losses. Unfortunately,

* The positions of these are shown on the Map.

during the South African War the experiment was never again repeated on a large scale, and the cavalry were only employed to turn positions, not to storm them. But the part played by cavalry in the main attack, when conditions of ground are favourable, is one that will grow in importance in the wars of the future, . . ."

So also thought, and perhaps still thinks, the German official historian, but to prophesy is unsafe. Nothing of the kind seems to have taken place in Manchuria. The German account says :—"This charge was one of the most remarkable phenomena of the war; it was the first and last occasion during the entire campaign that infantry was attacked by so large a body of cavalry, and its staggering success shows that, in future wars, the charge of great masses of cavalry will be by no means a hopeless undertaking even against modern rifles, *although it must not be forgotten that there is a difference between charging strong infantry in front and breaking through small and isolated groups of skirmishers.*"

The italics are not in the original; and even the qualification there suggested does not go to the root of the matter. The Boers, on many occasions such as this, were not only in a thin line, but in a single line; how would it have fared with the cavalry if there had been a supporting line, and a reserve in rear of that; if any one of these three had had a barbed wire obstruction in front of it?

This is not said with the aim of detracting in any way from the merit or this charge—there *were* no supporting lines, and there *were* no obstructive preparations. With a splendid flash of genius the cavalry commander saw that the Boers would least expect a charge between the two lines of kopjes, held by their riflemen, and that the col would therefore be least defended, either by men or artifices.

French was now in rear of Cronje's head-laager, and had it at his mercy, but he was quite unaware of its proximity, and his orders were to go on to Kimberley.

Meantime the British main body was massing towards the Drifts of the Modder, and Cronje was trying to understand the meaning of French's ride. Fresh information soon set all his doubts at rest, and he broke up from Magersfontein on the same day. Next morning, 16th February, at 4.30 a.m., the M.I. Division, 2,000 strong, advanced on the line French had taken, in time to see a great cloud of travelling dust away to the right front. This was Cronje's march; he had slipped through between French and the main body, discovered the advance of the latter, posted a strong rear guard on the kopjes, and escaped.

Kitchener urged on the pursuit, and French was instructed to head the Boers off, a feat which he performed with great skill. Cronje was "held up" just after he had passed the Paardeberg; and we are now concerned with the attack which Lord Kitchener made on his position along the river-bed.

PAARDEBERG, 18TH FEBRUARY, 1900.

v. Plate XXVI.*

Cronje reached Wolvekraal on the 17th February at 8 a.m., and there established his laager on the river bank. The VI. Division followed him at 3 a.m. on that day, with a brigade on each bank. The M.I. Division, with reprehensible tardiness, did not get under way till 7.30, and only regained touch with the hostile rear guard at the conical hill called Paardeberg. Cronje was now so confident of being able to escape that he

* The distance from Kimberley to Wolvekraal, omitted in the map, is about 28 miles.

determined to postpone any further advance until next morning, or at least for some hours, when shells began to fall among the wagons, not from the rear, but from the north-east. These were fired by French's horse-artillery; and the sound of his guns had the incidental effect of frightening off for the time a Free State commandant who was coming down from the north to Cronje's help with a force double the strength of French's single brigade.

Cronje now set to work to prepare his peculiar position for defence, such a position as has seldom, if ever, been occupied by a defending force. Narrow and sinuous and commanded on all sides by artillery positions, it had nevertheless some peculiar advantages. The Boer riflemen had a field of fire of the same quality as that at the Modder River battle, and their dispositions were completely hidden by the fringe of bush along the river banks. The river here was about 60 yards broad, crossable at many points at this season, the water being some 40 feet below the level of the plain. Opposite "Signal Hill," a narrow and deep watercourse, dry at this season and thickly covered with bush, dipped into the river from the north. This was the west end of the position, and it added considerably to the strength of this flank. To the east of the laager, a number of similar gullies existed. On both sides there was gently sloping ground for a minimum of 1,100 yards, on which the only cover was some small ant-hills. Movements within the position could not be perceived from any point outside, while the keen eyes among the bushes could see every preparation for serious attack, the moment it was put in hand.

The VI. Division was plodding along during the night of the 17th, and it arrived and bivouacked where shown, without quite knowing where it had got to, or where the enemy was. Two thousand yards in front of it was Hannay's M.I., with which was Lord Kitchener. In the rear the IX. Division was toiling up, having accomplished 31 miles in 24 hours, arriving between 11 p.m. and 4 a.m.

Daybreak, at 5.30 a.m., showed the commanders that the prize lay close before them. Under instructions from Lord Roberts, Kitchener took the command, and at once determined to "rush" the laager.

At 6.30, two batteries were firing from "Gun Hill," and these succeeded in a very short time in silencing the three guns that Cronje had near his laager, and in blowing up some ammunition wagons; infantry of the VI. Division were deploying for attack, five battalions preparing for a frontal attack from between "Gun Hill" and "Kitchener's Kopje," and two moving further east towards the other flank of the position, to assist some M.I. who had, an hour earlier, become entangled unexpectedly in a fight within 500 yards of the Boer laager. Four of the five battalions were put into 1st line, only one being kept back; and it was split up into detachments to cover the guns, to protect the baggage and to occupy "Kitchener's Kopje." The four battalions had a front of nearly two miles, being deployed in the usual three lines, almost exactly as the Highland Brigade were to deploy at Magersfontein. The enemy opened fire at well over 2,000 yards, but without effect, the British skirmishers not replying till their brisk advance brought them to within 1,500 yards. The firing at first was by volleys, and by word of command.

After a short exchange, mutually supported rushes of 100 yards by sections began, and the supports began to reach the firing line. The supports were suffering almost more than the firing line, and presently the former, when the firing line had been brought to a halt, began to reach it by crawling. As 500 yards from the enemy was reached, all started crawling, a company at a time, the rest redoubling their fire. Two of the battalions reached to within a quarter of a mile, and there checked; the other two lay down at 750 yards—one account says, "conformably to the express orders of the divisional commander," but this is almost incredible.

At 7 a.m., Kitchener had ordered the IX. Division to put in its Highland Brigade, and to push its batteries into action, the infantry prolonging the left of the VI. Division. One of the batteries on "Gun Hill" moved to a point north of "Kitchener's Kopje," to help to the east against any attempt of the enemy to break out, and was replaced by a howitzer battery. By 8.30, it was evident that the VI. Division's attack was going to be no easy success, and the Highland Brigade was by that time committed beyond recall to the attack on the left bank. It had advanced in one single line, eastwards from its bivouac, men at four paces interval, until the head of the line was behind the left of the VI. Division. The immense thin column then faced to the left, and advanced towards the river bank just as it was, without supports or reserves. The right moved forward, without halt and without firing, till it joined the VI. Division at 800 yards from the enemy; the left got to within 500 yards, then lay down and commenced firing.

The left of the Highlanders far overlapped the Boer position; finding itself on the bank with no enemy in front, it splashed across the stream, and joined the 19th Brigade (IX. Division) in the attack ordered by Kitchener on the other side. The Highlanders lost men freely in their silent advance across the open; the unsupported line made no further advance worth speaking of during the day; but the left, across the river, having support, reached to within 300 yards of the enemy.

Meanwhile, French had been shelling the position from north-east; but it seems that the natural and artificial shelter of the Boers rendered the fire of all these batteries almost innocuous against their riflemen. French signalled that he was being threatened by Ferreira from the north-east and east, and Kitchener directed him to turn his attention that way.

We saw that two battalions of the VI. Division had gone east at the outset; and the M.I. had been ordered in the same direction, the intention being to attack on both banks from the east. But the battalions, insufficiently impressed with their objective, and finding the M.I. skirmishing with a thin line of Boers to the east, turned away from the laager and joined in the skirmishing towards the south-east. A section of M.I. had taken partial possession of "Kitchener's Kopje." Hannay, commanding the M.I., was now ordered to join the infantry in the proposed attack, but his men were much scattered, and most of the infantry, with a field battery, continued to carry on the stationary action with their backs to the laager.

It thus happened that the attack on the laager from the east was made with insufficient numbers; even so, it reached to within very short range. Later, urgent pressure from Kitchener drove Hannay to the desperate expedient of mounting with a handful of his men and charging to within a few yards of the laager. Hannay and most of the party were killed.

The day was wearing on, and everything was at a standstill. About 3 p.m., the party on "Kitchener's Kopje" moved down to a farm to water their horses; De Wet, who had been hovering about with 500 men and 2 guns to the south, promptly occupied it, and the VI. Division found themselves fired at from the rear as well as the front. Energetic action on the part of De Wet would now have afforded Cronje a good chance of escape, but De Wet had not yet developed the skill and confidence he showed later.

A staff officer hurriedly collected some M.I., many of whom were rambling about rather aimlessly, formed a semi-circle round De Wet's position, and to some extent diverted his fire ; but from this moment our attack, begun so vigorously, dwindled "into a toilsome and dragging defensive action. Everybody was lying down close to the bare plain, anxiously awaiting the approach of darkness, in order to seek for cover further to the rear. Late in the afternoon the fire began to slacken, there being no more troops to reinforce the firing line and to bring up ammunition. . . . The infantry had to face this continuous and intense fire for nearly 12 hours without food and water. The men suffered terribly from thirst. Many British officers consider that to provide troops in action with water is . . . almost as necessary as bringing up ammunition."

The 19th Brigade, led by Colvile, commander of the IX. Division, began well. Making good use of the ground, it performed a wide flanking movement beyond the watercourse before mentioned, and then advanced in a single line, just as the Highlanders had done, to envelope the river-bend, getting to within 850 yards without opening fire. The 82nd Battery fired for some time at the same point, but diverted its attention to the laager under the impression that its shells were interfering with the advance of the left wing of the Highlanders.

The "interior lines" which the Boers enjoyed enabled them quickly to change this flank attack into a purely frontal one, and by 2.30 p.m. the impulsion was expended. To Lord Kitchener, watching from a distance, it seemed that a resolute bayonet charge of the whole should succeed, and he ordered Colvile accordingly. Colvile demurred, the distance being still 700 yards ; Kitchener repeated the order peremptorily, but even now Colvile did not carry it out in the true spirit. Having only half a battalion of fresh troops, he communicated the order to it alone. The four companies reached the Royal Canadians by rushes and crawling, the latter without orders joined in a further crawl to 500 yards, the 12 companies fixed bayonets, rose and charged with cheers. But the line became thinner and thinner, and the survivors threw themselves down short of the objective. In these few moments the half-battalion had lost 22 per cent. of its men, and the Canadians almost as heavily.

The attempt to charge may have been premature, but the commander on the spot who allowed the thing to be done, but held back more than half his men, cannot be exculpated.

The troops were all withdrawn after dark, and things were *in statu quo;* the British had lost over 1,200, 67 being officers, making 8 per cent. of the 15,000 men on the field ; the Boer loss was about 300.

Comments on Kitchener's Attack.

This attack by Lord Kitchener has been keenly criticized from two points—(1), whether the attack was necessary at all ; (2), whether it was reasonably well planned.

As to (1), those who object to the attack *in toto,* say that the object might as well have been attained by occupying defensive positions such that Cronje would be hemmed in ; that the object *was* attained quite comfortably in this manner. This seems rather like judging after the whole event ; on the other hand, it is true that both Kelly Kenny and Colvile wished from the outset to act in this way.

But Kitchener's available information must be taken into account. He, and others, supposed Cronje's force, in reality 4,500, to be much stronger than it was; and he knew that 3,000 Boers were outside, and within reach. Thus, he had to deal with a force equal to half his own and French's, and he thought he had to deal with one a good deal larger than this. Of his enemy he had more than half cooped up in the river bed, and after the events of the previous days he may well be excused if he thought this portion was demoralized.

The German Official Account is of opinion that the determination to attack on the 18th was perfectly sound. The *Times'* History says that "as far as available estimates went, he might have to reckon, even at the end of a few hours, with 12,000—15,000 Boers in all." Even if this is an exaggeration, the desire to make a rapid end of the surrounded enemy was quite natural, and might quite well seem possible.

But as to (2), the room for criticism is great, and the criticism is almost universal, and it is mostly directed towards the same points.

"Careful preparations should have been made," "the divisions should have been told off to definite points of attack," "the objectives of each should have been clearly defined," "the morning could have been employed in bringing the divisions on to their ground, and in letting them thoroughly reconnoitre that portion of the position each was to attack," "only by such reconnaissance could the guns produce their full effect by knowing where to fire."

Such is the gist of the German criticisms; but the writers do not say how they propose to reconnoitre under such circumstances as existed. To push forward men here and there to draw fire would tell very little, for the short and completely hidden communications of the Boers enabled them to shift their firing line rapidly and without molestation.

It was a case where any part of the fringe of bushes must be held to contain the enemy; unless, then, some part of it could be seen to afford any special advantage, either in the approach to it, or in itself when occupied, there seems little reason for attacking at one part rather than at another. Some point near the laager naturally suggests itself.

But the real error lay in the extent of front in the attack. This was such that all the men were used up in front line, and no adequate force remained in reserve, to be used where success seemed likely. This matter should be clearly understood, for it is more complex than appears at first sight.

In a case were reconnaissance can be effective, and a perfectly definite part of the hostile position is therefrom selected for the main effort, the reserve is naturally kept in rear of that part of the attacking line. But where no such definite part is discovered, the natural process would be to treat the general advance of a front line as a reconnaissance in force, watch carefully which part of the line made best progress, and throw in the reserve there. On a widely-extending attack, the thing could not be done quickly unless several bodies of reserve were in hand; the numbers required would thus militate against wide extension, and a mean would have to be carefully thought out and planned. This is exactly what does not seem to have been done on this 18th of February.

To one who was not there, and who only takes his knowledge from maps and many descriptions, it seems as if one or other of the narrow ends of the position should have proved the easiest objective—and not only the easiest to penetrate, but the most effective when penetrated. A semi-circle of firing line, supports and reserves, starting at six paces *per* man when a mile distant, would have contained (say) 3,000 men, or a brigade; such a line would have converged its fire on the end of the position, and in advancing would have become ever stronger and denser. A second brigade in general reserve would have been ready to advance in precisely the same manner, to carry on the impulse if necessary. Once the whole edge of the bushes was cleared, every man of the two brigades could have been rushed in helter-skelter, with no slanting or flank fire to fear. The other division and the M.I. and the guns could have been defensively posted to prevent breaking out, but would at the same time keep threatening, in order to tie down as many as possible of the defenders at a distance from the attacked end.

It was noticed as a curious fact that casualties were less when short ranges like 300 and 400 yards were reached than at 700 or 800 yards. This was apparently due to the fact that, when our men got close, the Boers were afraid to show their heads long enough to take aim, and even fired blindly without putting the head above the parapet at all.

The German lays great stress on the lack of "carefully prepared fire tactics." "There was no instance," he says, "of a strong and determined effort being made to acquire the ascendancy over the enemy's fire." And later—"wrong and fatal endeavours to rush the enemy's position as quickly as possible without first overwhelming him with fire."

No doubt these are excellent precepts; but the critic seems again to forget the peculiar conditions. At Gravelotte or Sedan the German attackers, when within effective range, saw the Frenchmen almost as well as the latter saw them. At Paardeberg and at Modder River the affair was very different; and it was the most natural thing in the world for brave men, ordered to push home, able to see nothing of the effect of their own fire, and naturally suspecting that it was having little or no result, to push on in an effort to use the bayonet.

The complete withdrawal at dark is also criticized; "the positions gained should have been entrenched, as points to start from in the morning." This is all very well, if fresh troops had been available; but a man is made of flesh and blood and not of iron.

CHAPTER XXI.

GENERAL COMMENTS ON THE BOER WAR.

THIS was the first war in which the belligerents, both armed with magazine rifles of great range and flat trajectory and using smokeless powder, fought in a country eminently adapted for the full development of the power of the weapons. If it be added to this that our enemy was an army of mounted infantry of very remarkable individual capacity as such, it will be expected that great developments of the method of fighting were bound to ensue. We have often before had to deal with an enemy whose individual mobility was far greater than our own; so have the French in Algeria and the Russians in Central Asia. But in those cases, as a rule, the enemy was indifferently armed, quickly realized that long-range fighting was no good, and adhered to his natural tactics of getting with all possible speed to close quarters, compelling the civilized troops on the other side to abandon the open order of their drill-books, and to fight in close order.

But in South Africa the enemy added to his mobility a weapon probably rather better than our own, saw that hand-to-hand fighting would be his weakest point, and consistently employed a tactics that aimed at victory long before the combatants could "see the whites of each other's eyes." So much was this the deliberate system of the Boers that their riflemen had no bayonets.

This war lasted so long, strained our resources to such an extent, and bulked so largely in our eyes, that there is a danger lest we forget the peculiarities of the case, and begin to assume that all we learnt there will be applicable to all future war. So much is this the case that some authorities are already saying that the Boer war was, in the matter of tactics, at a far more advanced stage of development than the war in Manchuria, thereby implying that the latter shows, as it were, a retrogression, and that we should model our fighting on the South African system. Such statements ignore the peculiarities of the case; at the same time, some of those peculiarities will crop up again, and valuable lessons for such occasions have been provided.

Common System of Training for Commanders.

The great extent of ground covered by a modern battle has brought into prominence a matter that has always been important as a factor towards success; the lack of it was painfully visible in several of our battles in South Africa, and its importance was consequently accentuated. I refer to the necessity of a common system of training for commanders. No men are perfect, no system is perfect and final; but it is of vast importance in war that battalion commanders, brigadiers, divisional generals, should all naturally strive to do the same thing in the same way in given circumstances. A certain situation being envisaged before them, their training should be such as to lead them individually to find something mutually similar as the proper way of meeting the situation. The almost unvarying success of the Germans in battle in 1870 was to a great extent due to this sort of uniform training of commanders. As various writers have said in different words, the Prussian commanders seemed to have all been at school together and to have all learnt out of the same book, and at the mouth of the same teacher. The book may not be perfect, the teacher may have some wrong ideas; but if two brigadiers have a joint operation before them, it is better that they should agree as to the natural way of doing the thing, even if the way be the second-best, than that one should act as if three hours of artillery preparation were necessary, while the other goes in after half-an-hour of it.

The difficulty of transmitting timely orders on a modern field has brought this matter into great prominence; but any approach to a narrow rigidity of training is just as much to be avoided. In the Prussian army of 1870 there was possibly rather too much of this rigid narrowness; to attack the enemy with the utmost fury, the moment your command finds itself in touch with him, often results in taking the whole planning of the operations out of the hands of the C.-in-C.—in the future this will probably be more pernicious than ever, on account of the extreme difficulty, the frequent impossibility, of withdrawing a corps once it is engaged; but the harm done will usually be lessened, if neighbouring corps, arriving, take the same view and plunge into the fight,—not blindly, of course. but with a due regard to the tactical and strategical conditions.

What is coming to be wanted, then, is a broad and careful teaching, and a teaching so thorough that commanders under similar circumstances will take similar action. Conspicuous examples of success under these conditions are to be found in the best period of the Napoleonic battles. A clear case of failure is to be found in the abortive charge of a fraction of the 19th Brigade in the already described attack on Cronje at Paardeberg. On the south side of the river eight or nine battalions are lying within 500 to 800 yards of the Boer position, "containing" a great number of the enemy, but unable to advance further, and approaching exhaustion. The 19th Brigade, on the north side, have gone later into action, and are lying down about the same distance from the enemy. Whether the whole attempt was foolish, or was badly planned, is not the question; one or other of the two attacks must push in, if success is to be achieved. The C.-in-C., surveying the whole field, orders the 19th Brigade to do it. Its general has now no proper say in the matter, and he should act as if he took it for granted that the thing can be done, if the whole brigade is pushed in. Would a Prussian brigadier of 1870 have ordered in half a battalion?

Generals to keep out of Details.

Another matter that belongs to all war emerged in an acute form on more than one occasion in this war, namely, the breach of the good rule that a superior, neither in written orders nor by personal action during an operation, should interfere with work that is within the proper sphere of a subordinate. On a particular occasion no harm may be done—even a temporary good may be achieved—but the effect is vicious of a system in which this sort of thing becomes at all common.

Thus Pole-Carew at Magersfontein had a definite written order not to engage closely; the supreme commander could not know the night before what the exact situation would be at daylight when the attack developed; when it did, he became so immersed in the Highland Brigade's mishap that Pole-Carew was left with the paralyzing, and too definite, order of the previous night. The latter should have had "instructions" that "the Highland attack is to be the chief attack, in which you will co-operate according to circumstances." As things turned out, the Boers soon saw that Pole-Carew had no mischief in hand, and they were able to send round men from opposite him to extend and strengthen their extreme left. Probably by mid-day Pole-Carew was "containing" not more than a fifth part of his own force. Had he been left to use his 2,000 good men "according to circumstances," a general of his vigour and tactical skill would probably have been among the kopjes near the railway within an hour of the time when he was aware that the main attack was at a standstill.

This was an instance of an unnecessary, and therefore harmful, stringency of formal orders. A case of interference in a local matter by the C.-in-C., and his complete immersion in a subordinate's business, to the extent of producing neglect of that general aspect of the operation which a commander should keep steadily in view, is one that has been mentioned in a previous chapter, namely, what happened on the occasion of the overwhelming of Long's ten guns at the Battle of Colenso. That a third of his field artillery should be out of action was certainly no trifling detail, but the remainder of his guns and the fortunes of a large force were the general's charge.

The less the C.-in-C. goes into the thick of the fight the better.

Difficulty of Reconnaissance.

The difficulty of reconnaissance under modern conditions, and especially in the clear atmosphere and open spaces of South Africa, has been already referred to. The increased difficulty is due to the long range of the rifle and the use of smokeless powder. Even a reconnaissance in force may find out little, if the enemy are well concealed, and make use of their magazines to produce the illusion of greater strength at some point than the actual. At the same time the necessity of reconnaissance is probably greater than ever, on account of the difficulty before-mentioned of extricating a force that has blundered into a trap. Thorough reconnaissance, therefore, must be done, and it will require in future specially trained men, the ordinary cavalry officer and trooper, as at present trained, being often quite unequal to the task.

When the work was not so difficult, when the scout could approach much nearer without the certainty of being hit, when the smoke of shots told him the exact locality of the shooter, when the careful concealment of gun-positions was less useful and was therefore not carried out, the cavalryman was naturally told off to reconnoitre, simply because of his swiftness. But his real business is to fight cavalry, to raid, to screen, to catch guns on the move, and perchance, if the German theorists are on the right track, even to charge infantry.

I have recommended elsewhere, as a thing I am sure will come soon, a special corps of scouts, fully trained in all the arts of the stalker of game—these arts being grafted on to a sound preliminary military training.

It is not intended that then the cavalry would never again reconnoitre; but that whenever circumstances prevented their doing any good, as in the failure to locate the Boers at Modder River, and the failure to find the low trench at Magersfontein, there would be men at hand who would at least have a better chance of discovering what was wanted.

The use of balloons for reconnaissance was no new thing; on several occasions they proved their use. Their utilization is part of the greater question of the introduction into the field and the fortress of the technical, a matter that will be discussed in the chapter on the war in Manchuria.

Locating of Guns.

The locating by scouting of the hostile guns proved quite impossible in South Africa; and as the Boers in most cases wisely declined the preliminary artillery duel, on account of the superiority of the number of our pieces, their discovery even after they had been firing for some time proved extraordinarily difficult. This was due to the great range, to the retired position that could consequently be occupied, to the smokeless powder—considerations that now apply to all cases. But there were other reasons peculiar to the war in question—the paucity of the pieces, and often the nature of the hilly country, strewn with huge boulders which gave cover to all the accessories of the gun, and threw shadows that puzzled the eyesight. A few guns can be easily hidden, a great number cannot all be hidden.

The Boers having still some smoky powder, and occasionally using it, were able to employ a cunning ruse. A big gun having fired with smokeless, a party ensconced hundreds of yards to the front would burn a bag of smoky powder simultaneously; this smoke attracted the hostile gun-fire. In future this ruse will be of no avail; but the Japanese are said to have sometimes deceived the Russians in an analogous manner, by throwing up sand or dust, at a distance from the true position of a battery or big gun, with the intention of making the enemy take it for the dust thrown up by the concussion of discharge—the ground round the battery itself being kept well soaked to prevent any dust rising there.

Indirect Fire.

The frequent use of indirect fire in this war in the open field was something of a development, such fire having previously been almost confined to siege-work. This method of fire by howitzers and the naval guns naturally aided the effort at concealing gun-positions, but brought into play the necessity for a systematic method of observing and signalling the effect of the fire. The matter is also intimately connected with the remarkable development of the size of guns taken into the field. The Boers began it, and we promptly found ourselves hopelessly outranged. At Ladysmith, when the enemy first began to sit down about it, their heavy Creusot guns were able to bombard from anything up to 9,000 yards. Placed thus well behind the investment line, we were powerless to engage them, till the 4·7-inch naval guns arrived. From this moment our forces were provided with an ever-increasing number of these long-range weapons, and no considerable engagement occurred without their employment.

Heavy Guns.

The older Boers, when war was imminent, looked askance at what they supposed would be the encumbrance of artillery; their perfect faith in their rifles and their horses made them sceptical of the use of guns, and to have to escort the artillery they looked upon as a nuisance. So the latter kept to retired positions, were left to look after themselves, and were consequently always ready to escape in good time. Their guns, few and precious, were therefore not expected, in European fashion, to be the chief instruments in covering a retreat. Retiring commandos saved themselves by the mobility afforded them by their mounts, and against cavalry pursuit by dismounted work in the mounted infantry style.

Lack of Aggressive Action on the part of the Boers.

While these proceedings saved artillery escorts, and kept the guns safe at the same time, they militated against the aggressive artillery action which stood the Germans in such good stead in 1870. In fact, during the first two phases of the war, aggressive action at all comparable to that of 1870 was notably absent.

The Boers certainly invaded, and they attacked; but as soon as they brought us to a standstill, they appeared to be content. The abortive attempt of Joubert beyond the Tugela was a feeble performance; the repulse of Buller's attacks was never followed by a counter-stroke; the *débacle* at Magersfontein left the Boers contentedly ensconced in their trenches; the siege of Ladysmith was an easy-going performance for the most part; the investment of Kimberley was a most unvigorous proceeding; only at Mafeking was there active work. The reasons for all this have been variously stated—the Boer character, the lack of trained generals, the democratic nature of their armies, causing plans of operations to be at the mercy of individual opinion, an ingrained dislike of dangerous ventures, lack of discipline in a national force controlled by no vigorous military law, lack of a trained staff.

Quite probably all of these things came into play; when it was too late, the survivors had been gradually hammered into disciplined troops, and under De Wet, De la Rey and Botha, performed feats of war that were remarkable.

British Artillery.

To return to the guns—the British artillery on most occasions did its work conspicuously well, keeping the guns safe when nothing was to be gained by risking them, pushing them forward boldly when an emergency arose. The case of Long's guns at Colenso was an instance of a complete misconception of what guns can do.

Prussian guns in 1870, British guns at Magersfontein, pushed in to rifle range; but these were occasions when the artillery either deliberately risked its existence to save its infantry in great jeopardy, or saw that the hostile infantry was fully occupied in its struggle with its immediate opponents.

Observation of Fire.

When an enemy is in deep and narrow trenches, shrapnel fire should be more effective when it descends at a considerable angle than when it flies horizontally; therefore a gun firing shrapnel may do more by firing at 3,500 yards, and obtaining an angle of descent of 1 in 7, than by advancing to 1,500 yards and getting hardly any angle of descent at all. But the longer range renders more difficult the hitting off of the exact elevation, and the observation of the effect from the battery becomes harder. Two developments are consequently in process—the provision of a great telescope for each battery, and the systematic arrangement for signalling back from the firing line instructions as to where to shoot. It has even been suggested that an officer of each battery should have charge of a great telescope mounted on a carriage like a machine-gun carriage; the French are said to be experimenting with these; and that an officer from each battery, or at least one from each artillery brigade, should as a matter of course accompany the C.O. of the front line where it is approaching the objective of the battery's fire, and by prearranged signals communicate with the battery. That this duty should be performed by an artillery officer, and not be left to the infantry, seems quite sound; and that the former should have plenty of practice at it in peace time is imperative. A well-arranged and well-practised plan of communicating would minimize the risk also of guns firing at their own infantry, a thing that happened at Talana Hill and at other places, and pretty often more recently in Manchuria. It has also been suggested that a flag should be carried in the firing line, but this might easily be used by the enemy to save his retreating troops. It has also been suggested that a few men of each infantry company should have a conspicuous back to their coats, such as white; but this would render these men a fine target to the enemy, if they had to retreat, and this again is a device that might be used by the enemy.

Guns Firing at their own Infantry.

Lyddite.

Great hopes were entertained of the probable effect of high-explosive shells, but the reports from the Boer side show that in well-planned field-entrenchments the effect was often small. In many cases the old-fashioned common shell would have been better; but our great fault often was in wasting ammunition by firing at trenches without first pushing forward infantry far enough to compel the enemy to man his trenches. Thus at Magersfontein, on the afternoon before the night-attack, the whole of our guns bombarded the kopjes for hours, the total effect being the wounding of three Boers. The pushing forward of infantry would not only have given the guns a human target, but the landing of the shells could have been continuously signalled, and the barren bombardment of acres of hillside avoided.

Pom-Poms and Machine Guns.

An effective weapon that emerged in this war was the so-called "pom-pom," a quick-firing piece that threw a continuous stream of 1½-inch shells, exploding them on impact. These guns were very effective against such targets as gun-teams, and their projectiles would penetrate, and explode within, corrugated iron or thin brick buildings. Our army was provided with them when we experienced their effect in the hands of the Boers.

Machine-guns were not of much use on the veldt; they will be fully discussed in connection with the war in Manchuria.

The Fortress of Pretoria.

Though on the open hillsides the Boers soon found that the lyddite bombardments were not of much consequence, they wisely recognized that to be shut up in a fort, with a ring of lyddite-firing batteries round about, would be an unpleasant experience; and no doubt from this tactical forecast, as well as from a sound strategic instinct, they properly declined to allow themselves to be shut up in Pretoria. The great questions of the use and abuse of permanent fortresses are so much a broad matter of strategy that we are not called upon to discuss them here; but we may note that for the Boers, whose strength lay in the rifle and mobility, to betake themselves to fortresses that could be surrounded would have been to throw away their chief advantages. Hence their extreme solicitude whenever any movement of ours seemed to threaten to surround them. Until Lord Roberts came on the scene, we did not make half enough use of this fear of theirs; it is on record that, some days before Magersfontein, the Boers were on the point of trekking from the place when they thought a certain movement of Methuen's seemed to portend the enveloping of their left flank.

Mounted Troops.

That we should have had a much larger mounted force from the outset has become a commonplace of criticism. The frequent impotence of cavalry in reconnoitring under modern conditions has been touched upon. The charge of French at Klip Drift has been described, and we have found it doubtful whether such things can be done except under the peculiar conditions of that case; it has been noted also that German military opinion seems to still think that effective cavalry charging against infantry may be often possible.

All cavalry, to a great extent as the result of this war, are now armed with the full-range rifle. The great defensive power of such a force was shown on the Boer side throughout the war, and most conspicuously by French's cavalry when they headed off Cronje, who outnumbered them by four to one, and then held off Ferreira, who had more than double French's numbers.

Two important points emerged in respect to mounted troops—(1) the ease with which a force can be destroyed by fatigue, and (2) the absolute necessity of the individuals of the force being thoroughly well-trained, if they are to be dependable.

Both points have emphasized the necessity of a thorough knowledge of horsemastership, so that the most may be made of the animals. A cavalry commander can hardly be too careful of inflicting unnecessary fatigue on his horses, for it is of the very essence of the cavalry *rôle* in war that it should expect sudden great calls on its energies.

General French went to Kimberley with four brigades of about 1,500 horses each, seven batteries of horse artillery and one ammunition column. The great exertions of this ride, and the next day's work against some parts of the Boer investing force, reduced, by sheer fatigue, to 1,200 men and two batteries the force that could be mounted to co-operate in the pursuit of Cronje.

The second point was painfully exemplified on more than one occasion. To put a man on a horse and give him a rifle does not necessarily provide a mounted infantry man. Thus, on 16th February, the day after French's charge, the *Times*' History says, "Hannay took his column (mounted infantry) forward towards the gap on the right. When about half a mile away the Boers opened fire. . . . The column began trotting forward, then suddenly halted" (for no apparent reason). "The Boers, bringing forward a pom-pom and a gun, redoubled their fire on the splendid stationary target now presented to them. A moment later a great part of the mounted infantry bolted headlong in a broken mass for the cover of the river bank in rear. Charging over the steep bank, they were thrown into yet more inextricable confusion, and the river-bed for half a mile was a seething *mêlée* of men and horses struggling in the deep current, hopelessly bogged in the treacherous edge of the 'Mud' River, or scrambling about under the bank."

The fact was that the mounted infantry could not yet be trusted to work for long in the scattered formation proper for it when in proximity to an enemy. It was felt that it could only be "kept in hand" by being kept together. When did we ever catch 1,000 Boers in a mass, on their horses and within easy range?

Infantry and Entrenching.

But, among all the more obvious developments, the plainest lessons, and the most important, are connected with the infantry in attack and defence, and with the use and arrangement of entrenchments.

To take the last point first. After a few trials, the Boers discovered the advantage of the trench on the level over the trench on the heights, when the field of fire is flat and open. The former gave to the flat trajectory its full effect, as a moment's thought will show; afforded a far less conspicuous mark for the hostile artillery to range on, but on the other hand rendered reinforcement of that firing line more difficult to conceal, and retreat from it more hazardous.

The reinforcement and retreating difficulties were overcome by the ready individuality of the Boer system, the men stealing forward or back singly, darting from cover to cover.

As a direct corollary to the long ranges at which effective fire now commences, there has come the utility of rendering entrenchments as inconspicuous as possible, this being effected partly by the sites chosen and partly by careful modifications of the profile.

The Infantry Attack.

Over the question of the formation in which infantry can best attack under these new conditions, a controversy rages; also over the "grand tactical" question as to whether a frontal attack can hope to succeed by itself.

Flank Attack.

Lord Roberts's instructions to the army in South Africa pretty plainly indicate the view that the frontal attack *per se* cannot, without unjustifiable loss. On the other hand, German authorities, and some of our own writers, assert that, with good formations and good leading, the thing is substantially no more difficult than it always was. To us it seems that the flank attack has become of necessity quite a different affair from what it once was. If the enemy has learned the elementary requirement of a deep formation, especially towards the flanks, a flank attack has no longer the thin end of a line to assail and roll up, but becomes virtually another frontal attack from a fresh direction;* it may, of course, hit off more favourable ground for attacking over, and will often have a profound strategical effect, while on the other hand the ground may present such difficulties on the flank that the attack there may have far less chance of succeeding against a staunch enemy than it might have on a skilfully chosen part of the front. (In these days, long smooth slopes, well seen by the defender, have become the best defence, while the *terrain accidenté* is the kind to be chosen by the assailant).

While, therefore, advocating combined frontal and flank attacks as the rule for the reasons given, the frontal attack pure and simple is sometimes necessary, and cannot be looked upon as impossible by good troops. The choice is governed by the nature of the ground, and by strategical considerations. No doubt, against an enemy who could escape by speed, the frontal attack *per se* is apt to prove an expensively barren success; but few enemies will be all mounted, like the Boers.

Attack Formations.

Two schools of military thought contend over the formation question, one advocating advance by a thin firing line, which is to be gradually strengthened as approach to decisive distances is made, by the rushing into it of handfuls of men from the support line and eventually from the reserve line—for all, of course, agree that the firing line must develop all possible intensity of fire as the crisis approaches; the other, among which the Germans are included, advocating a dense firing line from the very beginning.

Our commanders in South Africa inclined plainly to the former view, the instructions for attack gradually moving from the Drill Book "two or three paces extension" to 8, 10 and 15 paces. Against this the second school say that in the end these reinforcements for the firing line run just the same risk as if they were in that line at first, having precisely the same ground to cover under fire in the one case as in the other; that in the meantime the weak firing line has been able to do little towards the great aim of dominating the defenders' fire, the next line in rear having often been unable to fire from fear of hitting their comrades in front; that a weak firing line is apt to be prematurely checked from the fact that each man of them is subjected to the fire of a multiplicity of rifles;† that dependence on very frequent reinforcement is unsafe, and that there is a danger of the firing line, or parts of it, being so long checked that it loses heart.

* The Boers seldom used depth from the outset, but obtained from their mobility the same effect, as far as hostile outflanking operations were concerned.

† In some experiments carried on at Hythe, this point seemed to be settled once for all. A number of men, say 84, were ranged at close order, from 400 to 500 yards from 12 targets, one foot square, spaced at seven yards; side by side with these, 12 picked shots, at seven yards intervals, were ranged opposite 84 equal targets, at one yard intervals. Firing commenced simultaneously, and men were withdrawn from each party, as the corresponding target opposite the other party was hit. The invariable result was the annihilation of the smaller party long before the larger party; but when the range was increased to 1,000 yards, the result was the other way. The inference seems obvious; but a mistake was made in the experiments; the scattered men would be able to find cover in a manner practically forbidden to the close-order men. It would be interesting to see what the result would be if the open-order targets displayed only (say) half the height of the close-order ones; and still more, if a complete experiment were made in the following manner:—(1). Assume that the 12 are advancing, as firing line, to attack the 84; the 84, having presumably some artificial or natural cover, to show one foot broad by six inches high; the 12, having freedom of movement, to show also a diminished target. (2). Assume the 84 to be in motion against the 12 in position; the 84 would now be full-sized targets, the 12 diminished. (3). Assume the 84 to be attacking 84, only 42 of the assailants to fire for the first ten minutes and then the whole; the defending 84 are now represented by half targets, half the assailants are represented by targets at two yards intervals for ten minutes, and then the remaining half are erected in the intervals.

To the writer it seems that neither side goes quite to the root of the matter. The substantial point is the flatness of the trajectory of the defender's rifle, and the fact that this flatness in a certain extreme case is nugatory. The extreme case is when the defender is firing down perpendicularly to the plane of the slope on which the assailant's lines are moving. At the other extreme, as at Modder River, the level rifle sends its projectile almost parallel to the plane for a great distance. Between these two extremes come the usual gradations.

Then there are the two requirements to be sought by the assailant in selecting his formation for attack, namely, to deliver as much good fire as possible, and to avoid loss as far as possible—two requirements that are in their very nature antagonistic, and between which the best compromise must be made.

Take first the case of the flat plain—only the front rank of the attacker can fire, and a hostile bullet that passes over the shoulder of an attacker in the firing line will get a man of the next line, though far in rear. In this case, have your firing line dense, on the German system.

Take the case where the general slope from the beginning of the attack to the defender's position is concave—either flat first, and then a steep rise to the defender, or downhill and then uphill to the defender. Here, the second and third of the advancing lines can fire, and the defender's bullet that misses a man in the firing line strikes the ground harmlessly a short distance further on. If the defender keeps his fire for the front line, the shooting of the other lines will be all the cooler; if he distributes his fire over the lines, the important leading line will be helped.

This case includes the case where the advance is over undulating folds of ground.

The effect of artillery fire on these lines must be considered; the shrapnel bullets will usually be falling at a fair angle of descent. If the defender's artillery is numerous and good, and cannot be dominated to some extent, the attack will of course be made more difficult; but it remains in doubt whether shrapnel fire would be less effective against a lot of thin lines than against a single dense line.* And, in any case, the chief enemy is usually the defender's rifle.

The summary of these ideas is—against a low trench and a flat field of fire, make your firing line dense, and keep the next line far off as long as possible—against a high trench and a plunging fire, use a large number of thin lines in the advance.

In this short exposition, not much insistence has been made upon the important matter of dominating the defender's rifle-fire, a point that is conspicuous in every paragraph of German arguments. But it is not a matter that needs any insisting upon. In both the ways above proposed, all possible rifles are brought to bear, and no one can do more than that. The soldier should be trained to shoot steadily all he can, from the moment individual firing is ordered, and to understand that the longer he takes to get there, the more chance he has of being killed.

Volley Firing.

The mention of individual firing brings up the question of the almost discarded volley. At the risk of being thought old-fashioned, the writer holds it most useful in many cases. At long ranges, it helps to keep discipline, and the simultaneous splash on

* Some light might be thrown on this by experiments on the artillery range. Put 1,000 dummies in an irregular line a mile long; let a brigade of field artillery fire at them from 3,500 yards for ten minutes, at 2,750 yards for ten minutes, and at 2,500 yards for ten minutes. Then try the same gun-fire on 1,000 dummies in three lines each a mile long, the lines being 300 yards apart.

the ground helps to find the range. It was used for this purpose successfully more than once in South Africa. As the range shortens and the noise increases, it becomes practically impossible in the attack. In defence, it may often be used with good effect to steady the men.

Long-Range Firing.

Very long range rifle-firing was sometimes done, great effects being sometimes produced at 1,500 yards, and even more, by Boers on the defensive; but the excessive clearness of the South African atmosphere must be remembered, and also that on these occasions the firers were not much troubled by shrapnel. The Boers seldom had the nerve to wait, as the Germans did often in 1870, till the very first blast was bound to produce a great effect; but they did it sometimes, notably at Colenso, and the moral effect was immense.

Creeping.

Creeping or crawling on the way to the attack was at times done by both sides. A very remarkable case of its successful use by the Boers is given by a German eye-witness, who says that the steady crawling advance of a Boer line, with never a halt and apparently never a casualty, completely demoralized the British defender on the occasion. The Boers worked it in pairs; while one grovelled along for about his own length, without raising head or shoulder an inch, the other fired; then the groveller took up the firing, while the other crawled. But it is to be observed that there is here again no shower of shrapnel. Recent German writers, notably von Boguslawski, heap scorn on all creeping and crawling, and particularly on officers lying down; but why the limited number of officers should offer themselves for easy demolition is not understood. The objection to the crawling for a line of men is that hostile shrapnel get in more shots effectively at the same range, whereas rushes suddenly alter the range.

Night-Attacks.

Night-attacks, by which is meant the approaching of the hostile position under cover of darkness with the intention of attacking at the first streak of light, were often undertaken, and their inherent difficulty was shown again and again, while the advantage gained when they were successful was amply proved. Apart from the difficulty of reaching the exact spot in the dark and in the proper formation, the draw-back to the soldier of the loss of a night's rest is very great. Medical reports show that more men break down from loss of sleep in a campaign than from physical strain; and if the attack is not quickly successful, the loss of the previous night's sleep induces such serious and rapid fatigue that the *moral* is apt to deteriorate rapidly.

It emerged, as a general thing, that the night-march should be short; if over a very few miles, that the men should be eased of carrying their packs, or should themselves be carried part of the way in carts; and, of course, that all the precautions that have always been advisable in such operations should be observed. At Stormberg the force lost its way, was made to march much further than was prudent, did not know where it was when day broke. In the course of a dragging action 600 tired men of the force, getting stuck on a precipitous hillside where the cliffs prevented them from advancing, and the Boer rifles, commanding the valley at the foot, prevented them from retreating, lay down and slept the sleep of exhaustion, only to wake up to find the battle over and themselves dazed prisoners.

At Magersfontein the column attained the right point, but want of previous scouting exposed it to the fire of an unexpected trench, and there was apparently no common understanding as to what should be done if any such *contretemps* occurred.

Change of Formations during a Campaign.

It must have occurred to many people to ask why we had to make such changes in our formations to suit the Boers, instead of compelling them to change theirs to suit ours. We went out with men and officers trained to extend to two or three paces, and were very soon increasing these figures sevenfold.

The writer does not remember having seen the reason anywhere stated. It was the superior tactical mobility of the Boers that compelled us to resort to their methods. This is a matter that should be thoroughly understood.

When we went into the field against the Dervishes or the Zulus we found an enemy who, man for man, could get over the ground more quickly than our soldiers, and whose physique for a hand-to-hand encounter was superior on the whole to ours. These two things, but the former chiefly, compelled us to abandon our usual formation; and, the enemy having no firearms to count, we naturally took to close order. If such an enemy took to advancing only to spear-throwing distance, and thence hurling his weapons, we should have found our advantage in keeping our open order, to lessen the effect of his missiles, to show a broader front, and to bring a converging fire on his masses. Here we should not be altering our tactics, simply because the enemy was abandoning his advantage of superior individual mobility.

When we faced the Boers, we found again an enemy more individually mobile than ourselves, and so armed that close order was no solution of the difficulty. To obviate converging fire and perpetual outflanking, we had to follow their suit, and this was because of their mobility. Had they had no horses close in rear, they would never have dared to present a single thin line to our attack; a dense attack, pushed into their centre, might have suffered in the advance, but their wings would have been an easy prey thereafter. Being mounted as they were, the expensive victory over their centre would have been won too dear. In Manchuria, where the two sides may be credited with approximately equal tactical, or at least individual, mobility, we see no extravagant extensions.

The summary of this matter, then, seems to be that, given equal weapons, the less mobile force is compelled to follow the enemy's formations.

The rapid extension to a flank during a fight, possible to the mobile belligerent, has a most disconcerting effect on the enemy, as we found to our cost at Magersfontein. Its effectiveness may be so great that it at once suggests the idea of keeping one's mounted infantry as an integral part of the commander's main reserve, and not allowing it to be frittered away in fragmentary operations. Being thus kept, the commander has his infantry reserve to attend to developments on the primary or original fronts of attack, and his mobile reserve to attend to exterior developments.

Perpetual entrenching on the part of the Boers became a monotonous commonplace. It may be expected in future that the organically inferior army will always be thus acting; but, as the nature of man has not changed, and as the *moral* of masses of men is affected, as ever, by what they do themselves, as well as by what the adversary does, perpetual entrenching is apt to diminish that dash and confidence that are so important in war. But under the power of modern weapons entrenching is often necessary; and it has now come about that the troops must be prepared to entrench themselves, without waiting for the Sappers. The Boer trenches, on many occasions, were perfect models of strict adaptation to a purpose; there is a lesson in this, for all entrenching should be the handmaid of tactics. The Boers made little use of artificial obstacles, but these may be expected in the future to be an important feature of the defence.

In deep positions, and when large armies are fighting, the attacker will entrench himself from day to day as he progresses, the object being to release as many troops as possible for the continuance of the attack next day; but many authorities, Germans in particular, say there is a tendency to do too much of this, and that the extra fatigue occasioned by digging half the night after fighting all day may do more harm than the risk of not entrenching may justify.

Finally, it has become more than ever essential that the training of the soldier be such that he has an intelligent appreciation of what he is doing; that he take cover at all ranges, as long as doing so does not interfere with the steady progress of the advance, that the sooner, using such precautions, the line reaches the enemy, the less chance has he of being killed; that to show his back in retreat means death. The sense of duty must be as strong as ever it was, but there must be an intelligent variety in carrying it out; discipline must be perfect, but not of a strangling quality.

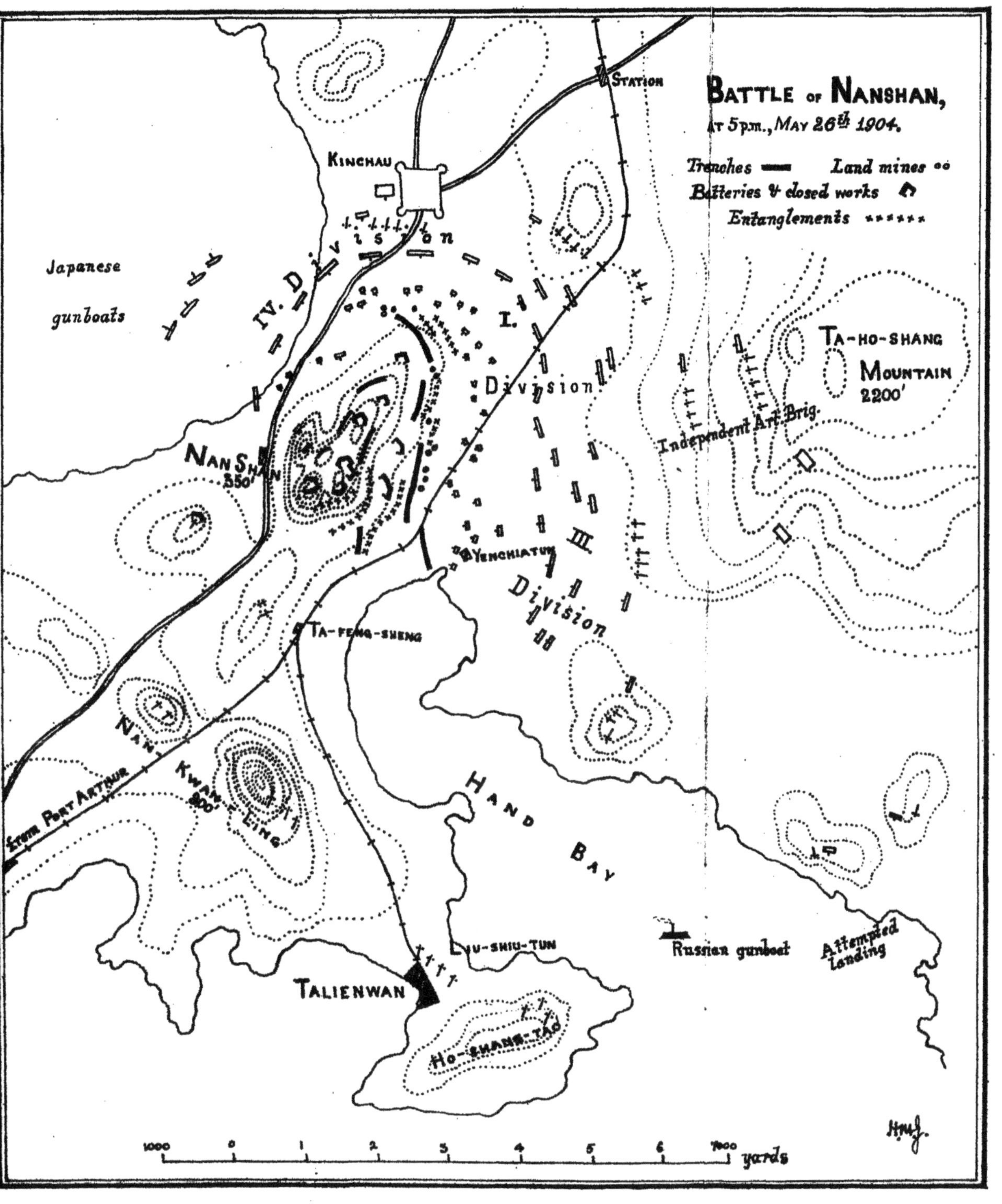
BATTLE OF NANSHAN,
AT 5 p.m., MAY 26th 1904.
Trenches
Land mines
Batteries & closed works
Entanglements
STATION
KINCHAU
Japanese
gunboats
IV. Division
I.
Division
III
Division
TA-HO-SHANG
MOUNTAIN
2200'
Independent Art. Brig.
NAN SHAN
350'
YENCHIATUN
TA-FENG-SHENG
NAN KWAN LING
From PORT ARTHUR
HAND BAY
LIU-SHIU-TUN
Russian gunboat
Attempted landing
TALIENWAN
HO-SHANG-TAO
1000
0
1
2
3
4
5
6
7000
yards

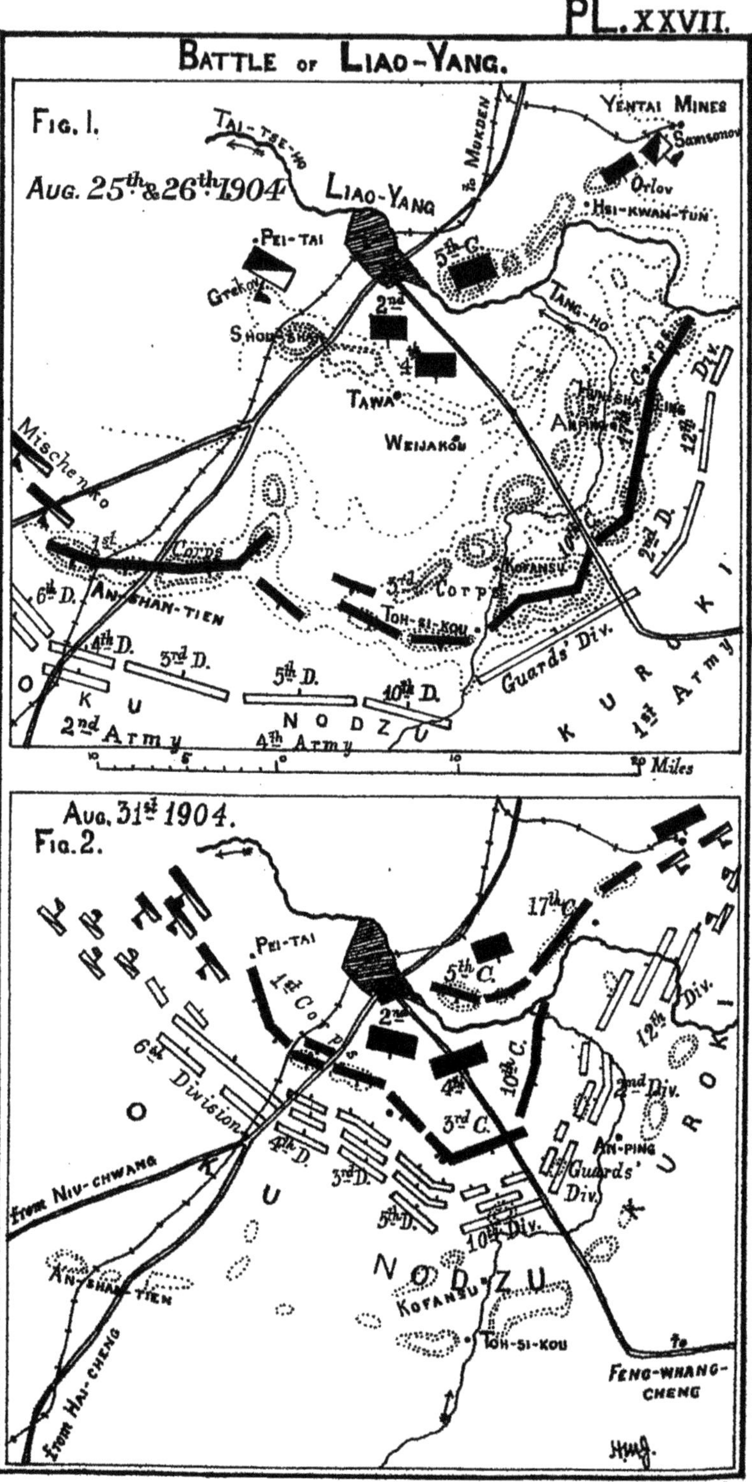

W. & A. K. Johnston Limited Edinburgh & London.

CHAPTER XXII.

THE RUSSO-JAPANESE WAR.

THE last great military drama with which we have to deal is the gigantic struggle in the Far East.

Battle of the Yalu.

The first land battle, fought on the banks of the Yalu, while giving an indication of the wonderful quality of the Japanese and the toughness on occasion of the Russian, was so one-sided an affair in the matter of numbers that it is not proposed to describe it in detail. The Japanese had 48 battalions, 108 field guns, 20 heavy guns, 1,800 cavalry—say 45,000 men and 128 guns. Considerable assistance was also rendered by 2 gunboats and some torpedo boats and armed launches.

The Russians had 14,000 men, 40 guns and 8 machine guns.

The Japanese showed their remarkable ingenuity and thoroughness in the manner in which they made the enemy pay attention to his right flank while the real objective was to be the left flank, in the skill with which they erected artificial screens to prevent the Russians from guessing even at the movements on foot, in the careful entrenching (strong as they were in numbers) of their whole front on the river, in the vigour of the frontal demonstration while the outflanking movement was on foot. So overwhelming, however, was the Japanese force, and so very poor the Russian shooting, that the frontal attack succeeded by itself. The idea is consequently suggested whether the Japanese made anything like adequate use of their great numbers—whether the flank attack from the north-east might not have been made into an enveloping attack from north and north-west—a procedure that might have brought about the capture of the whole Russian force.

The Russians had had plenty of time, but their entrenchments were flimsy, as compared with those we find them constructing later.

The battle ended in the defeat of the Russians, who lost nearly 3,000 in killed, wounded, and prisoners, 21 Q.F. guns and the 8 machine guns, against a loss of over 1,000 for the Japanese.

BATTLE OF NANSHAN.

Plate XXVII.

The next fight of importance was that at Nanshan, waged on 26th May between General Stössel, commanding IV. Siberian Army Corps, and General Oku, commanding the 2nd Japanese Army.

The Russians had 11,000 infantry, 2 field batteries, nearly 70 guns in the works, 1,500 sailors, the total *personnel* amounting to about 13,000; and they had substantial help during the fight from a gunboat and some destroyers and launches in Hand Bay.

The Japanese had three divisions, and an independent artillery brigade of 36 guns. The number of combatants amounted to about 45,000, and there were 144 guns available, but none of these reached the calibre of the five or six 6-inch guns in the Russian works.

Four gunboats joined in the fight from Kinchau Bay, and had a very great influence on the result of the battle.

The neck of land between the two bays is less than two miles wide. "The ground near the sea on each side is low, but along the centre there rises a ridge of higher ground, with a general bearing north-east and south-west," the culminating point at the north end being Nanshan, 350 feet in height. This is over 3,000 yards south of Kinchau, which the Russians held, for no sound reason, with 600 men. The town, standing in a level plain, is dominated by heights within easy range, and its strong brick *enceinte* was of little value as a defence against modern weapons.* It seems likely that the Russians, confident of holding their own, thought it might prove a good jumping-off place for eventual counter-attack; otherwise, the 600 men and 4 guns wasted in it seem a piece of folly.

The chief position was on Nanshan, which consists of four or five peaks; on and about these there were 45 guns and four 6-inch mortars, in solidly constructed works of ordinary design; but some of the guns were too conspicuous, and the more numerous, but lighter, artillery of the Japanese succeeded, with the help of the four gunboats, in partially silencing them before the day was over.

Connecting the batteries, and in several tiers lower down, were lines of infantry trenches, following the contours of the ground and of low command, and blinded in parts; there were good covered communications except in the low ground towards Hand Bay, where all the men holding the foremost trench were slaughtered by the hostile shells, owing to the impossibility of getting the wounded away.

The chief batteries were connected by telegraph and telephone, 10 balloons were in use, and search-lights were in working order.

In front of the trenches a great deal of wire-entanglement was constructed, and several land-mines sunk.

By May 22nd the Russian outposts had been driven in, and they now held the Nanshan position, Kinchau on the north of it, Liu-shiu-tun on Hand Bay; and there was a small work on the second line of heights called Nan-kwan-ling. This second line would, in some respects, have made a much stronger position than the one chosen; rising to 800 feet, it dominated everything in front, and though accessible by the shells of the gunboats in Kinchau Bay, the range for these would have been long. The line requiring to be held would have been six miles long, it is true, but where was the rest of Stössel's Army Corps of 30,000 men? He knew Oku's numbers. Whatever be the reason, only one division was on the spot; the Russian regulations laid down one and a-half miles as the natural front for one division, and so the 2nd line was neglected.

The left flank of the position of Nanshan was going to prove the weak point, and it is a remarkable fact that no trenches were made on the low ground on this side. In front of no other part of the position was there any cover for an attacker. The plain was flat and sandy, the slopes of the hills gentle and smooth; but on the north-west face of the hill there was a patch of dead ground 500 yards from the principal battery on that side.

The first touch with this formidable position gave pause to Oku, who proceeded to a most thorough reconnaissance. By drawing fire from various points, he was soon able not only to locate the guns, but to estimate their calibre by examination of the projectiles that arrived.

* The Japanese field artillery carry a proportion of common shell.

By the 24th May the plan of attack was formed, and it was perforce to be a frontal attack from north and east—the IV. Division from the north, the I. in the centre, the III. on the left. The independent artillery brigade to take up a position at 4,000 yards, due east of Nanshan, on the lower slopes of Ta-ho-shang; but there was a moving forward of much of the artillery during the fight. When the attack closed in, Nanshan would be assailed on three sides, and it was recognized that the right flank of the IV. Division would have the best chance of success, having the gunboats to help it.

Rough weather on May 25th caused postponement of the assault, as it would have been unsafe for the gunboats to venture into the shallow bay. The artillery, however, bombarded the position for some hours.

On the night of the 26th, the weather continuing very bad, the IV. Division made an attempt on Kinchau, but effected nothing during the darkness.

The dawn came with thick fog, and the artillery duel could not begin for an hour after the appointed time, 4.30. In the interval, a detachment of sappers of the I. Division blew in one of the gates of Kinchau, and the place was rushed, the 600 Russians in it were cut off from Nanshan, driven into the sea, and all but 12 killed.

From about 6 a.m. till 9.30, the Russians kept up the artillery duel well, against the combined fire of the four gunboats and the whole of the Japanese field artillery. They then appeared to slacken; some of their accounts say that this was due more to lack of ammunition than to their guns being "silenced"; that their heavy artillery was chiefly Chinese from Port Arthur, and that the suitable ammunition was scanty. As the slackening became apparent, the Japanese infantry advanced by rushes, and presently reached to within 300 to 500 yards of the entanglements; but on the perfectly open ground they could get no further. Meantime the Russian field guns had retired to the second line of hills, and thence kept up a galling fire on the Japanese III. Division. The latter was further tried, about 10 a.m., by the appearance of the Russian gunboat, which enfiladed its advance from Hand Bay, until well on in the afternoon.

Seeing the attack was well checked, the Russian commander now attempted to land a party across the Bay opposite Liu-shiu-tun; but the Japanese had calculated upon this and repulsed the landing.

Right through the middle of the day till 5 p.m., the Japanese infantry made no progress, except on their extreme right, where a part of the IV. Division was now wading in the shallows west of the Nanshan. Here and there on the east, small portions of the firing line made desperate rushes, but there was no glimmer of success, and the position of the III. Division must have become a source of anxiety to General Oku, who now had the whole of his troops engaged.

It was the psychological moment for a Russian counter-attack, but nothing of the kind occurred.

At the hour named, 5 p.m., the Japanese artillery reported their ammunition running out. Oku replied by ordering the infantry to rush the place at any cost. The III. Division got across the first trench line, the I. to within 100 yards, but there the impulse again gave out. But the IV. Division, well covered by the gunboat fire, and most intelligently helped by their own divisional batteries, which fired steadily on the works that bore upon the beach, managed to put several companies into the piece of "dead ground" previously referred to. Waiting here to be breathed and to re-form, during which period every available gun from sea and land poured a rapid fire on the works 500 yards up the hill, the companies gathered themselves for the final spring and won their object. The I. Division now made another effort and broke into the Russian centre. The right fought with the courage of despair, but, taken in front and on the left flank, were finally beaten by 7 p.m. The Liu-shiu-tun position was perforce abandoned, and the battle was over.

The Russians had lost about 4,000 men, or about 30 per cent., along with 68 guns and 10 machine guns; the Japanese loss was over 4,000, or nearly 10 per cent.

The general opinion is that the battle shows the impossibility of success for a purely frontal attack on a well-prepared position over open ground, even when the assailant outnumbers the defender by more than three to one—provided, of course, that the latter's line is not too extended. And there is little doubt that success came through the direction of attack of the right wing of the IV. Division and the fire of the gunboats in the same quarter. But it is a question whether the calling of this "a flank attack" is not a mere form of words.

The Russian errors were manifold. However strong a position may seem to be, you should have your whole strength at hand when battle is imminent, unless the rest of your troops are engaged on more important work elsewhere. Napoleon said, in effect, "Do not engage in battle, unless you are prepared to do everything possible to produce a decisive result." To prevent an enemy from driving you out—that, and no more—produces no decisive result in such a case as this. Granted that 13,000 were required to hold the ground safely—and they probably would have done it but for the gunboats in Kinchau Bay—then there should have been a reserve in hand, free from the duties of defence; and Stössel certainly had troops available for this in his command. Another division, waiting comfortably out of sight near Ta-feng-sheng Station till 4 p.m., could have struck the hard-pressed III. Division on its left flank when it was already almost broken; and the direction of the attack would almost certainly have caused the collapse of the rest of Oku's assault. Stössel's other division would have been much better employed here than in digging on the way to Port Arthur; the Russian commander, starting quite correctly on the defensive, would have been all along aiming at the "something decisive."

Oku was committed by circumstances to a frontal attack, and, his force being numerous enough for attack all along, he rightly took that method; but his careful reconnaissance told him much, and his prudence in delaying the assault till the navy could co-operate was first-rate. From sheer lack of space the attacking force was dense—40,000 men on a 3-mile curve gives 8 men to a yard. To have kept back half of them for any pedantic reasons would have been absurd, for he had plainly made up his mind that he had no serious counter-attack to fear. On the other hand, during the days of reconnaissance his troops had been preparing positions against the eventuality of failure.

It is not possible, as yet, from want of authentic information, to say much about the "fighting tactics" of these battles, as distinct from the "grand tactics"; but we know that the Russians fought, whether in attack or defence, in very dense formations according to our modern ideas, while the Japanese spread more freely, but to nothing like the extent that we found advisable in South Africa. Their enemy proved themselves very poor rifle-shots, and this may have been partly the cause of the density of the Japanese as compared with South African methods; but it was also due to the necessity, having no greater mobility than the Russians, of opposing at least equal numbers at the points where breaking-in was to be accomplished.

We have no information as to what amount of reserve Oku kept in hand at Nanshan, either in the way of "general reserve" or local reserves for his three divisions; but a general survey of all the offensive battles seems to indicate that the Japanese commanders did not keep large reserves, but made up for it by careful entrenching of a position to be held in case of reverse and hostile counter-attack.

Nor do we know the arrangement in depth and front of units in attack—whether, for instance, a regiment sent one battalion entire into the firing line, and supported it with the second battalion, keeping the third for reserve, or distributed each battalion in depth, as we do.

As to the artillery, the field guns were inferior to the Russian in range and power, but they were much better handled in the way of concentrating fire intelligently and in the way of inconspicuousness. Now and then we read of a battery closing in to rifle range, in the manner beloved by the Germans, but whether this was a feature of the dogged attacks of the Japanese, we do not know.

Battle of Liao-Yang.

On the 23rd August, 1904, General Kouropatkin had the Russian Army, about 200,000 strong, covering Liao-Yang. Three positions had been prepared; the advanced one was from Anshantien through Toh-si-kou to Hunshaling on the Tang-ho, and thence to the Tai-tse-ho.* Anshantien is 20 miles down the railway from Liao-Yang; the whole front from west of the railway at this place to the Tai-tse-ho is 40 miles, and the left of it is 15 miles distant from the city.

Plate XXVI.

The Japanese, under Marshal Oyama, were in three armies; on the right, Kuroki's or the 1st Army, about 60,000 strong, had the 12th, 2nd and Guards Divisions from right to left, and an uncertain number of "detached brigades"; in the centre, and entirely to the right of the railway, Nodzu's or the 4th Army, about 60,000 strong, had the 10th and 5th Divisions and some extra troops; on the left, on both sides of the railway, Oku's or the 2nd Army had the 3rd, 4th and 6th Divisions, probably about 90,000 strong, and most of the Japanese cavalry was on this flank. It did some useful work as mounted infantry, but on neither side was cavalry used as such to any appreciable extent.†

Kuroki and Nodzu seem to have had about 200 guns each, field and mountain only, as their lines of attack were to be in the hills; during the week's fighting that ensued, these two armies seem to have found great difficulty in using their guns to advantage, owing to the difficulty of finding gun-positions. Oku, on the other hand, with 200 guns, had many batteries of 5-inch howitzers, and used also the 6-inch guns he had captured at Nanshan.

* On the Russian right the three positions were Anshantien, Shou-shan, and the great redoubts north of that, close to the town. On the left, the 1st line was from a mile south of Toh-si-kou along the right bank of the Tang-ho through Hunshaling; the 2nd line was from the heights north and west of Kofansu along the left bank and across the neck of the peninsula of the two rivers; when the Russian right had retired to its second position (Shou-shan), the left had been forced back to its third position, Tawa, Weijakou, confluence of Tang-ho and Tai-tse-ho, and the line of heights on the other side of the latter to Yentai Coal Mines.

† The 3rd Army, Nogi's, was besieging Port Arthur.

The attack was commenced by Kuroki, who on the 24th and 25th and during the following night assailed the 10th and 17th Army Corps in their position in front of the Tang-ho. Up to the night of the 26th, both sides fought with stubborn courage, the Russian artillery was unsubdued, and the powerful works above Kofansu (to the south-east), aided by strong reinforcements from the reserve corps, baffled all the efforts of the Guards' Division. Kuroki had all his troops employed, and was making but little headway.

What Oyama's original plan was is matter for conjecture; did he intend that Kuroki's attack was to be the chief operation, while his other two armies "contained" the enemy's front?—or was Kuroki allowed to hammer away for days on the Russian left flank, in order to attract Kouropatkin's masses in that direction and thus make an easier job for Nodzu and Oku in their frontal advance? As Kuroki was later moved still further round the Russian left, it may be supposed that the latter is the correct idea, for the 1st Army was not reinforced in the way one would have expected if everything was to depend upon it.

Kuroki's left Division, the Guards, drawn by the fortified strength of the Toh-si-kou salient, and the frequent threats of counter-attack from it, kept edging to its left, and his army was already getting dangerously extended. On the night of the 26th a determined, but unsuccessful, Russian attack from Anping was followed by the capture of the Hunshaling position, by the 12th Japanese Division, from Kouropatkin's 17th Corps.

The Russian resistance at this part now weakened badly, and by evening of the 27th August the Japanese had cleared the enemy from the right bank of the Tang-ho, except in the peninsula between the Tang-ho and the Tai-tse-ho; but the right of the 10th Corps, with the 3rd Corps in support, still held on to the strong works north and west of Kofansu.

When Oyama saw that Kuroki had a hard task, he had put in the 4th and 2nd Armies to attack the south front. Kouropatkin stood for a day at Anshantien, and then abandoned the strong position there on the night of the 27th, losing during the retreat on the morning of the 28th a battery and a number of ammunition wagons.

Fig. 2, Plate XXVII.

On the 28th August, therefore, the Russian right was falling back to the second position, on the preparation of which for defence months of work had been spent. Kuroki was so hard pressed and in such difficulties on his long front, and Nodzu's right was so immediately checked, that Oyama held it essential to attack at once on the south front without the usual careful and prolonged reconnaissance that such formidable positions call for. A correspondent writes:—"We are told that but for the difficulties of the 1st and 4th Armies, we in the 2nd would have waited for two days more before beginning to attack the Shou-shan."

The crescent of hills four miles from the town, extending from Pai-tai on the west to near the junction of the Tang-ho and the Tai-tse-ho on the east, and having its culminating point at Shou-shan, 425* feet high, in the centre, bristled with every conceivable defensive work. The wide plain in front was, at the time of the battle, covered with a crop of kao-liang, which grows close and ten feet high; but this had been levelled into a natural entanglement for hundreds of yards in front of all trenches,

* One account puts it at 900 feet.

and the Russians had improved upon their former methods by locating trenches at the forward foot of the slopes. On this occasion much of their artillery used indirect fire from the reverse slopes.

When, on the 29th, Nodzu and Oku commenced their attack on this position, Stackelberg, with 1st Siberian Corps, held Shou-shan and about three miles of hills to the leftward; the other corps were as shown in *Fig.* 2, the 17th Corps having retired across the Tai-tse-ho, and having its left extended to the Yentai Mines by a scratch division under Orloff and a numerous Cossack Cavalry Division. The rest of the Russian cavalry covered Stackelberg's right, and two and a-half Army Corps were being held in reserve.

The "second position," from Pei-tai on the right, through Shou-shan and the Weijakou salient, to the left of the 17th Corps, measures about 25 miles, giving nearly five men per yard of front. The Japanese had now much better artillery positions available than they had had earlier, except opposite the "salient."

The salient held by the 3rd Corps was enveloped on the 29th and 30th by Nodzu's Army, but without being forced.* Oku hammered away at Shou-shan and the works to the west of it, but found them unexpectedly formidable; on the night of the 30th, Kuroki's 12th Division received orders to cross the Tai-tse-ho to its right, and attempt the complete outflanking of the enemy. That this was a most risky proceeding is at once apparent. Oyama had practically all his men fighting, his enemy had still some corps in reserve; from Nodzu's right (Kuroki's left) to the Tai-tse-ho was over ten miles, a front amply extended for an exhausted army of three divisions. The 12th Division by itself was little likely to effect anything decisive across the river, but Oyama plainly counted upon its giving Kouropatkin a fright.

The 2nd and 4th Armies continued their attacks on August 31st and September 1st without effecting anything decisive, and the 12th Division, with the "detached brigades," gained some ground towards Yentai Coal Mines.

Kouropatkin had news on August 31st that the 12th Division was across the river, and saw that there was little activity on Kuroki's old front; he jumped to the conclusion that the bulk of the Japanese 1st Army was engaged in the movement towards Yentai, and he took a most extraordinary means of meeting the threat.

On the evening of August 31st he ordered the 1st, 3rd and 10th Corps to withdraw from the formidable positions they had held so stubbornly, to cross the river by the numerous bridges that had been provided, and to march north-east to face Kuroki. The second position having been thus abandoned, the 2nd and 4th Corps, in reserve until now, were to occupy and hold the third position of strong redoubts in the plain close round the town.

* A correspondent, describing this part of the fight, says :—"From the volume of fire you could hear the Russian infantry discharge against the 4th Army, Kouropatkin was still striving hard (on the 30th) to break in between Kuroki and Nodzu." This assumes that the Russian C.-in-C. had a definite offensive plan; if this was the plan, it was a first-rate one, and it is pretty certain that he did use great portions of his reserve in counter-attacks at this salient. It will be remembered that, in our discussion of "Sickles' salient" at Gettysburg, it was said that such a configuration in a line certainly afforded a good jumping-off place for counter-attack, but that it must be strong by nature or art, or by both. Kouropatkin's earlier salient at Kofansu was so in both ways; if he had at any time a definite plan of separating Nodzu and Kuroki at this point, he must have weakened on it. In fact, the whole of Kouropatkin's proceedings seem to have been vitiated by want of that persistence which led the Japanese to victory. The Russian general sometimes planned well, and started the plan well, but allowed the initiative to slip from him under the moral stress exerted by events elsewhere.

Why the strong line of the second position should be abandoned is not apparent, nor is the utility of this shifting of defenders. If Kouropatkin thought that Kuroki in bulk was attacking towards Yentai, why did he not counter-attack from behind the 3rd Corps' salient on Kuroki's old front, which must now be weak? Or if he thought Nodzu had probably extended to his right to strengthen Kuroki's old front, and drawn some of Oku eastwards in doing so, then why abandon the Shou-shan front, if the enemy was thus weakening himself there? The Russian game now was to find out where the weak point was between the railway and the Tai-tse-ho, put in the two Army Corps there with the greatest vigour, and cut the Japanese front in two. As this day of the 31st wore to a close, the Japanese Staff reckoned the position so critical that they began to take the precaution of sending reserve ammunition to the rear. Oyama, in fact, felt that the forward momentum was waning and would not suffice for victory, if the Russian resistance should continue as stubborn as it had been. And Kouropatkin failed at the critical moment to divine the plight of his enemy.

He seemed now to have little idea beyond defence. The withdrawn corps and much cavalry marched towards Yentai Coal Mines; either they had no clear orders, or the C.-in-C. had no clear conception of the situation; Kuroki drew across nearly the whole of the 2nd Division and made a brave show on a 6-mile front, with every man practically in the firing line. Three Russian Army Corps, as the *Times'* correspondent says, "pirouetted about for two days" in front of two weak divisions; and meanwhile Oku had his guns firing within range of the railway station of Liao-Yang.

The Russian C.-in-C., learning that his left was making no progress, lost his head, and the order for retreat all round was issued. The operations pursuant to that order seem to have been well done; all the bridges near the city were destroyed by the rear guards, and the pursuit was checked.

The Russian casualties in the seven days' fighting are reckoned at 25,000 as a minimum; the Japanese lost nearly 18,000, the 2nd Army losing about 8,000.

Details of the "fighting tactics" are not yet obtainable in authentic form; but a few remarks from intelligent eye-witnesses are worthy of note.

Thus one reads: "The Japanese used all the cover as they advanced, while the Russian tactics of preserving close company formations sacrificed much of the natural advantages of the ground. That stiffness and precision of the German school, which some foreign officers noting in the home (Japanese) manœuvres had thought might handicap the Japanese in the field, was entirely abandoned. When a battalion went into action the companies worked most independently."

"Between the methods of shooting of the two armies there was as great a contrast as in other essentials. Russian firing is almost all by volleys, with quite mechanical regularity. Even in the use of artillery, they follow the plan of discharging one gun after another rapidly and with precision. . . . Occasionally the Japanese use volleys, but most of the time they fire at will."

"Japanese officers explain that it is not necessary for the generals to go to the firing line and supervise in person, because they have confidence that every subordinate officer knows and is competent to do his part in the work." When this is the case, it shows that the leading maxim of organization has been understood and acted upon, and that the training has been thorough.

"The active Japanese infantry were, with practised activity, making rows of trenches and rifle-pits in front of those from which the Russians had just been driven out."

"When the darkness set in, . . . all of them hastily dug shelter trenches for the gunners and the infantry. . . . In the same way, whether in front or far in the rear, they took especial care to guard against surprises, and to protect their ammunition columns."

"Many of the Japanese pioneers and infantrymen through that long night pressed on stealthily, seeking to gain ground inch by inch, and taking precaution to dig . . . as cover for themselves on the morrow." "From first to last they always strove in this manner to avoid the necessity of yielding an inch of the ground they had won."

In a case where the Russians were making a counter-attack,—"It was well that the Japanese had taken great pains to secure the ground they had gained, by numerous trenches. . . ."

"The Japanese . . . had prepared for their spring (on Shou-shan). Single line behind single line of men, *6 or 8 deep*, as far forward as they could get, they lay ready for half an hour." This attack, from 900 yards' distance, failed.

The 34th Regiment (Japanese) had just captured an important work on a conical hill on the skirts of Shou-shan; "the 15th Artillery, seeing numbers of men outside the Russian works, and taking notice of the furious cannonading that was proceeding, made sure the enemy were preparing to deliver some counter-attack. They thereupon turned all their guns upon the conical hill. . . . Nothing remained for the remains of the 34th, that had carried the works, but to lie down and try to escape destruction from the hands of their countrymen." The Russians quickly realized what was happening, returned, and retook the works.

Accidents of this kind are likely to be frequent in future wars, unless the most perfect communication is kept up between the firing line and the distant artillery.

CHAPTER XXIII.

THE PRESENT DAY.

HAVING in the previous chapters traversed briefly, in respect to the development of Tactics, the period from Frederick the Great to the war in Manchuria, it remains to see where we stand.

The most recent editions of the official text-books on Tactics and on Training are a good guide as to what the military authorities of the Great Powers have learned, or fancy they have learned, from the experience of the campaigns of the generation that has seen the latest developments in weapons.

One naturally turns first to the opinions of the General Staff in Germany.

After minute study and much discussion, this great authority seems to have come to the opinion that our South African War is far from providing adequate reason for revolution in methods of attack or defence; that our difficulties there were due to a great extent to our own deficiencies in tactical knowledge and training, to our notable lack of understanding the necessity of using all our arms and all our strength in due combination, to the exaggerated fear on the part of our commanders of losing men. The Germans therefore hold to the establishing, in the attack, of a dense firing line from the outset, laying great stress, as they do, on the paramount importance of establishing superiority of fire as early as possible. These ideas are no doubt to a great extent due to their supreme confidence in the tactical and shooting capacity instilled into their men by a most businesslike system of training. The more careful defensive-offensive system approved by Clausewitz, and in his later days by von Moltke, they look upon rather as interesting speculations than as opinions in accordance with the real spirit of these great teachers.

The General Staff breathes "Attack, attack!"—attack everywhere; when strategy has done what it can, attack all along the line. A battle like Austerlitz, they say, needs a Napoleon; in default of such genius as his, envelopment and attack is the way to victory. Threatening of communications is only one of the ways of attaining the object, which is the destruction of the enemy's military force. To envelope one of his wings—that will be usually the first aim; but, with the great extent of front in these times, superiority may become possible at some point which the supreme command had not foreseen. So the attack should be vigorous everywhere, and should not be abandoned as hopeless because the enemy has laid himself out to bar the original plan.

These are virile ideas; their steady inculcation into generation after generation of an armed nation produces that spirit that leads to victory. But the bullet of the magazine rifle does not spare high spirit; and these German soldiers, "in a dense firing line from the outset," may suffer beyond endurance if they meet, on particular ground, a staunch enemy who has been taught to shoot.

Great stress is laid on the necessity for officers to be ready to shoulder responsibility. The Japanese have a custom, in connection with their peace-manœuvres and training, that might well be imitated more freely; it is the custom of frequently removing, in the midst of a sham-fight, some of the commanding officers, and judging the juniors by the way they carry on the operation.

The Japanese attacked either in a dense line from the outset or by groups of a dozen working along more or less independently. Accurately observed details are wanting as yet; but it seems that in open ground the firing line *was* a line, while in broken ground the firing line was in irregular groups. This seems like sound sense. In open ground, where cover is scanty or non-existent, stick to the greater formality of the skirmishing *line*, advances being made by rushes here and there, quite irregularly as to locality and intervals of time, of portions of the line; in broken ground, leave the advance more to the discretion of squad commanders, pushing on their groups from cover to cover till decisive range is reached.*

In all the latest drill regulations for the attack, two matters of a minor kind are noteworthy—the order to a portion of the firing line to "Rise" is omitted, and the arrangement for the rushes is not to be of any stereotyped kind, such as "by half-companies from right to left," or "by alternate sections." A cool and steady defender behind his loopholed parapet will prepare a blast of fire towards the direction whence the cry of "Rise" comes; and if he sees that the half-companies rise in regular rotation, he will be ready for them.

The latest ideas on fire-discipline in the attack form a subject on which a whole book might be written. It may be here said in short that very little of it (fire-control) can be enforced when battle is once joined, that the men must be left to do what they have been taught to do, and that therefore fire-discipline will in future date from the training-ground. As one item of this it may be suggested that a man advancing should only shoot from behind some sort of cover, and that when his rush leaves him actually in the open, he will be much better to take his short rest quite prone, leaving the shooting for the moment to others more fortunately placed.

The French regulations have some peculiarities, among them the system of pushing forward independent detachments of all arms far ahead of the advanced guards; this intention is strongly combated by several leading French generals, who object to subordinating concentrated and decisive tactics to mere attempts at annoying the enemy. Great stress is laid upon fights for localities; the decision, when on the defensive, is to be expected only by offensive action on the part of the Reserve. The group system is favoured in the attack over open ground; it is a lineal descendant of the ideas of Folard and Ménil-Durand.

* A remarkable instance of success under the group system is afforded by Japanese action at Penlin on July 31st, 1904. A frontal attack by a skirmishing *line* was being made against the Russian main body, amounting to about a brigade; the Japanese thought they saw their chance in pressing forward their right in a direction just clear of the apparent left of the enemy, but very soon this movement was brought to a standstill by the fire of half a battalion of well-hidden Russians, whom the Japanese mountain-guns could not locate. The right portion of the Japanese line, being in broken ground, began to break up into groups; one of these, consisting of an officer and seven men, worked its way unnoticed by the enemy to a point covered with scrub, where they suddenly became aware that they were within 400 yards of the outer end of the troublesome Russian trench, which they were ably promptly to enfilade. This they did at once with magazine fire, giving the illusion of a strong attack, while their small numbers and the scrub prevented the Russians from ever locating or estimating them. In a few minutes the trench was evacuated. Had this outflanking been made by a line of skirmishers, the Russians would have seen it and prepared against it.

France and Great Britain are willing to allow unrestricted activity to the individual skirmisher—almost, it may be said, to throw upon him the bringing-about of the decision; Germany is to trust more to drill and discipline, education and thorough training in shooting, and to keeping the men more under the control of their officers.

In all drill-books every cut-and-dried method of attack has disappeared, except to a slight extent in the United States, where also volley-firing has still a vogue. There has been all round a still further diminution in the number of drill manœuvres taught. In the German book, for instance, square and double-column movements no longer exist; the actual instruction for attack talks of "skirmishers five to ten paces apart advancing by irregular rushes of 30 or 40 paces; lines of skirmishers following one another at considerable distance apart to the nearest cover in the line of advance—failing this, to within effective and decisive ranges." This is the method while under the hostile artillery fire only, or perhaps very long range rifle fire; but by the time effective range is reached, the front skirmish line will have been so strongly reinforced as to become quite a dense firing line.

The British Army having the prospect of a continuance of "small wars," our Regulations still include the square and some échelon formations.

In the view of Continental officers who have seen our recent manœuvres, our infantry showed great skill in the use of the ground; no close order within 3,000 yards; no frontal attacks; when attempted against entrenchments, the attack was judged to have failed unless the numerical superiority was as six to one. The opinion is that we extend too much, and that attack therefore can never be really powerful; that enveloping often failed on account of the feeble "containing" action along the front.

No Power except ourselves seems to be adopting permanent bodies of mounted infantry.

Pure frontal attack is on the whole condemned; but a school of French military thought, deeply impressed by study of the tactics of Napoleon, holds to the idea of bringing overwhelming masses suddenly against a chosen point of the hostile front. How such a thing is going to be done in practice is not clear, unless the enemy allows you first to choose his position for him.

The fact seems to be that the frightfully destructive power of the modern firearms is bringing back the War of Positions; our own "Combined Training" tacitly assumes that battles will all be attacks or defences of Positions, and that Napoleonic battles, with both sides on the move, are not worthy of discussion. German writers look upon this as a deterioration of the Art of War; knowing their own strength, they expect always to be the attackers, and they do not like the idea of hammering away at carefully prepared positions such as Kouropatkin's at Liao-Yang. But it seems quite likely that the wars of the future will be just such wars. All troops must be prepared to entrench themselves at short or long notice, and it may be expected that great armies on the defensive will have, not one line of defences, but two or three. This system of *depth* will make flank attack simply an attack on a lateral front; and it has brought again the use of the closed redoubt as a tactical *point d'appui*.

Until quite recently text-books on Military Engineering stated that closed redoubts would be seldom used, "except for isolated posts on a line of communications, or in order to provide security for a small force liable to be attacked by overwhelming numbers of savages." This is no longer the case, and the reason should be understood.

As long as field armies were only expected to make use, in their active operations, of mobile field-guns, firing for the most part only shrapnel or case-shot, redoubts very naturally fell into disfavour. They afforded a good target for artillery fire, their construction was laborious. The same labour expended on tiers of musketry-trenches was found to be more efficacious; the redoubt did not give a return in fire proportionate to the labour expended upon it, nor did it give more cover from shrapnel than could be got from properly constructed rifle-trenches.

But the Boers brought heavy guns into the field, and both Russians and Japanese have brought them heavier still. Against these, you must either hide from the enemy's view, behind hills, whatever you want to save, or you must provide enormous earth-cover; and it is natural to provide this in a closed work rather than in patches in the open. The patchwork system, on the other hand, would have the undoubted advantage of scattering the enemy's fire, and the existence of carefully made bomb-proofs at certain points might escape the enemy's notice altogether.

This, then, is not quite a sufficient reason for the recrudescence of the redoubt, and a reason must be sought in another direction. The Boers hardly used redoubts, and they did most of their earth-moving well, and on sound defensive principles. The reason is in connection with the huge size of modern armies, and the great depth of defensive positions.

A great army may take up a position like the French at Gravelotte, strong in front but essentially shallow. If pierced to such an extent that the assailant establishes himself where he has broken in, such a position is lost to the defender, unless the counter-attack by reserves is adequate in force and very immediate. If not rapidly successful in ousting the intruder, the defender finds his wings hopelessly severed as far as that battle-field is concerned.

But where a position is of the nature of that at Liao-Yang, where line behind line was prepared for defence, and the *depth* of the position was therefore great, a thrust by the enemy into the middle of a line did not to the same extent sever the defender into two halves; there was more chance, in fact, of retrieving the position, even if this could not be done instantaneously. Therefore it was indicated that it was worth while to prevent the intruding corps of the enemy from rolling up the line to right and left as soon as it had established itself. The trenches constructed for frontal fire would give no help for this; there must be works perpendicular to the front, and these, too, must be safe on their rear against a lapping-round by the intruded enemy. Hence the Russian redoubts. Though they attracted fire, they came to be reckoned necessary; they were powerful flanks for the inner ends of the two portions of a pierced line; the enemy must take them before he could advance, even at this point only, against the 2nd line, and before he could begin any "rolling-up" of the 1st line; the delay in effecting this afforded time for the bringing up of supports from other points of the vast theatre of operations.

But these redoubts attracted fire, and the fire was the formidable fire of heavy fortress and naval guns, discharging high-explosive shells of terrific potency. Therefore the redoubts had to be of huge construction, with parapets of enormous thickness and underground shelters covered by mountains of earth. Such a redoubt, threatened by a flank attack of victorious infantry and a target of the best kind for many guns, must have great obstacles provided—a deep ditch with steep scarps, flank defence from points not affected by the immediate bombardment (*e.g.*, from caponiers and counterscarp galleries), entanglements, and every conceivable obstacle; for a few good shots in succession from the big guns may denude the threatened face for the moment of all its parapet defenders and pave the way for a successful infantry rush.

Hence, in my opinion, the recrudescence of the great redoubt in the line of battle. But that there is a lurking snare in all this elaborate fortifying admits of no doubt. Man's nature induces him to stay where he feels safe, when there are deadly dangers around; and this applies to the commanding officer as well as to the man. It is necessary, therefore, that we do not allow the art of the engineer to stifle the spirit of the bold strategist. Keep the fortification in its proper place—*i.e.*, use the art simply in order that you may hold a given front with fewer men. Keep a large part of your force free of the defence of the works, watch for the critical moment, then put in your free force in a vigorous counterstroke.

The enemy will attack your formidably prepared front, simply to keep immobile as many of your men and guns as possible; he will inevitably use his superior numbers to lap round your flank or flanks, and it is as true now as it was in the time of Jomini, and as it has always been, that the only effective way of dealing with outflanking manœuvres is by active and vigorous countermoves. Usually these will be best made in the direction of the enemy's operation, but on occasion a commander will do great things by a strenuous offensive on the other flank or along the front or a portion of it. The question of time will rule; if you get early information that the enemy is beginning a movement round (say) your left, and you have reason to believe it will take him two or three days to develop that movement into a serious threat, you may well consider the advisability of putting in against his other flank, there and then, every man and gun you have kept in hand.

The method would be audacious, but it is the audacious that wins decisively; your action should at least have the result of compelling the enemy to abandon his outflanking movement—*i.e.*, you have now taken the initiative from him, and are forcing him to follow your lead.

Early information is essential; reconnaissance is insufficient at such a stage; *to the full extent* that the situation in the theatre of war allows you must be served by a host of spies. A bag full of gold paid to a single man, who brought you such information as that in the case just supposed, would be a cheap purchase.

The first idea produced by the South African fighting was, not unnaturally, that the day of the bayonet was gone; but soon the very difficulty of reaching a well-concealed, good-shooting enemy over open ground suggested the resuscitation of the plan of night-attack—*i.e.*, the endeavour to reach, under cover of darkness, a point from which you might hope at dawn to rush the position without firing.

In Manchuria the bayonet has been still more in evidence, and it seems likely to continue in the future to be so. There it was used freely by the Japanese during the night, not waiting for dawn. Fortified localities, almost but not quite captured during the day, the bullet-swept glacis having proved impassable—such points, whose capture you see will give you a good footing in the hostile line, are the natural objectives of such attacks. But now, more than ever, there must be nothing haphazard in the planning of such attempts; the troops who are to do it must be accompanied by pioneers for the passage of obstacles, and be followed closely by entrenching parties and by an intact body; this last will pass rapidly round or through the captured site, to ward off the inevitable *retour offensif** and cover the entrenchers at their work; and above all, every single commander of any unit, however small, must have a perfectly definite task, which should be pointed out to him by daylight as far as possible.

Work of this sort is likely to be so common in future war that practice at it, *in the dark*, should occupy a substantial portion of the soldier's training time; and officers should be practiced frequently at leading their commands across country at midnight to points they have only been able to reconnoitre by day at three miles' range, and even to points they have never seen except on a map. Orders of this kind appear, more or less urgently, in most of the latest official training instructions of the Powers, along with increased stress upon physical training, gymnastics, etc.

A deadly weapon, whose use takes one's mind back to the sieges of the Peninsula and to Sebastopol, is the hand-grenade. This has been much used, both in front of Port Arthur and in the attack and defence of the great fortified positions in Manchuria. Very close work is the occasion for their use—a well-sheltered defender confronted at a few yards' range by an assailant who dare not mount that last parapet till he feels there is dire confusion among the staunch adversary on the other side of it.

A similar revival is the little mortar, sometimes improvised by the Japanese out of bamboos and iron hoops, for use when the distance is too great for hand-throwing, but still far less than the range of any ordinary artillery.

Of technicalities now in use in the field, there is no end—searchlights, star-shells, telegraphs, telephones, wireless telegraphy, balloons—all the inventions are to take part in the battles of the future. While the great principles of strategy and tactics remain as ever, these adjuncts have to be understood and efficiently worked; every army except our own is continuously adding to the numbers and variety of its technical troops. The working of these elaborate helps to warfare cannot be improvised, if they are to be really efficient; commanding officers and staffs must be practised in their use. Witness the following wail from a Russian Staff Officer, who concludes upon the "impossibility" of having a great mass of artillery directed by one commander; but the alleged impossibility was due to lack of *practice* at it.

"The case of the employment of a mass of artillery is really exceptional, and therefore it is not surprising that when the opportune moment for its employment came it was forgotten. Gradually the desirability of having one responsible person for directing a large quantity of artillery, even that of several Army Corps, was realized. All the necessary preparations were made, orders laid down, and checkered maps distributed.†

"I recollect that on 28th February and 4th March an attempt was made to follow this method of procedure, which however did not succeed, and miscarried before the orders were issued to the executive, and clearly demonstrated the impossibility of confiding to one person alone the fire direction of such a great number of guns. Therefore to judiciously direct fire, the entire field of battle must be visible, which,

* But the French authorities say that you cannot depend on any body for this purpose other than the body that has won the position.

† These are maps divided into small numbered squares, each square being a point on which it may be desirable to concentrate fire from many directions by telephone or other order.

given the extent of the positions occupied, is most difficult for the sector of a single Army Corps. If one is content with information sent by telephone, the dispositions will not be taken or communicated in time necessary for the corps which are in a position to carry them out, and, above all, they will no longer correspond to the actual facts of the case. The desire to have the whole of the artillery directed by a single commander entails the separation of the artillery and infantry; the necessary bond between the two arms and their community of action disappears."

All of this simply shows the necessity of combined practice of all the arms. It is quite essential that the O.C. Artillery should have the power of directing the whole of his weapons when, in concert with the C.-in-C., some great decisive operation is afoot; but the O.C. Artillery who meddles with (say) a Divisional Artillery in action, simply because he *is* O.C. Artillery, is violating one of the fundamental principles of command and of organization.

In most of the latest regulations there is a leaning towards strong advanced guards, who practically make a reconnaissance in force, and who enable the main body of the artillery to come forward early. Thus our "Combined Training" says that the artillery shall establish itself on preparatory positions under the protection of the advanced guard, but will not enter into action till a good objective is presented to it; and that at any moment of the fight its objective shall be that part of the enemy (infantry or artillery) which is of the greatest tactical importance at the moment. In discussing the *rôle* of the different classes of guns, it gives, during the preliminaries of a fight, a somewhat roving commission to the horse-artillery, and calls upon it for great boldness of action in supporting the infantry attack, by flanking or enfilade fire if possible.

The field artillery, less mobile but more powerful, uses its guns and light howitzers to bombard the hostile guns and then the point of attack, the howitzers turning their special attention to deep trenches and to reverse slopes. The heavy guns, finally, have the task of silencing from afar any hostile weapons that can range upon the proposed lines of advance, to destroy villages or constructions occupied by the enemy, to enfilade if possible positions which the shorter range of the other weapons only enables them to attack in front, and finally to join in the final bombardment of the point of assault. Howitzers, owing to their high-angle fire, can continue firing till the last moment at this point.

It seems that a great deal of the artillery of the future will carry portable shields.

After 1870 the cry was all for "massing the guns," but smokeless powder, the consequent value of concealment, and the ever-increasing destructive power of the weapons have caused a revulsion towards the method of dispersion. A single battery, so placed as to be able to fire for a quarter of an hour before being located by the enemy, may often do more effective work than several batteries that immediately betray their presence.

All regulations now lay down that artillery can call upon the nearest infantry for its protection; at the beginning of a fight, a particular body of infantry may have been allotted as support to some guns, but mobile guns must on no account be hampered in their movements by having to take their original escort along with them—hence the regulation just referred to.

Many writers are remarking on the small percentage of casualties due to gun-fire, as compared with those due to infantry fire—in Manchuria one to twelve seems to have been the usual ratio—and these writers seem to consider the fact remarkable, seeing the great power of modern Q.F. guns. To us it seems quite natural. During great part of the artillery bombardment of the hostile infantry, the latter are not firing and can hide; when the guns are engaged with the hostile guns, any gun-detachment of the latter which finds itself being overpowered can cease fire temporarily and take cover, leaving their guns *in situ*—shrapnel does no harm to the *matériel*. The Japanese often did this, and very wisely—the same when ammunition supply gives out for the time. It is not to be supposed that the continued artillery fire is therefore wasted in either of these cases, for your own infantry can then advance more easily.

The Japanese usually withheld their gun-fire till the enemy began his; their field-guns were inferior to the Russian.

Most armies now have a regulation forbidding a battery to leave its position in order to refill. The Q.F. guns of the future will use so much ammunition on occasion that the nine wagons per battery of six guns will probably have to be much increased. The problem of bringing up the shells to the guns when the position is an exposed one is very difficult. The Russians on more than one occasion fitted up a sort of cable tramway, up to as much as 600 yards in length.

Every effort will in future be made to reduce to a minimum the number of men who have to *stand up* about a gun in action.

The question of high-velocity *versus* high-angle gun-fire is exercising some minds. The Germans seem to favour the former; but if they are going to attack, and the enemy entrenches deeply, they will find the flat trajectory not very effective for their shrapnel. It seems to us that a force on the defensive could better dispense with howitzer fire than a force attacking an entrenched enemy. But high-angle shrapnel fire with small bursting charge has the disadvantage of imparting very little penetrating power to the bullets, and the lyddite system has been found rather disappointing against *personnel*; probably the solution will be found in high-angle shrapnel fire with a large bursting charge, the weapons being brought up as near the defender's position as the folds of the ground will allow. The "observation of fire" would be much easier in this case than when the steep "angle of descent" is attempted to be obtained, with a high-velocity gun, by firing at long range.

All regulations recommend the turning of a portion of the artillery at times against unseen ground where reserves, etc., may be supposed to be. The Japanese did this in the early part of the war, but seem to have learned through their Chinese spies that little good was done in return for a heavy expenditure of ammunition; so that later they did little of this fancy shooting. But they adhered with good results to the showering of shrapnel over the ground in rear of a battery or batteries that had been some time in action, and which would therefore be likely to be bringing up more rounds.

The discussion of artillery naturally leads on to the question of the *rôle* of the machine-gun. The Russians and Japanese both multiplied their numbers of these weapons as the war progressed; but it was always clear that such guns, firing from high travelling carriages, were too conspicuous, and were at once silenced as soon as they were located by the hostile cannon. Experiments are consequently being made in many countries with an automatic gun, such as the new "Rexer," that shall be fired by a man lying down. All such guns are to be used in pairs, in case of an accident to one of them. Whether they should be formed into batteries of six, or kept in pairs as regimental weapons, is exercising the minds of authorities. The bulk of evidence seems to point to these weapons being essentially weapons for occasions, and that therefore they should not be brigaded into batteries. Their great usefulness on occasion was shown at Liao-Yang, for instance, where, at 1,200 yards, the Russians say they stood them in better stead than a line of skirmishers. They caught a Japanese mountain battery in motion, and in a few seconds of firing put it out of action for the rest of the day. Again, the same guns absolutely prevented Japanese skirmishers from establishing themselves along the railway embankment where their presence would have been fatal at the moment.

The Russians used eight Maxims in a battery; their weak point showed whenever they were pitted against sharpshooting snipers, or against artillery.

The German publication, *Kriegstechnische Zeitschrift*, has a very interesting article on the machine-gun. It acknowledges its utility in small wars, but cautiously says that this is no criterion for the War of Masses, while noting that both Russians and Japanese laid great store by them. Holding that from 600 yards inwards is the zone of effective infantry fire, and from 1,200 yards outwards is the artillery range, it asks whether the intermediate zone might not be, as it were, handed over to the machine-gun.

The matter should be discussed in relation to the three arms in turn. Infantry fire at medium and long ranges, to be of much use, demands the collective fire of a large number of rifles; this may often have to be done, when the artillery is for any reason not helping at the point where infantry have to advance. The requisite number of rifles, all firing at once, is not by any means always available; for, where 500 on a given front might be ample for the work at 600 yards, five times as many might be necessary to produce a similar adequate effect at 1,400 yards. A too-small number, in their effort, would be apt to degenerate into a kind of long-range magazine-fire, with a demoralizing effect on the shooter; and a good infantry commander is always glad when his men do not need to fire till effective range is approached. Under such circumstances, and they will be common, a machine-gun per company would not be out of place.

I say advisedly "per company." The object of these weapons is not to engage in set duels, but to be used from moment to moment as opportunity offers, and to be preserved in order that they may add their quota of fire when the crisis arrives.

The cautious doubt, expressed in the article referred to, as to the "War of Masses," seems to originate in the mind of a writer envisaging to himself the employment of *batteries* of machine-guns; but if one holds them to be best used as battalion or company weapons, there seems no reason why the war being one of masses should affect their utility at all. But the officers in command of these guns must both study and practise their tactics.

In many situations a pair of Maxims might aptly form the escort of a battery, or even of a brigade of artillery in mass; and it has been suggested that one of these might often or always accompany the artillery staff when they go forward to find gun-positions—a work that is very apt to be interfered with by hostile cavalry; and also to cover the first artillery arrivals, when firing is to be postponed till the whole artillery force is in position.

The assistance of machine-guns to cavalry is very grateful to all that large class of cavalry officers who fear they are in danger of being called upon to "degenerate into mounted infantry." They would turn over to this weapon—one per squadron—the cavalry fire-fight; and there is little doubt that on many an occasion the being actually in the saddle might make all the difference in the lightning rapidity of the cavalry combat. If a cavalry machine-gun can be devised, to travel on a high mobile carriage and to be capable of rapid dismounting so as to be fired by two or three men lying down, great things may be done.

A single word may be here interpolated about a possible use of cavalry in a pitched battle. We shall suppose the crisis approaching, the attacker turning his whole attention to the decisive point, his reserves in motion towards that point—the defender's mass of cavalry waiting under cover on a flank, longing for an opportunity. The Japanese themselves have allowed that there were occasions—at Nanshan, for example—when a highly-trained division of real cavalry, imbued with the true cavalry spirit, might have done great things—might have ridden over the flank of their firing line and supports, and have reached the guns, now engaged in the final *rafale* that supports the pushing home of the infantry attack. The cavalry might lose 30 or 50 per cent. of their horses, but they might win the battle, especially if the defender had succeeded in keeping a portion of his reserve and in sending them to attack on the heels of the horsemen.

Is it for this sort of purpose that the German Kaiser fosters the "cavalry spirit," keeps a great force of horsemen and preaches the attack for them in mass?

We have seen, then, that the greatest European army is of opinion that no fundamental change has been produced in tactics by the latest weapons; that attacks across any kind of ground are as feasible as ever, but that along with the old discipline and patriotic spirit there must be a high level of intelligence and a mutual helpfulness between unit and unit, and between arm and arm.

The training to be aimed at is both intellectual and moral—the development of individual intelligence and ingenuity, and the fostering of individual and collective patriotism. The schoolmaster should have a large share in this. The Japanese soldier's Pocket Ledger contains the following words:—"Our soldiers to keep in mind the following virtues, viz.:—Loyalty, courtesy, bravery, uprightness and modesty." "Every inhabitant of Japan is rejoiced to perform an act which will benefit his country; but a soldier who has no such desire is of *no value*." "Soldiers must think before acting, neither despising an inferior enemy nor fearing a powerful one, and by so doing they will display the higher kind of bravery." In an order to his officers, General Oku wrote:—"You must devote the greatest attention to the surrounding details, especially to the movements of the troops to right and left of you; in action you are constantly responsible and must act by your own judgment. Soldiers who only think of their own interests, neglecting their comrades, understand nothing of the art of war."

The Russian instructions contain the following:—"Die, but save your comrade. Advance, even if those before you fall. . . . On the defensive, do not content yourself with parrying blows, strike; attack is the best defence. . . . In the fight, victory comes to him who is the most tenacious and the most bold, not to him who is the strongest. . . . A well-formed unit knows neither flank nor rear, it sees only the front, where the enemy is. . . . However unexpected and sudden may be the enemy's appearance, you have both fire and bayonet for him. The choice is easy, and the formation is a secondary matter. . . . There is no situation from which you cannot emerge with honour. In the fight, every unit once engaged ought to keep up the struggle ; it may be supported, never relieved."

With these quotations I close this chapter and this book. They serve to emphasize the fact that is so well expressed in the phrase, "Behind the gun there is the Man."

And *in* the Man there is the spirit that leads to victory, or that leads to defeat. That spirit is the outcome of his environment and of his education from the very cradle.

Battle of Sadowa, 3rd July 1866, at 2p.m.
Pl. XI.
Gerekwitz
Milowitz
from Horsitz 3 miles
Cernutek
Zelkowitz
from Burglitz 3 miles
Wrchownitz
from Jericek 1½ miles
Habrina
Trotinka Brook
Horicka Berg
from Welchow 2 miles
Racitz
Horenowes
Benatek
Hnewcowes
Sowetitz
Klenitz
Stracow
Dub
Res. Art.
Roskos B.
Skalka Wood
6th Div.
Sadowa
Mzan
from Poanek 4 miles
Lhota
Sugar Factory
5th Div.
7th Division
2nd Guard Div.
Rodow
Trotina Brook
1st Guard Div.
Maslowed
Cistowes
Brig. of 2nd Corps
8th Division
Zawadilka
Kopanina
Dohalitz
Dohalicka
3rd Division
Mokrowous
Neresow
Sucha
Johanneshof
Lipa
3rd Corps
No 1
No 2
No 3
No 4
No 5
4th Corps
Chlum
Sendrasitz
VI. Corps
Nedelist
Brigades of 2nd Corps
Lochenitz
to Josephstadt
10th Corps
Langenhof
Rosberitz
Tresowitz
R. Bistritz
Popowitz
14th Divn
Stresetitz
6th Corps
Wsestar
Sweti
Art. Reserve
Predmeritz
1st Corps
Rosnitz
Saxon Div.
2nd Saxon Div.
Problus
Komarow
Nechanitz
Lubno
from Bidsow 11 Miles
from Kobilitz 3 miles
Nieder Prim
Plotist

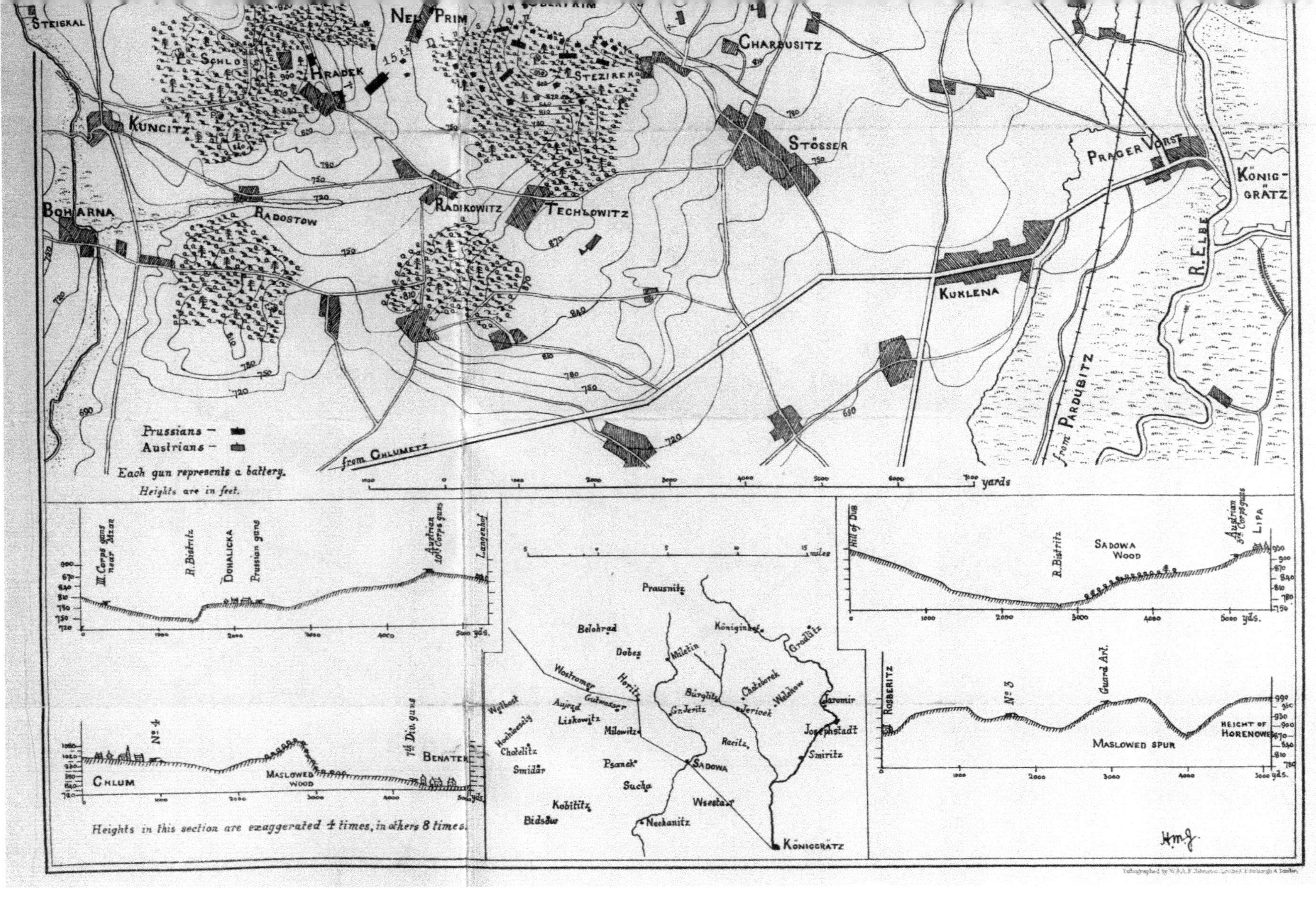
Steiskal
Schloss
Hradek
Neu Prim
Stezirek
Charbusitz
Kuncitz
Stösser
Prager Vorst
König-grätz
Boharna
Radostow
Radikowitz
Techlowitz
R. Elbe
Kuklena
from Pardubitz
from Chlumetz
Prussians –
Austrians –
Each gun represents a battery.
Heights are in feet.
yards
III. Corps guns near Mzan
R. Bistritz
Dohalicka
Prussian guns
Austrian 10th Corps guns
Langenhof
Hill of Dub
Sadowa Wood
Austrian 3rd Corps guns
Lipa
yds.
Prausnitz
Königinhof
Gradlitz
Belohrad
Dobes
Miletin
Wostromer
Horitz
Bürglitz
Chotěborek
Welchow
Jaromir
Josephstadt
Gutwasser
Aujezd
Liskowitz
Milowitz
Gr. Jeritz
Jericek
Racitz
Smiritz
Chotelitz
Smidar
Psanek
Sadowa
Sucha
Wsestar
Kobititz
Bidsow
Neckanitz
Königgrätz
miles
No. 4
7th Div. guns
Benatek
Chlum
Maslowed Wood
Heights in this section are exaggerated 4 times, in others 8 times.
Rosberitz
No. 3
Guard Art.
Height of Horenowes
Maslowed Spur

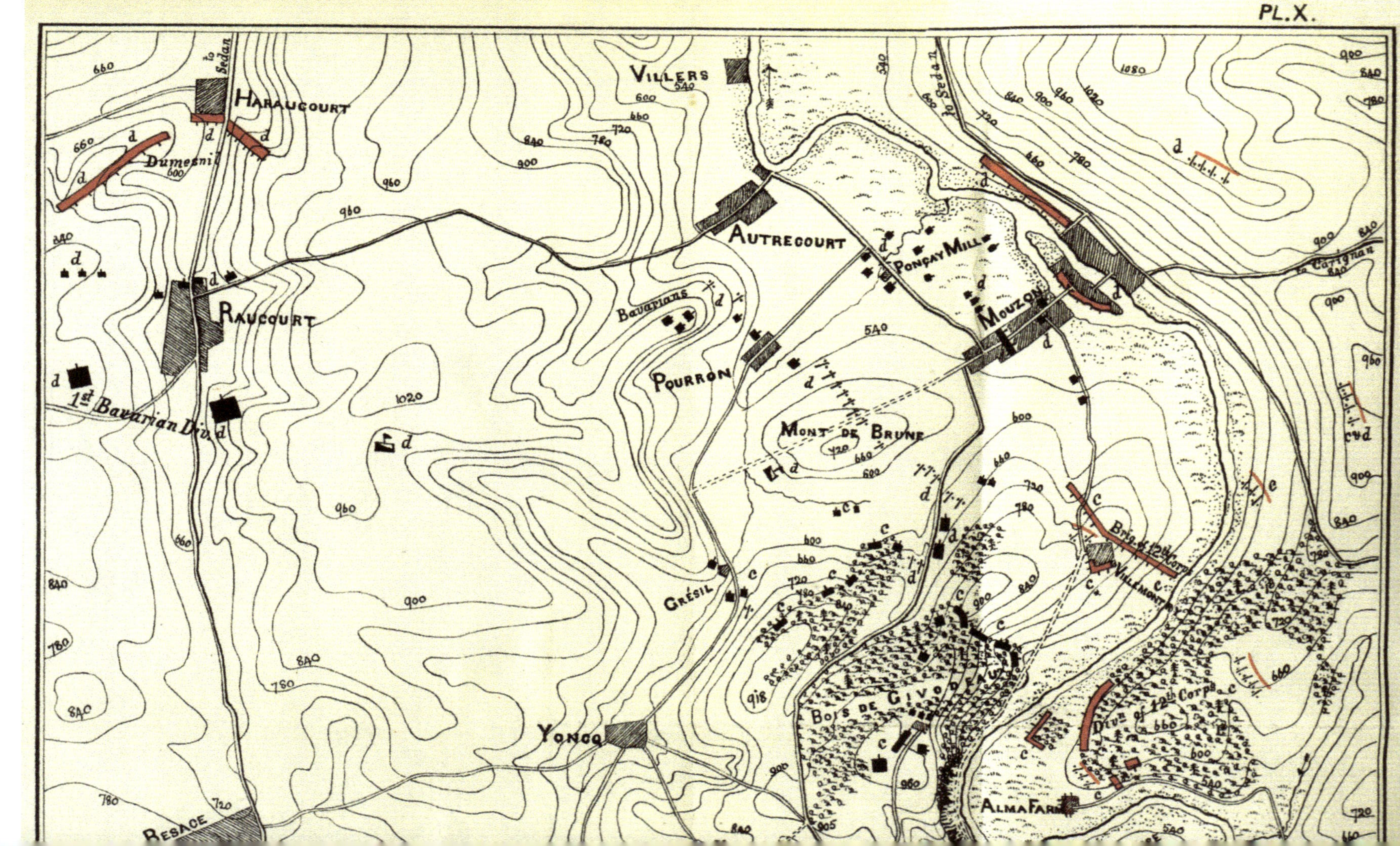
PL. X.
Villers
Haraucourt
to Sedan
Dumesnil
Raucourt
1st Bavarian Div.
Autrecourt
Ponçay Mill
Mouzon
to Sedan
to Carignan
Bavarians
Pourron
Mont de Brune
Grésil
Yoncq
Bois de Givodeau
Villemontry
Brig. of 12th Corps
Div. of 12th Corps
Alma Farm
Resace

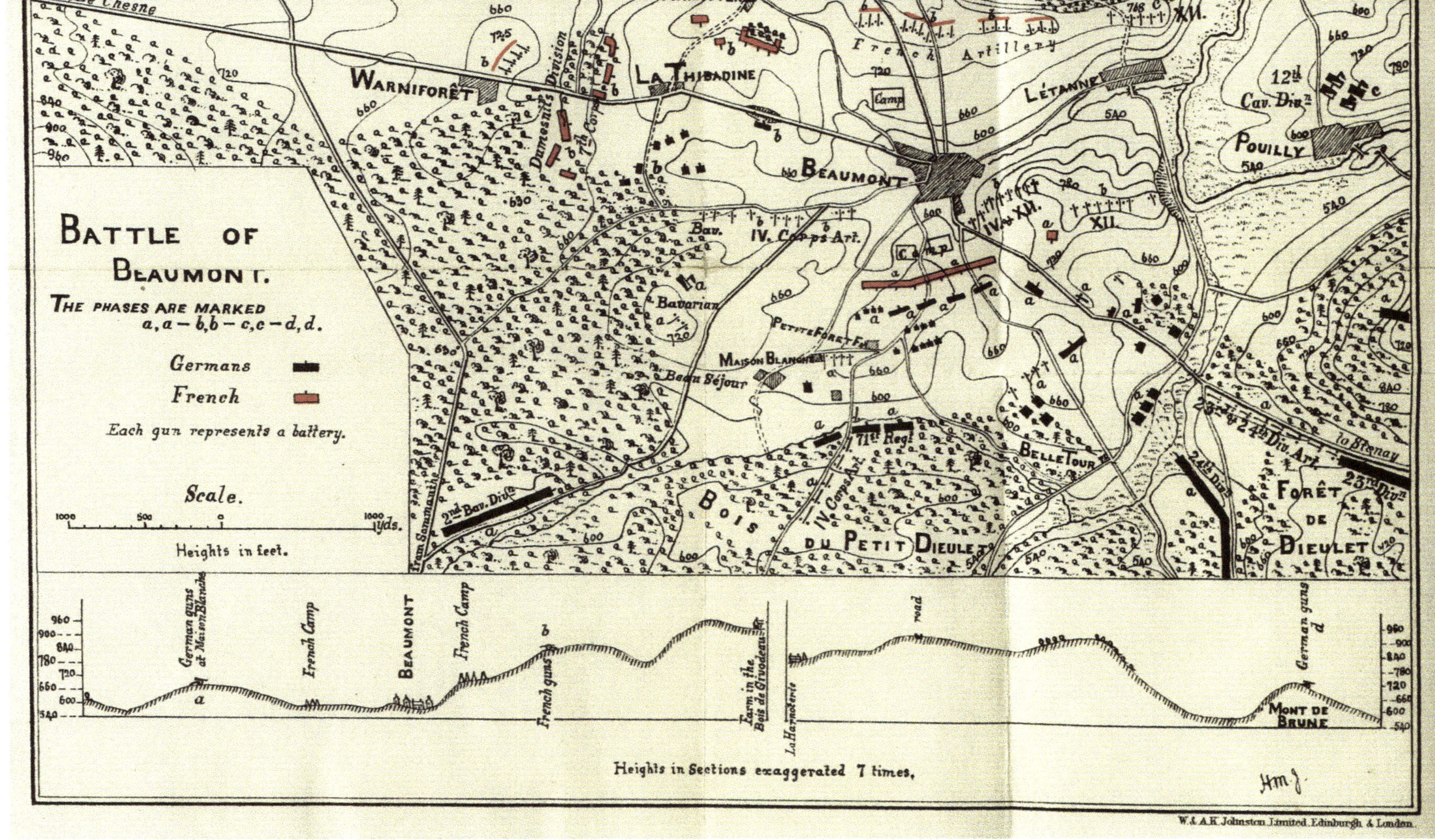
BATTLE OF BEAUMONT.
THE PHASES ARE MARKED a, a – b, b – c, c – d, d.
Germans
French
Each gun represents a battery.
Scale.
1000 500 0 1000 yds.
Heights in feet.
Le Chesne
Warniforêt
Dumesnils Division
7th Corps
La Thibadine
French Artillery
Camp
Létanne
Beaumont
XII.
IV. & XII.
12th Cav. Divn.
Pouilly
Bav.
IV. Corps Art.
Camp
Bavarian
Petite Forêt Fm.
Maison Blanche
Beau Séjour
71st Regt.
Belle Tour
IV. Corps Art.
23rd & 24th Div. Art.
to Stenay
24th Divn.
23rd Divn.
Forêt de Dieulet
Bois du Petit Dieulet
2nd Bav. Divn.
Tram Sommauthe
German guns at Maison Blanche
French Camp
Beaumont
French Camp
French guns
Farm in the Bois de Givodeau
La Harnoterie
road
German guns
Mont de Brune
Heights in Sections exaggerated 7 times.
W. & A.K. Johnston Limited, Edinburgh & London.

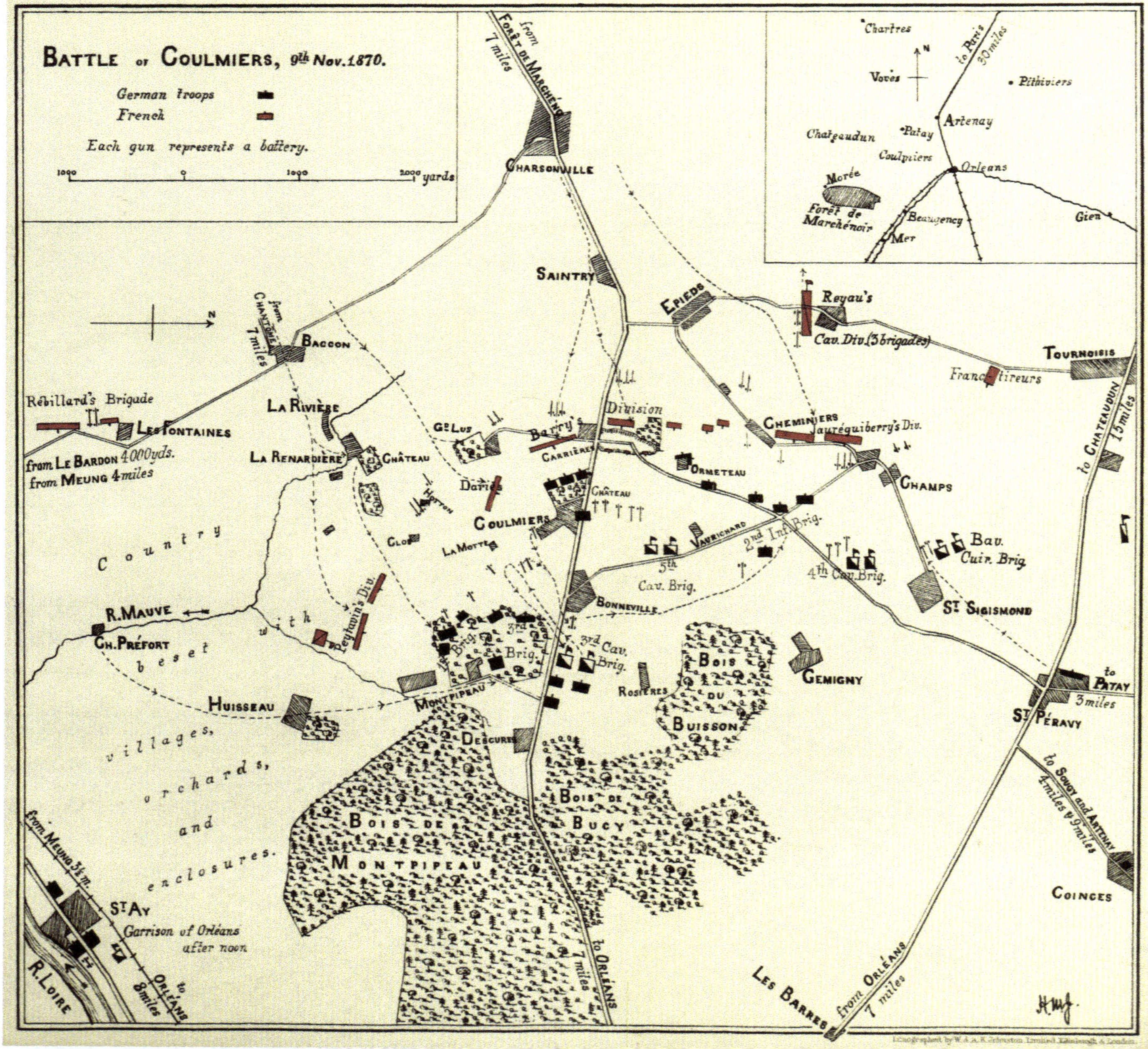

Battle of Coulmiers, 9th Nov. 1870.

German troops
French
Each gun represents a battery.
1000 0 1000 2000 yards

N

Chartres
Voves
to Paris 30 miles
Pithiviers
Artenay
Chateaudun
Patay
Coulmiers
Orleans
Morée
Forêt de Marchénoir
Beaugency
Mer
Gien

from Forêt de Marchénoir 7 miles
Charsonville
Saintry
Epieds
Reyau's Cav. Div. (3 brigades)
Tournoisis
Franc-tireurs
to Chateaudun 15 miles
from Chantome 7 miles
Baccon
Rebillard's Brigade
Les Fontaines
from Le Bardon 4000 yds.
from Meung 4 miles
La Rivière
La Renardière
Château
Gd Lus
Barry's Division
Carrières
Cheminiers
Jauréguiberry's Div.
Ormeteau
Champs
Dariès
Château
Coulmiers
Huetton
Clos
La Motte
Vaurichard
2nd Inf. Brig.
5th Cav. Brig.
4th Cav. Brig.
Bav. Cuir. Brig.
St Sigismond
Country beset with villages, orchards, and enclosures.
R. Mauve
Ch. Préfort
Peytavin's Div.
1st Brig.
3rd Brig.
3rd Cav. Brig.
Bonneville
Huisseau
Montpipeau
Rosières
Bois du Buisson
Gemigny
to Patay 3 miles
St Péravy
Descures
Bois de Bucy
Bois de Montpipeau
to Orléans 7 miles
to Sougy and Artenay 4 miles & 9 miles
Coinces
from Meung 3½ m.
St Ay
Garrison of Orléans after noon
to Orléans 8 miles
R. Loire
Les Barres
from Orléans 7 miles

Lithographed by W. & A. K. Johnston Limited, Edinburgh & London.

www.ingramcontent.com/pod-product-compliance
Ingram Content Group UK Ltd.
Pitfield, Milton Keynes, MK11 3LW, UK
UKHW051030290726
14058UKWH00012B/863